Rolls Royce
SILVER SHADOW
Gold Portfolio
1965-1980

Compiled by
R.M.Clarke

ISBN 1 85520 2298

Brooklands Books Ltd.
PO Box 146, Cobham, KT11 1LG
Surrey, England

Printed in Hong Kong

BROOKLANDS BOOKS

BROOKLANDS ROAD TEST SERIES

Abarth Gold Portfolio 1950-1971
AC Ace & Aceca 1953-1983
Alfa Romeo Giulietta Gold Portfolio 1954-1965
Alfa Romeo Giulia Berlinas 1962-1976
Alfa Romeo Giulia Coupés 1963-1976
Alfa Romeo Giulia Coupés Gold P. 1963-1976
Alfa Romeo Spider 1966-1990
Alfa Romeo Spider Gold Portfolio 1966-1991
Alfa Romeo Alfasud 1972-1984
Alfa Romeo Alfetta Gold Portfolio 1972-1987
Alfa Romeo Alfetta GTV6 1980-1987
Allard Gold Portfolio 1937-1959
Alvis Gold Portfolio 1919-1967
American Motors Muscle Cars 1966-1970
Armstrong Siddeley Gold Portfolio 1945-1960
Aston Martin Gold Portfolio 1972-1985
Austin Seven 1922-1982
Austin A30 & A35 1951-1962
Austin Healey 100 & 100/6 Gold P. 1952-1959
Austin Healey 3000 Gold Portfolio 1959-1967
Austin Healey Sprite 1958-1971
BMW Six Cyl. Coupés 1969-1975
BMW 1600 Collection No.1 1966-1981
BMW 2002 Gold Portfolio1968-1976
BMW 316, 318, 320 (4 cyl.) Gold P. 1975-1990
BMW 320, 323, 325 (6 cyl.) Gold P. 1977-1990
BMW 5 Series Gold Portfolio1981-1987
BMW M Series Performance Portfolio1976-1993
Bristol Cars Gold Portfolio 1946-1992
Buick Automobiles 1947-1960
Buick Muscle Cars 1965-1970
Cadillac Automobiles 1949-1959
Cadillac Automobiles 1960-1969
Chevrolet 1955-1957
Chevrolet Impala & SS 1958-1971
Chevrolet Corvair 1959-1969
Chevy El Camino & SS 1959-1987
Chevy II Nova & SS 1962-1973
Chevelle & SS Muscle Portfolio 1964-1972
Chevrolet Muscle Cars 1966-1971
Chevy Blazer 1969-1981
Chevrolet Corvette Gold Portfolio 1953-1962
Chevrolet Corvette Sting Ray Gold P. 1963-1967
Chevrolet Corvette Gold Portfolio 1968-1977
High Performance Corvettes 1983-1989
Camaro Muscle Portfolio 1967-1973
Chevrolet Camaro Z28 & SS 1966-1973
Chevrolet Camaro & Z28 1973-1981
High Performance Camaros 1982-1988
Chrysler 300 Gold Portfolio 1955-1970
Chrysler Valiant 1960-1962
Citroen Traction Avant Gold Portfolio 1934-1957
Citroen 2CV Gold Portfolio 1948-1989
Citroen DS & ID 1955-1975
Citroen DS & ID Gold Portfolio 1955-1975
Citroen SM 1970-1975
Cobras & Replicas 1962-1983
Shelby Cobra Gold Portfolio 1962-1969
Cobras & Cobra Replicas Gold P. 1962-1989
Cunningham Automobiles 1951-1955
Daimler SP250 Sports & V-8 250 Saloon Gold
Portfolio 1959-1969
Datsun Roadsters 1962-1971
Datsun 240Z 1970-1973
Datsun 280Z & ZX 1975-1983
The De Lorean 1977-1993
De Tomaso Collection No. 1 1962-1981
Dodge Charger 1966-1974
Dodge Muscle Cars 1967-1970
Dodge Viper on the Road
The De Lorean 1977-1993
Excalibur Collection No. 1 1952-1981
Facel Vega 1954-1964
Ferrari Cars 1946-1956
Ferrari Collection No. 1 1960-1970
Ferrari Dino 1965-1974
Ferrari Dino 308 1974-1979
Ferrari 308 & Mondial 1980-1984
Motor & T&CC Ferrari 1966-1976
Motor & T&CC Ferrari 1976-1984
Fiat Pininfarina 124 & 2000 Spider 1968-1985
Fiat-Bertone X1/9 1973-1988
Ford Consul, Zephyr, Zodiac Mk.I & II 1950-1962
Ford Zephyr, Zodiac, Executive, Mk.III & Mk.IV
1962-1971
Ford Cortina 1600E & GT 1967-1970
High Performance Capris Gold P. 1969-1987
Capri Muscle Portfolio 1974-1987
High Performance Fiestas 1979-1991
High Performance Escorts Mk.I 1968-1974
High Performance Escorts Mk.II 1975-1980
High Performance Escorts 1980-1985
High Performance Escorts 1985-1990
High Performance Sierras & Merkurs Gold
Portfolio 1983-1990
Ford Automobiles 1949-1959
Ford Fairlane 1955-1970
Ford Ranchero 1957-1959
Thunderbird 1955-1957
Thunderbird 1958-1963
Thunderbird 1964-1976
Ford Falcon 1960-1970
Ford GT40 Gold Portfolio 1964-1987
Ford Bronco 1966-1977
Ford Bronco 1978-1988
Holden 1948-1962

Honda CRX 1983-1987
Hudson & Railton 1936-1940
Isetta 1953-1964
Jaguar and SS Gold Portfolio 1931-1951
Jaguar XK120, 140, 150 Gold P. 1948-1960
Jaguar Mk.VII, VIII, IX, X, 420 Gold P.1950-1970
Jaguar 1957-1961
Jaguar Mk.2 1959-1969
Jaguar Cars 1961-1964
Jaguar E-Type Gold Portfolio 1961-1971
Jaguar E-Type 1966-1971
Jaguar E-Type V-12 1971-1975
Jaguar XJ12, XJ5.3, V12 Gold P. 1972-1990
Jaguar XJ6 Series II 1973-1979
Jaguar XJ6 Series III 1979-1986
Jaguar XJS Gold Portfolio 1975-1990
Jeep CJ5 & CJ6 1960-1976
Jeep CJ5 & CJ7 1976-1986
Jensen Cars 1946-1967
Jensen Cars 1967-1979
Jensen Interceptor Gold Portfolio 1966-1986
Jensen Healey 1972-1976
Lagonda Gold Portfolio 1919-1964
Lamborghini Cars 1964-1970
Lamborghini Countach & Urraco 1974-1980
Lamborghini Countach & Jalpa 1980-1985
Lancia Beta Gold Portfolio 1972-1984
Lancia Fulvia Gold Portfolio 1963-1976
Lancia Stratos 1972-1985
Land Rover Series I 1948-1958
Land Rover Series II & IIa 1958-1971
Land Rover Series III 1971-1985
Land Rover 90 & 110 1983-1989
Land Rover Discovery 1989-1994
Lincoln Gold Portfolio 1949-1960
Lincoln Continental 1961-1969
Lincoln Continental 1969-1976
Lotus & Caterham Seven Gold P. 1957-1993
Lotus Sports Racers Gold Portfolio 1953-1965
Lotus Elite 1957-1964
Lotus Elite & Eclat 1974-1982
Lotus Elan Gold Portfolio 1962-1974
Lotus Elan Collection No. 2 1963-1972
Lotus Elan 1989-1992
Lotus Cortina Gold Portfolio 1963-1970
Lotus Europa Gold Portfolio 1966-1975
Lotus Turbo Esprit 1980-1986
Motor & T&CC on Lotus 1979-1983
Marcos Cars 1960-1988
Maserati 1965-1970
Maserati 1970-1975
Mazda RX7 Collection No. 1 1978-1981
Mercedes Benz Cars 1949-1954
Mercedes Benz Competition Cars 1950-1957
Mercedes Benz Cars 1954-1957
Mercedes Benz Cars 1957-1961
Mercedes 190 & 300 SL 1954-1963
Mercedes 230/250/280SL 1963-1971
Mercedes Benz SLs & SLCs Gold P. 1971-1989
Mercedes S & 600 1965-1972
Mercedes S Class 1972-1979
Mercury Muscle Cars 1966-1971
Metropolitan 1954-1962
MG Gold Portfolio 1929-1939
MG TC 1945-1949
MG TD 1949-1953
MG TF 1953-1955
MGA & Twin Cam Gold Portfolio 1955-1962
MG Midget Gold Portfolio1961-1979
MGB Roadsters 1962-1980
MGB MGC & V8 Gold Portfolio 1962-1980
MGB GT 1965-1980
Mini Cooper Gold Portfolio 1961-1971
Mini Muscle Cars 1961-1979
Mini Moke Gold Portfolio1964-1994
Mopar Muscle Cars 1964-1967
Morgan Three-Wheeler Gold Portfolio 1910-1952
Morgan Plus 4 & Four 4 Gold P. 1936-1967
Morgan Cars 1960-1970
Morgan Cars Gold Portfolio 1968-1989
Morris Minor Collection No. 1 1948-1980
Shelby Mustang Muscle Portfolio 1965-1970
High Performance Mustang IIs 1974-1978
High Performance Mustangs 1982-1988
Oldsmobile Automobiles 1955-1963
Oldsmobile Cutlass & 4-4-2 1964-1972
Oldsmobile Muscle Cars 1964-1971
Oldsmobile Toronado 1966-1978
Opel GT 1968-1973
Packard Gold Portfolio 1946-1958
Pantera Gold Portfolio 1970-1989
Panther Gold Portfolio 1972-1990
Plymouth Barracuda 1964-1974
Plymouth Muscle Cars 1966-1971
Pontiac Tempest & GTO 1961-1965
Pontiac Muscle Cars 1966-1972
Pontiac Firebird & Trans-Am 1973-1981
High Performance Firebirds 1982-1988
Pontiac Fiero 1984-1988
Porsche 356 1952-1965
Porsche 911 1965-1969
Porsche 911 1970-1972
Porsche 911 1973-1977
Porsche 911 Carrera 1973-1977
Porsche 911 Turbo 1975-1984
Porsche 911 SC 1978-1983
Porsche 914 Collection No. 1 1969-1983
Porsche 914 Gold Portfolio 1969-1976
Porsche 924 Gold Portfolio 1975-1988
Porsche 928 1977-1989

Porsche 944 Gold P.1981-1991
Range Rover Gold Portfolio 1970-1992
Reliant Scimitar 1964-1986
Riley Gold Portfolio 1924-1939
Riley 1.5 & 2.5 Litre Gold Portfolio 1945-1955
Rolls Royce Silver Cloud & Bentley 'S' Series
Gold Portfolio 1955-1965
Rolls Royce Silver Shadow Gold P. 1965-1980
Rover P4 1949-1959
Rover P4 1955-1964
Rover 3 & 3.5 Litre Gold Portfolio 1958-1973
Rover 2000 & 2200 1963-1977
Rover 3500 1968-1977
Rover 3500 & Vitesse 1976-1986
Saab Sonett Collection No.1 1966-1974
Saab Turbo 1976-1983
Studebaker Gold Portfolio 1947-1966
Studebaker Hawks & Larks 1956-1963
Avanti 1962-1990
Sunbeam Tiger & Alpine Gold P. 1959-1967
Toyota MR2 1984-1988
Toyota Land Cruiser 1956-1984
Triumph TR2 & TR3 1952-1960
Triumph TR4, TR5, TR250 1961-1968
Triumph TR6 Gold Portfolio 1969-1976
Triumph TR7 & TR8 Gold Portfolio 1975-1982
Triumph Herald 1959-1971
Triumph Vitesse 1962-1971
Triumph Spitfire Gold Portfolio 1962-1980
Triumph 2000, 2.5, 2500 1963-1977
Triumph GT6 1966-1974
Triumph Stag 1970-1980
TVR Gold Portfolio 1959-1990
VW Beetle Gold Portfolio1935-1967
VW Beetle Gold Portfolio1968-1991
VW Beetle Collection No.1 1970-1982
VW Karmann Ghia 1955-1982
VW Bus, Camper, Van 1954-1967
VW Bus, Camper, Van 1968-1979
VW Bus, Camper, Van 1979-1989
VW Scirocco 1974-1981
VW Golf GTI 1976-1986
Volvo PV444 & PV544 1945-1965
Volvo Amazon-120 Gold Portfolio 1956-1970
Volvo 1800 Gold Portfolio 1960-1973

BROOKLANDS ROAD & TRACK SERIES

Road & Track on Alfa Romeo 1949-1963
Road & Track on Alfa Romeo 1964-1970
Road & Track on Alfa Romeo 1971-1976
Road & Track on Alfa Romeo 1977-1989
Road & Track on Aston Martin 1962-1990
R & T on Auburn Cord and Duesenburg 1952-84
Road & Track on Audi & Auto Union 1952-1980
Road & Track on Audi & Auto Union 1980-1986
Road & Track on Austin Healey 1953-1970
Road & Track on BMW Cars 1966-1974
Road & Track on BMW Cars 1975-1978
Road & Track on BMW Cars 1979-1983
R & T on Cobra, Shelby & Ford GT40 1962-1992
Road & Track on Corvette 1953-1967
Road & Track on Corvette 1968-1982
Road & Track on Corvette 1982-1986
Road & Track on Corvette 1986-1990
Road & Track on Datsun Z 1970-1983
Road & Track on Ferrari 1975-1981
Road & Track on Ferrari 1981-1984
Road & Track on Ferrari 1984-1988
Road & Track on Fiat Sports Cars 1968-1987
Road & Track on Jaguar 1950-1960
Road & Track on Jaguar 1961-1968
Road & Track on Jaguar 1968-1974
Road & Track on Jaguar 1974-1982
Road & Track on Jaguar 1983-1989
Road & Track on Lamborghini 1964-1985
Road & Track on Lotus 1972-1981
Road & Track on Maserati 1952-1974
Road & Track on Maserati 1975-1983
R & T on Mazda RX7 & MX5 Miata 1986-1991
Road & Track on Mercedes 1952-1962
Road & Track on Mercedes 1963-1970
Road & Track on Mercedes 1971-1979
Road & Track on Mercedes 1980-1987
Road & Track on MG Sports Cars 1949-1961
Road & Track on MG Sports Cars 1962-1980
Road & Track on Mustang 1964-1977
R & T on Nissan 300-ZX & Turbo 1984-1989
Road & Track on Peugeot 1955-1986
Road & Track on Pontiac 1960-1983
Road & Track on Porsche 1951-1967
Road & Track on Porsche 1968-1971
Road & Track on Porsche 1972-1975
Road & Track on Porsche 1975-1978
Road & Track on Porsche 1979-1982
Road & Track on Porsche 1982-1985
Road & Track on Porsche 1985-1988
R & T on Rolls Royce & Bentley 1950-1965
R & T on Rolls Royce & Bentley 1966-1984
Road & Track on Saab 1972-1992
R & T on Toyota Sports & GT Cars 1966-1984
R & T on Triumph Sports Cars 1953-1967
R & T on Triumph Sports Cars 1967-1974
R & T on Triumph Sports Cars 1974-1982
Road & Track on Volkswagen 1951-1968
Road & Track on Volkswagen 1968-1978
Road & Track on Volkswagen 1978-1985

Road & Track on Volvo 1957-1974
Road & Track on Volvo 1975-1985
R&T - Henry Manney at Large & Abroad

BROOKLANDS CAR AND DRIVER SERIES

Car and Driver on BMW 1955-1977
Car and Driver on BMW 1977-1985
C and D on Cobra, Shelby & Ford GT40 1963-84
Car and Driver on Corvette 1956-1967
Car and Driver on Corvette 1968-1977
Car and Driver on Corvette 1978-1982
Car and Driver on Corvette 1983-1988
C and D on Datsun Z 1600 & 2000 1966-1984
Car and Driver on Ferrari 1955-1962
Car and Driver on Ferrari 1963-1975
Car and Driver on Ferrari 1976-1983
Car and Driver on Mopar 1956-1967
Car and Driver on Mopar 1968-1975
Car and Driver on Mustang 1964-1972
Car and Driver on Pontiac 1961-1975
Car and Driver on Porsche 1955-1962
Car and Driver on Porsche 1963-1970
Car and Driver on Porsche 1970-1976
Car and Driver on Porsche 1977-1981
Car and Driver on Porsche 1982-1986
Car and Driver on Saab 1956-1985
Car and Driver on Volvo 1955-1986

BROOKLANDS PRACTICAL CLASSICS SERIES

PC on Austin A40 Restoration
PC on Land Rover Restoration
PC on Metalworking in Restoration
PC on Midget/Sprite Restoration
PC on Mini Cooper Restoration
PC on MGB Restoration
PC on Morris Minor Restoration
PC on Sunbeam Rapier Restoration
PC on Triumph Herald/Vitesse
PC on Spitfire Restoration
PC on Beetle Restoration
PC on 1930s Car Restoration

BROOKLANDS HOT ROD 'MUSCLECAR & HI-PO ENGINES' SERIES

Chevy 265 & 283
Chevy 302 & 327
Chevy 348 & 409
Chevy 350 & 400
Chevy 396 & 427
Chevy 454 thru 512
Chrysler Hemi
Chrysler 273, 318, 340 & 360
Chrysler 361, 383, 400, 413, 426, 440
Ford 289, 302, Boss 302 & 351W
Ford 351C & Boss 351
Ford Big Block

BROOKLANDS RESTORATION SERIES

Auto Restoration Tips & Techniques
Basic Bodywork Tips & Techniques
Basic Painting Tips & Techniques
Camaro Restoration Tips & Techniques
Chevrolet High Performance Tips & Techniques
Chevy Engine Swapping Tips & Techniques
Chevy-GMC Pickup Repair
Chrysler Engine Swapping Tips & Techniques
Custom Painting Tips & Techniques
Engine Swapping Tips & Techniques
Ford Pickup Repair
How to Build a Street Rod
Land Rover Restoration Tips & Techniques
MG 'T' Series Restoration Guide
Mustang Restoration Tips & Techniques
Performance Tuning - Chevrolets of the 60's
Performance Tuning - Pontiacs of the '60's

BROOKLANDS MILITARY VEHICLES SERIES

Allied Military Vehicles No.1 1942-1945
Allied Military Vehicles No.2 1941-1946
Complete WW2 Military Jeep Manual
Dodge Military Vehicles No.1 1940-1945
Hail To The Jeep
Land Rovers in Military Service
Off Road Jeeps: Civ. & Mil. 1944-1971
US Military Vehicles 1941-1945
US Army Military Vehicles WW2-TM9-2800
VW Kubelwagen Military Portfolio1940-1990
WW2 Jeep Military Portfolio 1941-1945

261033

CONTENTS

5	The New Rolls-Royce and Bentley Cars	Autosport	Oct. 8	1965
6	After the Cloud - The Shadow	Autocar	Oct. 8	1965
15	Silver Shadow two-door saloon	Car South Africa	June	1965
17	Revolution at RR	Modern Motor	Nov.	1965
18	The substance behind the Shadow	Motor	Oct. 9	1965
21	Rolls-Royce Silver Shadow Road Test	Autocar	Mar. 30	1967
27	The Rolls-Royce Silver Shadow Road Test	Autosport	Apr. 28	1967
30	Long Weekend with a Rolls-Royce Silver Shadow	Motor Sport	May	1968
34	Rolls-Royce Silver Shadow Road Test	Autosport	Apr. 11	1969
38	Rolls-Royce Silver Shadow Road Test	Road & Track	Aug.	1969
40	Rolls Royce Silver Shadow Road Test	World Car Guide	Oct.	1969
45	There and Back	Autocar	June 12	1969
48	West in Luxury	Autocar	July 30	1970
50	Rolls-Royce Silver Shadow Road Test	Autosport	Dec. 10	1970
57	For the man who has everything	Australian Motor Manual	Jan.	1971
58	Rolls-Royce	Road Test	Dec.	1970
60	Rolls-Royce Silver Shadow Road Test	Motor Trend	Apr.	1973
64	Rolls-Royce Silver Shadow	Motor	Apr. 21	1973
70	Taking Stock	Autocar	June 1	1972
73	Marseilles and back in a day	Autocar	June 5	1973
76	Rolls-Royce Silver Shadow: finer than ever	Autosport	Jan. 24	1974
78	Rolls-Royce Corniche Road Test	Autocar	Apr. 6	1974
84	Rolls-Royce Shadow and Corniche	Car South Africa	Apr.	1974
86	Cadillac Seville vs Rolls-Royce Silver Shadow	Car and Driver	July	1975
92	Rolls-Royce Silver Shadow Road Test	Autocar	May 1	1976
98	Rolls-Royce Silver Shadow Road Test	Road & Track	Sept	1976
103	The Silver Lining	Autosport	Feb. 24	1977
105	Newcomers: The Shadow's Second Coming	Car	Apr.	1977
106	Rolls-Royce Silver Shadow II Road Test	Motor	May 21	1977
112	The tool-kit is a telephone...	Modern Motor	Aug.	1977
114	Rich the treasure, sweet the pleasure	Motor	July 9	1977
119	In the Shadow of Rolls-Royce	Motor Sport	Oct.	1977
122	West Head is only a Silver Shadow away	Wheels	Oct.	1977
126	Incomparable engineering Road Test	Autosport	Nov. 2	1978
128	The Standard	Wheels	Dec.	1978
136	J2 to Le Mans	Autocar	June 9	1979
138	Bentley T2 Road Test	Autocar	Oct. 20	1979
144	Rolls-Royce Silver Shadow II Road Test	Road & Track	Nov.	1979
147	R-R's Silver Shadow II	Road Test	July	1980
150	High-Roller Ragtops Comparison Test	Car and Driver	June	1980
156	Just Another Roller	Car	Oct.	1980
162	Bargain or bankruptcy	Practical Classics	Feb.	1987
169	Shadow of doubt	Your Classic	Nov.	1990

ACKNOWLEDGEMENTS

When our regular Road Test book on the Silver Shadow went out of print, we searched through our archives to see if there was more material on these cars which might prove of interest to enthusiasts. There was, and so we have put it together with material from our first book to produce this new and enlarged Silver Shadow Gold Portfolio.

As always, we are indebted to a number of people for their help and cooperation. Motoring writer James Taylor, who ran a Shadow for six years, has kindly provided a few words of introduction; and of course these who hold the copyright to the original material we have reproduced here once again deserve our sincere thanks. They are the managements of *Australian Motor Manual, Autocar, Autosport, Car, Car and Driver, Car South Africa, Modern Motor, Motor, Motor, Motor Sport, Motor Trend, Practical Classics, Road Test, Road & Track, Wheels, World Car Guide and Your Classic.*

R M Clarke

Many people tut-tutted when the new Rolls-Royce was announced in 1965. No chassis, no sweeping wing-lines - just the Palladian radiator and the $6^{1}/_{2}$ Litre V8 from the final Silver Clouds to link it to previous models. Was this the final break with the great tradition?

It wasn't, of course, and in time the Silver Shadow and its Bentley T-type cousin became recognised as truly great cars just like their forebears. Style, durability, ride quality, and all the qualities expected of Crewe's products were present in abundance for those who were not too blinkered to see them, and today the Silver Shadow draws as many enthusiastic admirers as other established classics.

In one respect, however, the Shadow did break with the past. As a monocoque design, it could not be the basis of the wide variety of custombuilt bodyshells which had graced earlier cars from Crewe. James Young bravely built a few two-door derivatives, but these were not sufficiently different from the standard article to attract much custom. It was left to the in-house coachbuilders, Mulliner-Park Ward, to come with a attactive two-door body with more flowing wing lines and to offerit as both fixed-head coupé and convertible. From 1971, this was marketed as the Corniche, a model seen as quite separate from the Silver Shadow.

The Shadow was built during a period when Bentley cars were simply rebadged Rolls-Royce models, and there were those who wondered whether there was any real point in keeping the Bentley marque alive. Sales were slow - which makes Bentley-badged models rarer today - and the point was not lost on Crewe. After the Shadow had ceased production, the Bentley marque was given a more distinctive profile of its own.

Reading these reports on the Shadow is a pleasurable experience in itself, but there is no real substitute for ownership of one of these fine cars. Well-kept examples command lower prices than many sports cars of a similar vintage, and represent a sound and enjoyable investment for today's motoring enthusiasts.

James Taylor

The new ROLLS-ROYCE and BENTLEY cars

By JOHN BOLSTER

THE NEW CARS have many technical innovations greatly differing from predecessors. Below the monocoque construction can be seen

Cars may come and cars may go, but there can never be any doubt that the Rolls-Royce is the British make with the highest reputation. When Henry Royce (the "Sir" came later) introduced his Silver Ghost in 1906, a car appeared which combined all the latest engineering knowledge in one chassis, while ignoring many modern trends which proved to be false trails. Now, Rolls-Royce announce models which they describe as being "the most radically new Rolls-Royce cars for 59 years"—the Silver Shadow.

Such an event must be regarded as a very special occasion, and so I set off in my 1911 Silver Ghost for Crewe. Her easy cruising speed of 60 m.p.h. soon annihilated the motorways, and she was locked up for the night among many of her great-grand-children. Next day, I was able to sample the progress which has been made in more than half-a-century. It would be pointless to compare the latest car with its forbear, because the requirements of the wealthy owner have changed so much. The Silver Ghost has a vast array of cups which must be removed and charged with lubricant. It can be done in 1 hour and 53 mins., according to my instruction book, but I have never managed to approach this time even remotely. The new car has six nipples that need the grease gun every 12,000 miles, and all that splendid maze of polished copper pipes no longer await the polishing rag. "Must keep your chauffeur busy," I was told in my youth, "or he's bound to start hanging round the back door and courting the maids." The new Rolls is as easy to wash as any medium-sized car, and needs the minimum of skilled attention.

The theory behind the design of the new cars has been simple, though it has brought into being some complicated engineering. First of all, a much more compact size was needed for modern city traffic. Yet, even more space for the seats was stipulated and a bigger luggage boot was required. In addition, riding comfort of an exceptional standard had to be allied with the best possible controllability, but all this had to be tied up with a lightness of operation which demanded a negligible effort from the driver. Quite a series of problems!

The necessity to save space required the deletion of the chassis frame and the adoption, for the first time, of a combined body and chassis. All four wheels are independently suspended, with wishbones in front and trailing arms behind. Very soft suspension was required, with 12 ins. of static deflection both ends and 8 ins. of working travel. To accommodate so much movement would not present a serious problem, provided that the load was constant. However, with varying numbers of passengers, often accompanied by heavy luggage, a self-levelling suspension system was the only solution. In particular, the universal joints of the driving shafts had to run normally at a reasonably moderate angle.

In the Rolls-Royce car, an elaborate arrangement for the operation of the suspension, brakes, and steering has been designed. The possibility of failure has been virtually eliminated and the brakes have three separate but complementary methods of operation. It is not practicable to explain the various circuits in so short an article as this, but suffice it to say that the engine camshaft has two extra lobes to operate a pair of pumps, running at a pressure of 2,500 lbs. per square inch.

The suspension is by helical springs, with four hydraulic jacks above them to settle the level. Three sensor devices, one in front and two at the rear of the car, measure the height and apply the control. There is a very rapid, but silent, height correction when the doors are open or the gear lever is in neutral. Much slower correction on the road avoids false levelling on acute bumps or hump-back bridges.

The brakes are discs all round. A method of damping them has been developed which eliminates any chance of squeal. The independent rear suspension has an anti-lift angle and the front suspension incorporates an anti-drive angle, to avoid nose-down braking. The suspension at both ends is mounted on sub-frames, which are insulated from the body to avoid the transmission of road noise.

The 6,230 c.c. V8 engine is a development of the existing pushrod light-alloy unit. More power and greater torque have been found, while the sparking plugs are now much more accessible. The automatic gearbox has been re-designed, having free-wheels on first and second speeds, a clutch to give jerk-free engagement of the ratios, and electrical operation of the gear selection. The overall ratios are 3.08, 4.46, 8.10, and 11.75 to 1 which, with 8.45 ins. × 15 ins. low profile tyres, gives 26.2 m.p.h. per 1,000 r.p.m. on top gear.

The car, although having more room inside, is 5 ins. lower, 3½ ins. narrower, and 7 ins. shorter than the previous model. The wheelbase is 9 ft. 11½ ins., track 4 ft. 9¼ ins., and weight 41 cwts. 44 lbs. No performance figures are available, but this is the fastest Rolls yet.

I was able to drive the new Rolls-Royce on the road. Actually, mine was a Bentley T series, the price of which is £6,496, including P.T., compared with £6,556 for the senior make.

The car accelerates very strongly, obviously having immense torque in the middle ranges. It feels astonishingly small, being a car of reasonable size, with a small turning circle and exceptionally light steering. There is no pitching, very good stability, and a remarkably comfortable ride. Some roll is felt if one corners very fast, but this is not excessive. Because the sound level is so low, it is possible to hear some road noise on occasion, which later designs of tyres may eliminate.

We hope to reinforce these brief impressions with the results of a full road test later on, but it can be said now that the very advanced chassis design has many advantages. Appearance is always a matter of opinion, but I found the looks of the new Rolls-Royce to be very much to my liking. It's a new car for modern conditions.

Frontal aspect of the Series T Bentley, identical to the Silver Cloud except for its radiator and bonnet pressing

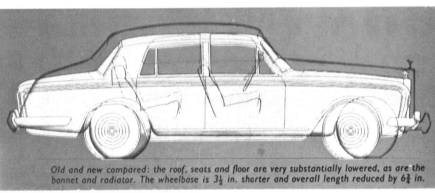

Old and new compared: the roof, seats and floor are very substantially lowered, as are the bonnet and radiator. The wheelbase is 3½ in. shorter and overall length reduced by 6¾ in.

AFTER

ROLLS-ROYCE ADOPT

A decade ago, when the six-cylinder Rolls-Royce Silver Cloud I and Bentley SI were safely hatched after the long incubation period that must lie behind any complex piece of machinery aspiring to near-perfection, the engineers at Crewe were already resetting their sights for the next generation of super cars.

Now at last we can see the fruits of their long labours and researches in the Rolls-Royce Silver Shadow—alias Bentley T Series—which conceals beneath an almost austerely orthodox new body a wealth of mechanical innovation. In fact, it possesses more individuality and advanced engineering than this company has ever displayed before in a new model.

It is, for instance, the first Rolls-Royce product to incorporate the following features:

A monocoque steel body shell, with separate front and rear sub-frames for engine and transmission, steering, suspension and wheel assemblies; all-independent suspension with automatic height control; four-wheel disc braking powered by dual high-pressure hydraulic systems; a recirculating ball steering gear with integral power assistance; electric actuation for the gear range selector, as well as for three-way front seat adjustment and window-lifts. In fact, the only direct legacies from the superseded car are its 6.3-litre vee-eight engine and four-speed automatic transmission, both with several interesting innovations.

To conform with current trends and the known preferences of many of the present Rolls-Royce and Bentley clientele, the new model is considerably more compact outwardly than its predecessors in every main dimension—4¼ in. lower, 6¾ in. shorter, and 3½ in. narrower, yet enveloping increased space for passengers and their baggage. The wheelbase has been reduced by 3½ in. from 10 ft 3 in. to 9 ft 11½ in., and front and rear tracks are now equal at 4 ft 9½ in.—compared with 4 ft 10½ in. front and 5 ft rear for the Cloud.

Several factors have contributed to the increased passenger space within an outwardly less bulky shell. First, dispensing with a separate chassis frame and having a fixed transmission line and final drive unit have allowed the floor and rear seat to be lowered several inches. The luggage boot has also gained from the static final drive assembly, as well as from a more boxy shape and having the spare wheel stowed in a hinged tray beneath it instead of inside. Moving the engine forward slightly has also played its part. Saving space has not meant saving much weight, the Shadow being about ¾ cwt lighter than the Cloud when carrying the same quantity of fuel; but, in fact, the total tank capacity has been increased from 18 to 24 gallons—very welcome for long-distance motorists.

Once again the classic Rolls-Royce radiator shell has been re-proportioned to match the greatly lowered bonnet line, and once again without appearing as a traditionalist anachronism. In fact it seems to have withstood the transformation perhaps more elegantly than the Bentley shell, which has lost some of its grandeur and distinction. No doubt Rolls-Royce archivists will be quick to remark upon a reversion to the proportions and dimensions of the early Edwardian era.

Still safe on her draughty perch is the Silver Lady mascot, now easy to detach when entering countries where her rigid posture contravenes road safety laws; with the bonnet open one can unscrew her from her pedestal with a special key. Those who live in such

Lighting switches and other minor controls are of traditional design and quality, but the instrument dials (apart from the clock) are now all in front of the driver, and the two-spoke steering-wheel is another innovation. At each side of the centre panel are adjustable outlets for cool air

'66 Models

New Look for The Best Car In The World, now with a lower, wider radiator shell and less flamboyant body contours

THE CLOUD — THE SHADOW

MONOCOQUE BODY, ALL INDEPENDENT SELF-LEVELLING SUSPENSION, DISC BRAKES

countries can now specify (at extra cost) a spring-heeled Lady.

Of all the Shadow's mechanical features, the most significant and intriguing is the high-pressure hydraulic system serving the suspension height control and braking system. There are, in fact, two independent circuits. Each is served by an identical single-plunger pump set in the vee between the cylinder banks and actuated through a short pushrod from the camshaft. Mounted low on the left side of the crankcase are two spherical accumulators with butyl separators, inflated with nitrogen to 1,000 p.s.i. The hydraulic pumps pressurize these to 2,500 p.s.i. (maximum), at which point pressure regulating valves (built under licence from Citroen) incorporated in the housing above each accumulator open a by-pass back to the reservoir. Accumulators not only store a reserve supply of fluid under pressure, but also damp down sharp fluctuations in pressure when any service is in demand.

One system—from the forward pump—is responsible only for a 47 per cent share of the total braking which is divided front and rear in the ratio 31 to 16. The other provides a further 31 per cent (of the total) for the front only, and also feeds the suspension's height control circuits. Each front disc has two independent two-cylinder calipers, and each back one a single, four-cylinder caliper. One pair of cylinders in each rear caliper is served by an ordinary hydraulic system with master cylinder and no servo assistance. This normally provides the remaining 22 per cent of total braking, but its main purpose—apart from the safety factor of having three distinct circuits—is to give the driver 'feel' through the pedal. In the master cylinder circuit is a 'G conscious' pressure limiting valve of the ball-and-ramp type to reduce the likelihood of rear wheel locking.

→

Left: In the back are picnic tables of a new pattern, with ashtrays beneath. Cigarette lighters are in the door armrests, which also contain red lamps, lit when the doors are open. Slots in the carpet are above safety harness anchorages. Right: Other luxuries in the back include a compartment let into the folding armrest and mirrors in each quarter

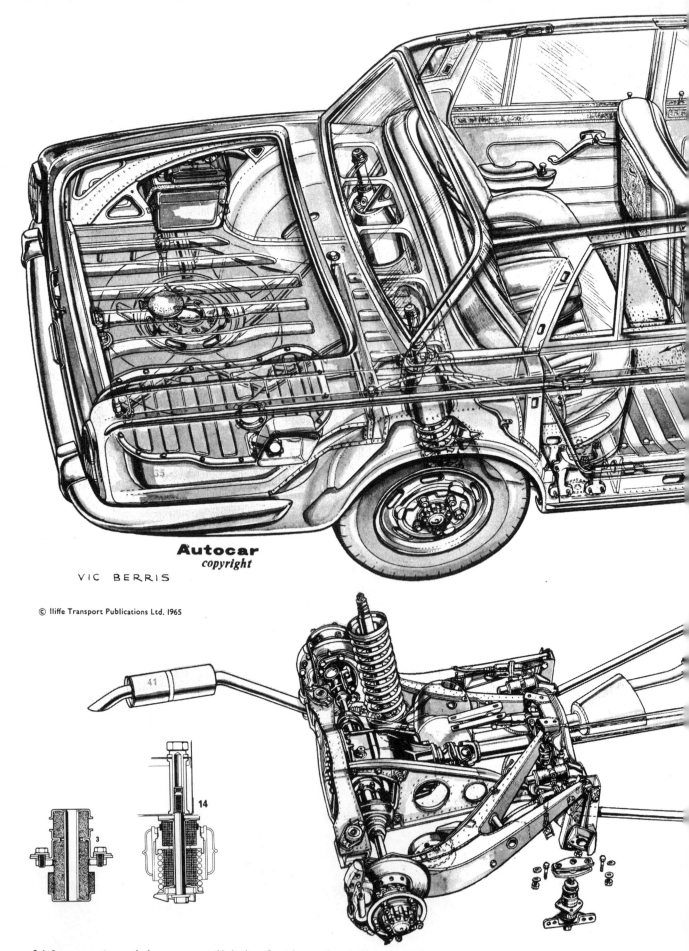

Sub-frame mountings embody new compressible bushes of stainless steel mesh. They provide closely controlled freedom of movement to absorb shock and road noise

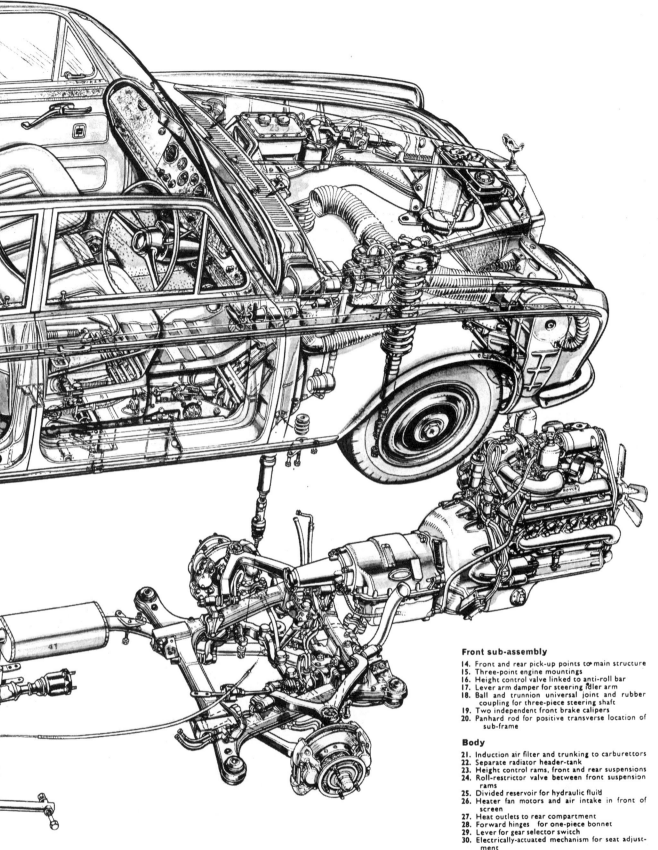

KEY:

Rear sub-assembly

1. Rear suspension main cross-member
2. Sub-frame and final drive cross member
3. Pick-up points to main structure
4. Horizontal fore-and-aft compliance mountings to main structure
5. Hydraulic damper supplementing compliance mountings
6. Torque arm bolted to final drive casing, coupled to rear suspension cross member
7. Ball-and-trunnion inboard universals
8. Suspension trailing arm pivots
9. Tubular tie bars between rear suspension cross member and underframe of body
10. Height control valves linked to trailing arms
11. Solenoid valve for quick-action levelling, front and rear
12. Single brake caliper incorporating two pairs of operating cylinders, and mechanical handbrake caliper
13. Mechanical linkage to handbrake caliper

Front sub-assembly

14. Front and rear pick-up points to main structure
15. Three-point engine mountings
16. Height control valve linked to anti-roll bar
17. Lever arm damper for steering idler arm
18. Ball and trunnion universal joint and rubber coupling for three-piece steering shaft
19. Two independent front brake calipers
20. Panhard rod for positive transverse location of sub-frame

Body

21. Induction air filter and trunking to carburettors
22. Separate radiator header-tank
23. Height control rams, front and rear suspensions
24. Roll-restrictor valve between front suspension rams
25. Divided reservoir for hydraulic fluid
26. Heater fan motors and air intake in front of screen
27. Heat outlets to rear compartment
28. Forward hinges for one-piece bonnet
29. Lever for gear selector switch
30. Electrically-actuated mechanism for seat adjustment
31. Seat adjusting switches
32. Footbrake unit operating three hydraulic systems
33. Twin S.U. electric fuel pumps
34. Fuel line filter
35. 24-gallon fuel tank integrated with boot-floor
36. Spare wheel in hinged carrier
37. Battery and tool kit
38. Rear window with electric demisting element
39. Monocoque pressed steel body structure
40. Cold air ventilation ducts from below headlamps
41. Exhaust system with four stainless steel silencers
42. Central jacking points

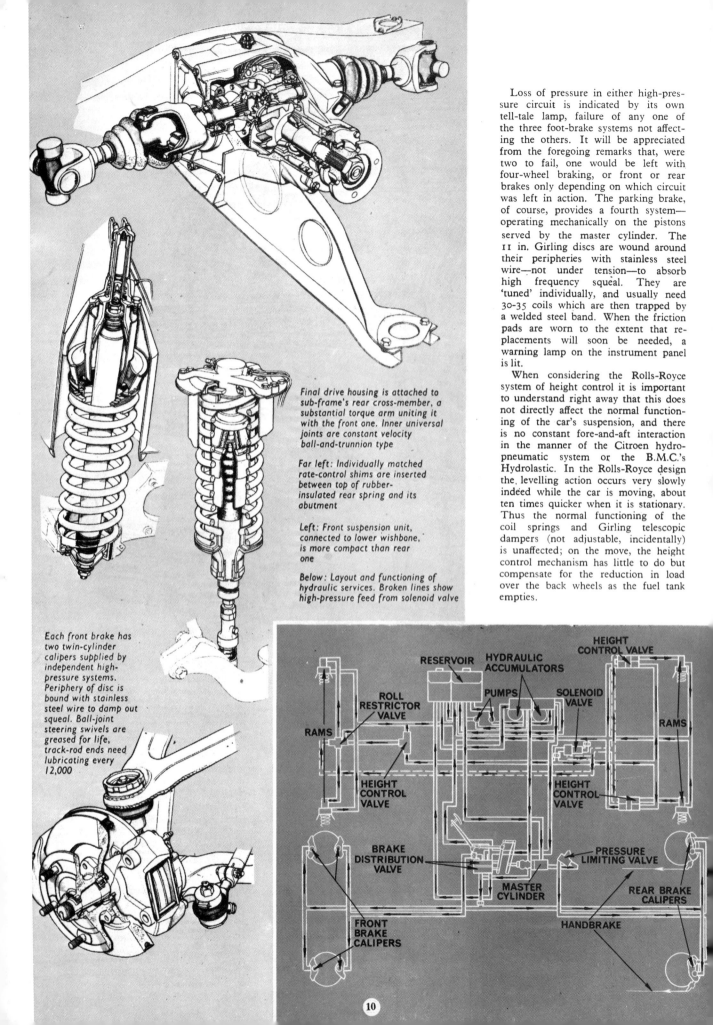

Final drive housing is attached to sub-frame's rear cross-member, a substantial torque arm uniting it with the front one. Inner universal joints are constant velocity ball-and-trunnion type

Far left: Individually matched rate-control shims are inserted between top of rubber-insulated rear spring and its abutment

Left: Front suspension unit, connected to lower wishbone, is more compact than rear one

Below: Layout and functioning of hydraulic services. Broken lines show high-pressure feed from solenoid valve

Each front brake has two twin-cylinder calipers supplied by independent high-pressure systems. Periphery of disc is bound with stainless steel wire to damp out squeal. Ball-joint steering swivels are greased for life, track-rod ends need lubricating every 12,000

Loss of pressure in either high-pressure circuit is indicated by its own tell-tale lamp, failure of any one of the three foot-brake systems not affecting the others. It will be appreciated from the foregoing remarks that, were two to fail, one would be left with four-wheel braking, or front or rear brakes only depending on which circuit was left in action. The parking brake, of course, provides a fourth system— operating mechanically on the pistons served by the master cylinder. The 11 in. Girling discs are wound around their peripheries with stainless steel wire—not under tension—to absorb high frequency squeal. They are 'tuned' individually, and usually need 30-35 coils which are then trapped by a welded steel band. When the friction pads are worn to the extent that replacements will soon be needed, a warning lamp on the instrument panel is lit.

When considering the Rolls-Royce system of height control it is important to understand right away that this does not directly affect the normal functioning of the car's suspension, and there is no constant fore-and-aft interaction in the manner of the Citroen hydro-pneumatic system or the B.M.C.'s Hydrolastic. In the Rolls-Royce design the levelling action occurs very slowly indeed while the car is moving, about ten times quicker when it is stationary. Thus the normal functioning of the coil springs and Girling telescopic dampers (not adjustable, incidentally) is unaffected; on the move, the height control mechanism has little to do but compensate for the reduction in load over the back wheels as the fuel tank empties.

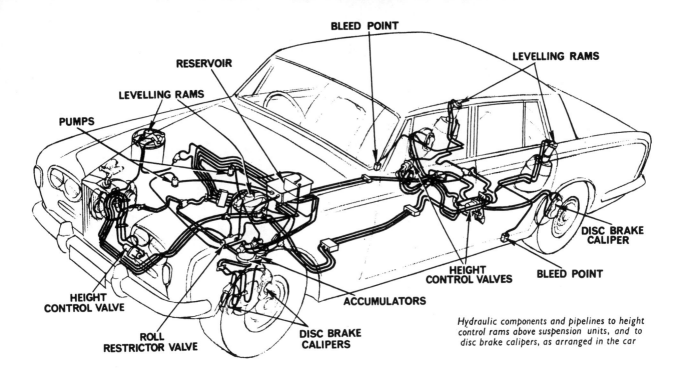

Hydraulic components and pipelines to height control rams above suspension units, and to disc brake calipers, as arranged in the car

In simple terms, the object is to keep the basic trim of the car constant regardless of the load carried or variations in weight distribution. The need for rapid levelling occurs only while the car is at rest, when immediate reactions are required to reset the trim as passengers enter or alight, or as the fuel tank is replenished. How this two-speed functioning is achieved will be explained a little later in the text, after considering the fundamental components and action of the system as a whole.

A hydraulic ram with 3 in. stroke is incorporated above each upper spring abutment; any static compression of the spring beyond the prearranged neutral setting for the height control (or levelling) valves causes this ram to extend by the same amount, so that the car, in effect, is then supported on a column of oil. If the car is heavily laden and carrying a boot full of luggage, for instance, then the rear springs will be much more compressed than the front ones, and the rams above them are extended that much further. Since the basic geometrical relationship between the suspension arms and the body structure thus does not vary with the load carried, the bump and rebound stop clearances also remain constant regardless of load, and much greater vertical wheel movements can be provided than with conventional suspensions.

There are three height control valves, one attending to both front rams, one for each at the back; apparently with an individual valve for each corner the system would never reach a state of equilibrium. The valves for the rear suspension are bolted to the sub-assembly's forward cross-member and linked to the trailing arms, whereas the single front one, bolted to that sub-frame, responds to rotary movements of the anti-roll bar. Between this valve and the levelling rams is a roll restriction valve. As well as dividing the flow between the two rams, it also restricts cross-flow between them, which otherwise would allow one ram to empty into the other and the car to roll.

When any door is opened, or if the gear selector is in neutral, a solenoid valve in circuit with the interior courtesy lamps and the selector is energized. It opens to pass fluid at high pressure to an overriding piston in each rear height control valve and in the front roll restrictor valve. In each of these there is a small, counterbored shuttle which normally provides a very restricted bleed path around it to the ports feeding the levelling rams. The solenoid-controlled piston pushes this shuttle against a spring loading, to bring a cross-drilling in the shuttle in line with the outlet ports, thereby permitting unrestricted flow—and hence rapid levelling.

With the car at rest and unladen it 'sits on its springs' with the rams unpressurized. The minimum system pressure for their operation is about 1,150 p.s.i. While the accompanying diagram should help to clarify the hydraulic circuits, to describe the precise design and operation of the individual components is beyond the scope of this article. After much experimenting with air springs, incidentally, the Rolls-Royce engineers decided that steel ones set fewer problems, particu-

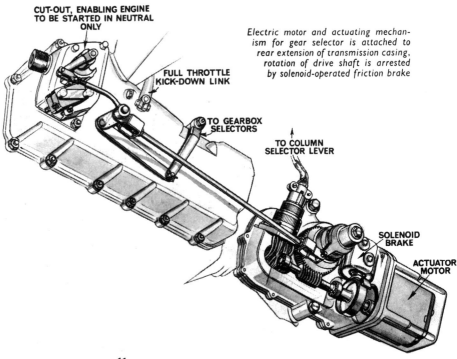

Electric motor and actuating mechanism for gear selector is attached to rear extension of transmission casing, rotation of drive shaft is arrested by solenoid-operated friction brake

ROLLS-ROYCE SILVER SHADOW...

Tail end of the new Rolls-Royce shows much larger rear window and squared-up boot

larly in very cold weather—and moreover cannot go flat overnight.

In attaching the front and rear subframes to the main structure, great pains have been taken to evolve mountings flexible enough to suppress road noise without giving the car spongy handling. The rear one is composite. It consists broadly of two vertical cross-members—one carrying the suspension trailing arms, the other the final drive unit, the only direct link between them being a sheet metal torque arm bolted to the right-hand side of the final drive casing. Its front end is attached through a flexible rubber coupling to a pick-up point on the forward cross-member.

A new type of resilient metal mounting is now used to attach these cross-members to the main structure. Developed by Delaney Gallay and called *Vibrashock*, it looks rather like a pan scrubber, embodying a stainless steel mesh claimed to have more closely controllable and constant characteristics than rubber. Supplementing the vertical mounts at each end of the forward cross-member is an arrangement to give it a measure of horizontal compliance—freedom of movement, that is. Below are two long tubular stays linking the cross-member to points well forward on the underside of the main structure, and bolted to the back of it are two upright forged steel horns of which the tops are sandwiched between two *Vibrashock* mountings. This movement is restricted by a small double-acting hydraulic damper.

Vic Berris's masterly cutaway drawing shows how the pivots of each trailing arm are angled to provide progressive changes in the wheel's attitude as it moves above or below the static mean. When the car is cornered fast, the outer rear wheel assumes a negative camber (that is, leans in at the top) which increases stability and cornering power. The rear-end geometry provides complete freedom from lift when the brakes are applied hard. Roll-centre is about 4 in. above ground level, near the parallel roll axis.

Constant velocity inboard universals on the half-axles are the ball-and-trunnion Detroit type, and the outers normal Hooke-type Hardy-Spicers.

Carrying the entire front assembly—engine, transmission, steering and suspension—is a rigid, box-section subframe welded from sheet steel pressings, with four-point attachment to the car body underframe. The *Vibrashock* mountings in this case are supplemented by heavy coil springs. A short Panhard rod anchored between a point on the front of the sub-frame and the main structure provides positive trans-

Driver's door has a map pocket, adjustable armrest, switches for all four electric window lifts

verse location. The upper suspension wishbones are one-piece fabrications from sheet steel, whereas the very long two-piece lower links are substantial forgings. Their trunnion axes are tilted to provide about 68 per cent anti-dive under heavy braking. Front wheel range of movement is given as 4 in. on both bump and rebound; at the back it is $3\frac{1}{2}$ in. on bump, $5\frac{1}{2}$ in. on rebound.

An American (Saginaw) recirculating ball steering gear, with constant ratio (19.3 to 1) and a built-in ram for the power assistance, is new to Rolls-Royce. It is fed, as on the Silver Cloud, by a belt-driven Hobourn-Eaton pump and compared with the earlier car the steering load—already light—is reduced still further, as is the turning circle. The column is divided and kinked for safety (and, of course, convenience of installation), having one rubber coupling and one ball-and-trunnion type. A lever-arm hydraulic damper is incorporated in the steering idler. The two-spoke steering-wheel defies tradition for a Rolls-Royce or Bentley; we must try to get used to it. At least it has the same slender rim section and diameter as the old three-spoke article.

Now that Harry Grylls, chief engineer at Crewe, no longer has to explain away drum brakes and a live rear axle, we might ask him: Why not fuel injection for the vee-8 engine? To shift the sparking-plugs

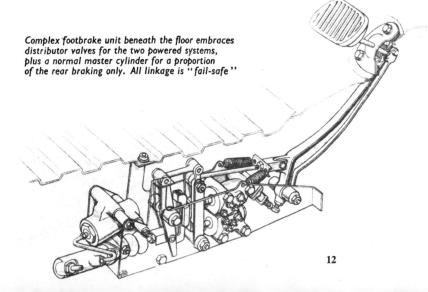

Complex footbrake unit beneath the floor embraces distributor valves for the two powered systems, plus a normal master cylinder for a proportion of the rear braking only. All linkage is "fail-safe"

Specification

ENGINE
Cylinders	8 in 90 deg vee
Cooling system	Water; pump, fan and thermostat
Bore	104·1 mm (4·10 in.)
Stroke	91·4 mm (3·60 in.)
Displacement	6,230 c.c. (380·5 cu. in.)
Valve gear	Overhead in-line, pushrods and rockers, hydraulic tappets
Compression ratio	9·0 to 1; option 8·0
Carburettors	2 diaphragm type S.U. HD8
Fuel pumps	2 S.U. electric
Oil filter	Full-flow, renewable paper element

TRANSMISSION
	Automatic 4-speed with fluid coupling and electric selection
Gear ratios	Top 1·00; Third 1·45; Second 2·63; First 3·82; Reverse 4·30
Final drive	Hypoid bevel, 3·08

CHASSIS AND BODY
Construction	Steel monocoque with separate front and rear sub-frames. Alloy doors, bonnet, boot lid

SUSPENSION
Front	Independent, double wishbone geometry, coil springs, Girling telescopic dampers, automatic hydraulic height control, anti-roll stabilizer
Rear	Independent, single trailing arms, coil springs, Girling telescopic dampers, automatic hydraulic height control

STEERING
Type	Saginaw recirculating ball, integral power assistance
Turns, lock-to-lock	4; steering-wheel dia. 17 in.

BRAKES
Make and type	Rolls-Royce-Girling, front discs each with two single calipers, rear discs each with one dual caliper, power assistance from two engine-driven hydraulic pumps, three independent foot-brake systems
Dimensions	F and R, 11 in. dia.
Swept area	F, 227 sq. in.: R, 286 sq. in. Total 513 sq. in.

WHEELS
Type	Pressed steel disc, five studs, 6in. wide rim

Tyres	8·45-15 in. low profile

EQUIPMENT
Battery	12-volt 64 amp. hr.
Generator	Lucas 35-amp
Fuel tank	24 Imp. gallons (109 litres) (warning lamp for 3-gal. reserve)
Cooling system	28 pints (16 litres)
Engine sump	14·5 pints (8 litres)
Gearbox and fluid coupling	24 pints (13·6 litres)
Final drive	4 pints (2·3 litres)

DIMENSIONS (manufacturer's figures)
Wheelbase	9 ft 11·5 in. (304 cm)
Track: front and rear	4 ft 9·5 in. (146 cm)
Overall length	16 ft 11·5 in. (517 cm)
Overall width	5 ft 11 in. (180 cm)
Overall height (unladen)	4 ft 11·75 in. (152 cm)
Ground clearance (laden)	6·5 in. (16·5 cm)
Turning circle	38 ft (11·6 m)
Kerb weight (with half-full fuel tank)	40·6 cwt, 4,546 lb. (2,062 kg)

PERFORMANCE DATA
Top gear m.p.h. per 1,000 r.p.m.	26·2

ROLLS-ROYCE SILVER SHADOW...

from their previously inaccessible position beneath the exhaust manifolds, and place them where they can be reached with ease above the manifolds, has meant a considerable redesign of the head. The new combustion chamber shape, shown in cross-section on these pages, allegedly adds about 2 per cent to a maximum output which the manufacturers will never reveal.

The engine is three-point mounted, the single front mounting being on a bridge piece bolted between the steering box on one side and idler arm casting on the other. Rear mountings are on the sub-frame itself. Three silencers are included in the exhaust system, all of them stainless steel, the front one being lagged with asbestos and sheathed in aluminium for heat and sound insulation.

Hydraulic accumulator has butyl diaphragm separator between the nitrogen chamber (below) and oil. Valves above return pump delivery to reservoir when pressure exceeds 2,500 p.s.i.

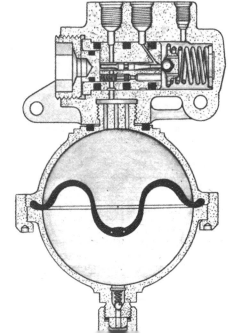

To achieve a low bonnet line the header tank has been divorced from the radiator; there is no overflow catch-tank. For the same reason the air cleaner has been buried down in the forepart of the right-hand wing beside the radiator, and leads to the twin S.U. diaphragm-type carburettors through flexible trunking.

For the first time, also, the bonnet is a one-piece aluminium pressing hinged on counterbalanced links just behind the radiator shell. It is lined with a heavy quilt for sound-deadening —another innovation for Rolls-Royce —and released or pulled shut positively by a substantial lever under the facia.

While retaining their 4-speed automatic transmission with simple fluid coupling, Rolls-Royce have incorporated several significant improvements. The main casing and some of the internal castings are now of aluminium instead of iron, the oil sealing is said to have been improved, and a free-wheel has been added for first and second, particularly to smooth out the big jump down from third to second. But with the second speed hold selected the free-wheel is inoperative and there is full engine braking.

A nice innovation is an electric motor to do the main physical labour of selecting the gears; the driver has a small lever, finger-light to move, which sends its instructions through a five-way switch. The electrical circuits are protected by an overload switch; if this should cut out for any reason, it can be reset by pressing a red button in the fuse box. For emergency get-you-home use there is a tommy bar in the tool-kit; for direct manual changes one inserts this in a hole in the top of the gearbox, after lifting the carpet.

The electric motor is bolted to the back of the aluminium casing containing the actuating mechanism, the whole unit being attached to the rear extension of the transmission casing. There are mechanical links to the selector levers.

Electrical services are supplied by a 12-volt battery with negative earth, charged by a 35-amp dynamo; cars equipped with the optional refrigeration unit have an alternator instead. Considerable use is made of printed circuits, these being employed behind the fuse box, under the seat switches, in the coolant level probe, for the relay box under the bonnet, and for the air-conditioning switches if fitted. Multi-pin sockets are also used where practical. Typical of the Rolls-Royce approach to motoring safety is a system of relays

Engine installation in the Bentley showing the cold-air ducting for the carburettors from a filter housed in the off-side wing. The tops of the suspension units can be seen each side

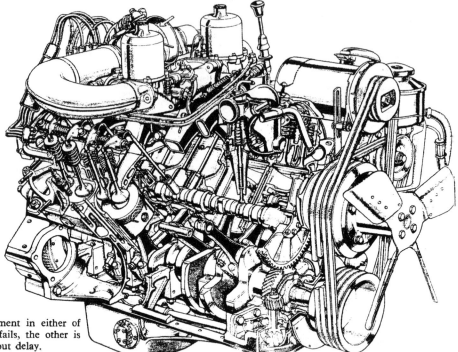

by which, if one filament in either of the outer headlamps fails, the other is lit automatically without delay.

As well as for the window lifts and seat adjustment, electric operation is used to raise and lower the radio aerial, demist the back window by built-in heater wires, and release the fuel filler trap-door. Separate switches on the dash illuminate right or left side and tail lamps for parking, and when these are lit the ignition cannot be switched on. An automatic red warning lamp for following traffic is built into the armrest on each back door.

It must have been a momentous decision to go over to a unitary structure for the chassis frame and body, made easier, of course, by the sad reduction in the number of specialist coachbuilders; and of the survivors, the biggest (H. J. Mulliner and Park Ward) belongs to Rolls-Royce anyhow. Pressed Steel, who have been responsible for the standard steel bodies for Rolls-Royce and Bentley cars for many years past, have been entrusted with the new composite structure. Without the subframes added, it has a torsional stiffness of about 10,000-11,000 lb per degree.

Although the Silver Cloud III and Bentley S3 will remain in production for the time being in standard and long wheelbase forms to carry the specialist bodies, there seems no reason why the Shadow structure should not be adapted for this purpose. Practically all the many Italian special bodies are based on integral structures of some sort.

Inevitably the Shadow will be now somewhat dwarfed beside most of the U.S. extravaganza, but the Rolls-Royce attitude is that, since they obviously cannot make their car in two sizes—one to suit their home customers and another for Americans—the home buyer must have preference. While an extra 2in. of shoulder room and the fact that the back seat is now ahead of the wheel arches have made the Shadow a bit more spacious than the standard Cloud, it is still essentially a four-seater, the cushions being rolled and bolstered to make a third person in the back feel little more welcome than a grub in a lettuce leaf. For four, the seating is, of

Revised engine showing new and enlarged porting with sparking-plugs now accessible above exhaust manifold. One of the two camshaft-driven hydraulic plunger pumps can be seen in the drawing above

course, extremely comfortable. There are the usual folding picnic tables behind the front seats, together with ashtrays and lighters, and small vanity mirrors in the quarters between side and rear windows. Footroom is generous, the small transmission tunnel being unobtrusive, and there are loose footrests.

All adjustments of the front seats, except that of varying the included angle between cushion and backrest, are now performed electrically. Back or front of the cushion can be lifted or lowered independently. Two simple switches—one for each seat—govern every movement; they are placed above the transmission tunnel between the front seats.

The Rolls-Royce and Bentley driver will no longer sit head and shoulders above most other motorists, but it is good to find that the Rolls-Royce pedal levers still disappear through the toe board, as distinct from the pendant type. On the restyled instrument panel, still with wood veneer facings, the main dials are now directly in front of the driver instead of grouped in the centre.

There are warning lamps or telltales for low level of fuel, handbrake left on (or stop-lamp bulb failure), lack of hydraulic pressure in either system, engine oil pressure, low coolant level and ignition.

Air intake for the standard heating and ventilation system is now just in front of the windscreen, supplemented by two fresh air pick-ups ducted from the front of the car to outlets each side of the scuttle and through the facia. There are independent blowers left and right, and one can divide the flow up and down in the usual way as well as having it delivered at different temperatures. Floor-level outlets are ducted to the rear compartment, but lack the means to close them. One

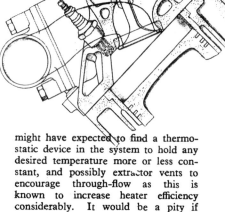

might have expected to find a thermostatic device in the system to hold any desired temperature more or less constant, and possibly extractor vents to encourage through-flow as this is known to increase heater efficiency considerably. It would be a pity if one needs to open a window to encourage this.

The layout of the refrigeration unit has been completely revised. On the former car the main components were housed in the right-hand front wing valance, but in the latest one space has been found for the evaporator behind the facia, as well as heater matrices which were also previously housed in the wing. As before, the compressor is belt-driven from the crankshaft and the condenser placed just ahead of the radiator.

Prices of these outstanding new models, including purchase tax, are increased by £924 in the case of the Rolls-Royce, £1,003 for the Bentley. The Silver Shadow is listed at £5,425 basic, £6,556 with tax, and the Bentley T-Series costs £5,375 basic, £6,496 with tax. ∎

SILVER SHADOW two-door saloon

ROLLS-ROYCE Limited announce today the introduction of a new Rolls-Royce Silver Shadow and Bentley T Series Two-Door Saloon, with coachwork by H. J. Mulliner, Park Ward. The Bentley T Series car was exhibited at the Geneva Motor Show in March.

The two new models replace the Rolls-Royce Silver Cloud III and Bentley S.3 Con-

NEW MODELS

tinental Two-Door Saloons. The modern, monocoque coachwork has graceful, dignified styling and is constructed of steel and light alloy, hand-built by craftsmen employing all the traditional skills of the coachbuilder. The new cars will appeal to those who require an individual car of particular distinction, in export and home markets.

The coachwork is designed to accommodate four persons in the utmost comfort and the interior is furnished with the finest materials available. Top grain English hide is used to upholster the seats. The two front bucket type seats are electrically adjustable for backward/forward movement, height and tilt and the backrest is reclinable. The door windows and rear quarter lights are electrically operated. Deep pile all wool carpets cover the floor and specially selected walnut veneers are used on the fascia panel and door garnish rails.

The new cars incorporate all the advanced mechanical features introduced in October 1965, including independent suspension and disc brakes on all four wheels, automatic height control, triplicated hydraulic brake system, new re-circulating ball and nut power assisted steering. It is powered by the highly successful 8-cylinder aluminium engine, but with redesigned cylinder heads for greater efficiency.

The cars with righthand steering are fitted with a 4-speed Rolls-Royce automatic gearbox, and lefthand cars with a 3-speed torque converter transmission. In both versions, the gear selection is made by an electric actuator on the gearbox, controlled by a lever mounted on the steering column which is finger light in operation. ●

SPECIFICATIONS

Engine:
Eight cylinder, 90° V unit, 6,230 c.c., of aluminium alloy. Bore 4·1 in., stroke 3·6 in. Compression ratio 9·0 : 1 (8·0 : 1 where 100 octane fuel not available). Twin S.U. H.D.8 side draught carburettors.

Final Drive:
Hypoid bevel gears. Ratio 3·08 : 1.

Tyres:
8·45 x 15 Low profile.

Heating and Ventilation:
Two separate fresh air systems, one for windscreen demisting and facia outlets, second for interior of car. Electrically demisted rear window. A Rolls-Royce refrigeration unit is available as an optional extra.

Lubrication:
Greasing required only at steering and height control ball joints every 12,000 miles. All other joints sealed for life.

Dimensions:
Track front	57·5 in.
Track rear	57·5 in.
Wheelbase	119·5 in.
Turning circle	38 ft.
Ground clearance	6·5 in.
Overall length	203·5 in.
Overall height	58·75 in.
Overall width	72·0 in.

(RECOMMENDED BRITISH RETAIL PRICE)
Rolls-Royce Silver Shadow Two-Door Saloon: by H. H. Mulliner, Park Ward: £8,150 basic, £1,699 9s. 7d. U.K. tax, total £9,849 9s. 7d. (R19,698·96).

Bentley T Series Two-Door Saloon: by H. J. Mulliner, Park Ward: £8,100 basic, £1,689 1s. 3d. U.K. tax, total £9,789 1s. 3d. (R19,578·12).

Rolls-Royce pull the surprise of the season — two new models with smart modernised lines, chassisless construction and a wealth of gadgetry. Special report by Harold Dvoretsky

ROLLS-ROYCE have gone independent, with disc brakes all-round and—monocoque construction!

Yes, it's true. After all these years the "best car in the world" has had to face up to the hard, cold economic and competitive facts of life and go the way of the rest of the hoi polloi. Well, almost.

For—boy, oh boy—when Rolls decide to change, they change in typical Rolls-Royce style—perfection personified.

The new car (it will be produced in both Rolls-Royce and Bentley versions) will be known as the Silver Shadow and the Bentley T series.

Under the bonnet of this sleek powerful-looking thoroughbred will be the V8 6.2-litre introduced in 1959, and now modified with a new head to give more horses.

How many horses, Rolls still aren't saying, though it must be up around the 300 mark. Top speed is around 117/119 m.p.h. by their own admission, "depending on the surface." By the time you've taken power off to drive ancillaries—and Rolls have ancillaries like other cars have rattles —you've got to have plenty left to push 41cwt. along at that speed.

No other performance figures have been released, but I would say they'd be sufficient to see most off at the lights.

The famous smooth Rolls-Royce automatic transmission, with its four forward speeds and complete over-riding control, has also been modified with the addition of another clutch and free-wheeling device that make it go up and down through the gears like a direct turbine drive.

But apart from those two contributions, and the luxury finish of this amazing marque, the rest is as different as model T's to T-birds.

On paper, this new product from the Crewe factory mechanically surpasses most, if not all, cars being produced anywhere in the world.

Yet the way Rolls have introduced it you'd never know the company's car division was about to embark on the biggest revolution in its history. I had to ask — nay, plead for details.

The four-foolscap-page typed specification I finally received is a masterful understatement.

Take what it says under body: "Steel monocoque construction, alloy doors, bonnet and bootlid. Boot capacity 22 cubic feet." That's all.

The bodies, incidentally, will be made by Pressed Steel (who have just been taken over by BMC, who also have a tie-up with Rolls with the Princess-R.).

BENTLEY version of new model is known as "T" series, shares body styling of the Silver Shadow.

CUTAWAY of Rolls Silver Shadow clearly shows monocoque design and fully-independent suspension.

SOME things never change... like that familiar square Rolls-Royce grille and winged figure.

What It's Got

The body is a four-five seater (though internally it is larger than the five-six seat S3 it replaces). It is luxuriously equipped with everything any self-respecting millionaire motorist would wish.

You get electric wind-up windows, heating and demisting with element-heated rear screen and individual fresh air vents for driver and front seat passenger as standard equipment —though you've got to pay extra for air-conditioning ("a Rolls - Royce refrigeration unit for full air-conditioning is available if required," murmur the specifications discreetly).

Compared with the S3, the new car is slightly longer—yet narrower. Overall length goes up almost five inches to 17ft. 11½in. Width drops from 6ft. 2¾in. to 5ft. 11in. Height drops from 5ft. 4in. to 4ft. 11¾in.

Front track drops one inch and rear track two and a half inches to a uniform 57½in. Turning circle goes down from 41ft. 8in. to 38ft., which is pretty good for a car of this size.

The old boot had 19 cubic feet of luggage space; this one holds 22 cubic feet, and the tank will take an extra gallon of tiger—or whatever you use in your tank—making 24 gallons in all.

Over the years the main worry of using monocoque for luxury bodies

(Continued on page 44)

Revolution at

The substance behind the

The people behind the new Rolls-Royce

AT Rolls-Royce they are rather against the personality cult. They practise engineering for its own sake and because they enjoy making The Best Car in the World, but they do it very much as a team. It is appropriate that Harry Grylls should be Chief Engineer. His father was one of the first people to buy a Rolls-Royce car, though he does not tell you this, any more than the fact that he went to Rugby and to Trinity College, Cambridge. He joined the company in 1930, when he was 21, and he used to go down to West Wittering while Royce was alive, a most formative experience; he became Chief Engineer in 1951. Today there is a strong physical resemblance between Grylls and the great man. The same commanding height (about six feet three), a similar stance, and a similarity too in the high domed forehead.

Baffling

One wonders whether Royce had the same ability to see the joke. Probably so, for one notices this so much about Rolls-Royce: just when a ghostly hush is expected, you discover a tongue in cheek or a throwaway gesture. Very baffling for the solemn. For the fullest illustration of this see "The History of a Dimension", a paper which Harry Grylls read to the Institution of Mechanical Engineers on becoming chairman of the Automotive Division. It is a triumphant alliance between irony and engineering.

The new Silver Shadow, described in this issue, is almost certainly the best and quite certainly the most sophisticated car in the world. It is full of ingenious devices to confound those who claim that the Silver Cloud was old-fashioned, although drivers may well find little change, apart from added surefootedness, better springing and a far wider view in all directions. It is also more practical in traffic being lower, shorter and three inches narrower. Quite something when it comes to a gap. The engine is little changed because Rolls-Royce do not believe in changing everything at once. If you remember, the Phantom I was a new engine in a Silver Ghost chassis, and the Phantom II was a new chassis for the Phantom I. Only in the V-12 Phantom III was everything new, and the development work was quite a headache.

The Silver Shadow, for those who have not yet read Charles Bulmer's description, uses roughly the same engine as the Silver Cloud III, but the chassis has independent rear suspension, a highly ingenious levelling device, a monocoque body and disc brakes, thus answering all the critics at once. The monocoque car has been running since 1958 and the disc brakes have been going now for 10 years, which is probably an answer not only to the critics but to any complacent competitors.

Design at Rolls-Royce is regarded as a philosophical problem. There is no doubt in anyone's mind what they want to achieve. They supply themselves with foreign cars to see what the Continental and U.S. opposition are doing, and adopt ideas where necessary. They never hesitate to pay a royalty although they do not have to. They avidly devour scientific papers, and they draw upon vast personal experiences. As an undergraduate Grylls had three Aston Martins—an International, a Le Mans and an Ulster. "No windscreen, no springs. I thoroughly enjoyed it; then I got married." He now uses a Rolls-Royce, tries out competitors' products, and drives the experimental cars at every stage, sometimes at Oulton Park which is quite close to Crewe and can be hired by the day for private testing. He also drives a good deal in America and has a great admiration for American cars, whose image in this country is still clouded by the memory of pre-war jellified heaps. Modern U.S. motorcars are lower built than most European sports cars and they no longer roll; what is more, they steer very well indeed, and should do even better in future, because Cornell University has just published a paper on handling problems which "explains in a few sentences what we have never known before".

The steering on the Silver Shadow will be "quicker" than that of the S.III, although only slightly so because they have the "sneeze factor" always in mind.

Incomprehensible steering

During their exploration of this question the Rolls-Royce engineers built experimental steering with zero movement—the ultimate in tillers, nudged by a mechanical linkage—and another with zero feel. They found the former quite incomprehensible at low speeds, lashing from lock to lock, but the latter proved manageable though unpleasant. They now have something they are dead sure will please customers on both sides of the Atlantic —even though the British *like* discomfort. A check at Rolls-Royce's London depot shows that 80 per cent keep their ride-control permanently on "hard", and many owners still mistrust power steering. "But", remarks Harry Grylls, "your bottom and eyes tell you what is happening on ice."

The new steering will have plenty of "feel". The aim is a car with the softest possible flat ride, which will swallow motorways at 110 m.p.h. in complete comfort, and be just as enjoyable on twisty English roads.

At one time it was pleasanter to ride in a Rolls-Royce than to drive it. Those days have gone. Everyone at Crewe, from Grylls

hadow

by D. B. Tubbs

down, seems to enjoy the act of driving, the physical sensation of handling a motorcar. They are proud of their new brakes. The old Rolls-Royce servo has gone. In its place are shrouded discs all round which are applied by two high-pressure hydraulic servos (as on the Citroen DS), plus a third hydraulic system working on half the back brakes. All this is coupled up to the brake pedal, and the brakes feel exactly like the brakes on any Rolls-Royce model since 1925. "We think we know what brake pedals should feel like, so we have built in artificial feel. This is due to the cunning way we have coupled up all the valves," says the Chief Engineer. Then comes that throwaway grin: "And at the heart of the feel is a squash ball".

The body is new in every way. Five inches lower, considerably shorter, three inches narrower outside but three inches wider inside (the Issigonis syndrome) and with a larger boot. The problem was to make a car that would look perfectly at home on both sides of the Atlantic: "Different, but right".

Aerodynamics have been taken care of; the car is quiet at speed with the windows closed. In fact they need never be opened. There is lots of ambient air, and the majority will be sold with air-conditioning, here as well as abroad. At one time cars were designed round the rear-view mirror and spare wheel.

The man and the machinery— Harry Grylls looks thoughtfully on the creation of his team at Rolls-Royce. It is something of which to be truly proud.

The substance behind the Shadow

Now they are built round the refrigerator evaporator.

Styling is done by John Blatchley, who is faithful to brown walnut although Grylls betrays a preference for sycamore, which Rolls-Royce have used in the past. But wood finish anyway, because the U.S. finds it "different" and the British still like their half-timbering. In the middle of the facia stands (on the insistence of the Chief Engineer) the time-honoured Rolls-Royce switchbox. "Nobody else can afford to put black paint on aluminium and then scratch it off to get the lettering." The dials and switches cluster symmetrically round it, and although Grylls whispered a rueful "No" when asked about ergonomics, he remarked later that it was part of the Rolls-Royce philosophy to avoid anything that infuriates the driver, like inaccessible switches, doors that you can't reach to open and pedals too close to the wheel.

Styling and crash-padding do not go together, but R-R have "padded to the limit of not ruining the styling". They are very concerned with safety. For example, should the headlamps fail for any reason a completely separate circuit is automatically energized, giving dipped beam, and if either stop-light fails the handbrake warning light comes on. This device was patented by Grylls after a circuit of London without stop-lights. "I should like to have a thing on the back to tell the man behind how fast I am stopping, because if you brake hard without warning, God has decided he will hit you." But the Construction and Use Regulations at present will not allow.

Awareness of motoring problems

Another thing Grylls feels strongly about is the barely mobile traffic jam between M1 and M6 that cuts off the North of England from the South. He has worked out a completely viable system of one-way roads to resolve this jam. One portion of it, Lichfield-Fazely-Harris Hall could be made a right-hand one-way circuit immediately, with no alterations at all except for widening a few islands. As he says, a motor manufacturer has to be aware of motoring problems if he wants to go on selling cars. Let us hope the Ministry will listen.

The best things in life are produced by versatile people. Harry Grylls is not narrowly an engineer. The best thing about going to New York, he says, is that one can visit the Whitney Museum and the Museum of Modern Art next door. He does not go much on the emptier and spikier schools of modern painting, but he admires the sculptors who explore pure form, making beautiful shapes for their own sake. He also likes architectural photography but is maddened by the lack of a rising front on 35 mm cameras. So he has designed one and is making it. It sounds very simple.

On our way to the experimental shops Grylls gave credit to his colleagues. Blatchley has already been mentioned; Frank Tarlton was in charge of design, with 30 designers working with him; Fisher is responsible for development, "and Fred Murray takes charge while I'm away. When I'm home, we share." As in F. H. Royce's day all big endurance testing is done on the Continent, for although the Company now has test rigs to ascertain the accelerated laboratory life of everything on the car, French roads and New York side-streets discover resonances that never show up on bump drums.

We reached the cars at last. Each prototype on its own mechanical hoist. Standing beneath such complexity we felt like a hick at the World's Fair. So little time, so much to see. We said as much while Grylls was having his photograph taken. A by-stander put us right. "I'm only a simple mechanic," he said "but it all looks to me like real engineering—and it's quite simple when you know your way about it." It was only then that we noticed the spanner in his hand. He was, of course, wearing a blue serge suit.

Driving the Silver Shadow

We will, of course, be road testing the new Rolls-Royce as soon as possible but a 15-minute run in Cheshire last week provided some initial impressions. Briefly, this car sets new standards of automobile behaviour.

The 10 years of development of its design features have brought about a complete revolution because while all the R-R virtues remain as before, there is now improved sensitivity of control and smoothness of ride that give the car a new character. Every control is feather light—the power steering almost *too* light for the first few miles, until the driver gets used to it—and the car is very deceptively fast. Acceleration is of the "instant" variety and the new driver can whoosh up to 90 m.p.h. and be under the impression that it is only 60. Cornering speeds can be extraordinarily high.

An eight-way, electrically-powered seat movement provides the ideal driving position for everyone, the view of the road and all around is outstanding because of the lower bonnet and boot line, and the smooth, powerful brakes complete a feeling of being 100 per cent in command of the situation. Without doubt, a world-beater.

R.A.B.C.

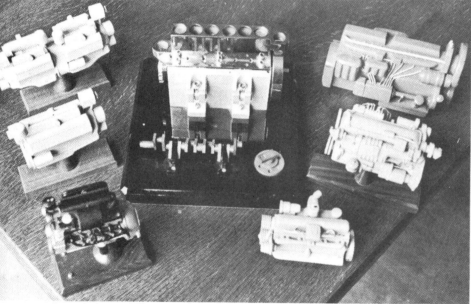

Beauty in the metal. The true engineer takes a delight in the sheer shape of a power unit or anything else mechanical. These engine shapes (above) could be classed as modern sculptures...

...and, below, more artistry, as the cars gradually take shape in the Rolls-Royce works under the hands of craftsmen.

Autocar ROAD TEST NUMBER 2125

Rolls-Royce Silver Shadow 6,230 c.c.

AT A GLANCE: An exceptionally comfortable and refined car, with soft springing and high, effortless cruising speed. Powerful and progressive brakes. Steering generally good but slow response when driving fast. Car quiet, tyres often relatively noisy. Lavishly equipped and important services duplicated. Has a finish to be proud of—inside and out.

MANUFACTURER:
Rolls-Royce Ltd., Pyms Lane, Crewe, Cheshire

PRICES:
Basic	£5,425	0s 0d
Purchase Tax	£1,244	19s 0d
Total (in G.B.)	£6,669	19s 0d

EXTRAS (inc. P.T.)
Refrigeration system	£178	4s 7d
Inertia reel seat belts—front (pair)	£19	19s 7d

PERFORMANCE SUMMARY
Mean maximum speed	115 m.p.h.
Standing start ¼-mile	17·6 sec
0-60 m.p.h.	10·9 sec
30-70 m.p.h. (through gears)	10·9 sec
Fuel consumption	13 m.p.g.
Miles per tankful	360

FOR so long accepted as the best car in the world, and probably still so, the Rolls-Royce is inevitably judged by the highest standards. This means that even minor criticism, acceptable in a lesser car, is likely to be regarded as very serious. Today such standards are no longer entirely realistic. The mechanical parts of ordinary inexpensive cars have become so much better; expensive cars can simply add top quality materials, luxurious equipment, a desirable elaboration of services, and a lot of personal and individual attention to details of design and finish. Relatively, but not actually of course, no expense is spared to please the customer; even so there is no reason to expect design miracles of a Rolls-Royce, though every reason to look for top quality in the broadest sense.

This Silver Shadow we have been testing is the latest of its kind, just nicely run in at 4,500 miles. It includes those modifications which are "part of the continuous process of development and improvement which Rolls-Royce carries out" discreetly and usually without mention.

If you look under the bonnet at the extremely full engine compartment, you will detect, from the necessary complication and the quality of detail design and finish around the aluminium vee-8 engine, where a fair proportion of the cost goes. The car was fully described in *Autocar* of 8 October 1965 but such is the extent of the equipment and the elaboration of systems and reserve systems that we have provided a page of refresher notes at the end of this test report.

First and foremost, the Silver Shadow is a superlative journey car for driver as well as for passengers. Those of us who took the car to France and Switzerland, as well as across England, were unanimous in our favourable comments on a complete absence of tiredness or tension at the end of a long day on the roads. The ride is considerably more level and comfortable than in the Silver Cloud this Shadow replaces, and the damping over humps or hollows is very good indeed. Cobbles, including the rough test sections at M.I.R.A., are traversed with the minimum of disturbance to passengers.

Suggestions that this car is less imposing or that you sit too low are no more than first impressions of traditionalists; the same kind of thing was said of previous models when they were new.

Contributing to the relaxed travel are the elaborate air conditioning system, the eight-way power adjusted front seats with additional recline control, the mechanical quietness and the excellent look-out all round. On the negative side is tyre noise, which, according to whether the car is on smooth tarmac or on coarse dressed or ribbed surfaces, varies from quiet to noisy. Passengers sitting back in the rear seats hear more from the tyres than those in front and there is additional rumble when cornering. In contrast, there is very little wind noise so you can talk normally or listen to the radio (balanced front and rear speakers) at 100 m.p.h. This Rolls-Royce is a gentle car by nature, and one wonders if it has not been developed with more than an eye on the American market where straight-line speeds seldom exceed 80 m.p.h., brisk acceleration is essential and a soft ride is demanded.

Our test car had the slightly stiffer home suspension but was still softly sprung by British standards. It became a bit flustered if hurried through a combination of bends. This is because there is quite a lot of initial roll in front as you go into, or alter line through, a bend; and of course a reversal left to right or vice versa doubles the effect. So winding roads are better taken sedately. Even so, when we hurried the car round the test road circuit at M.I.R.A. on both dry and wet occasions, the roll proved be to firmly restrained after the first lean-over, and the adhesion was outstanding. We never made the back wheels slide.

Several people have asked expect- ▶

Right: A comfortable, luxuriously fitted rear compartment for two people and with room for three. Note the red warning lamps in the backs of the arm rests. Left: Large, deeply, upholstered seats with eight-way electric adjustment and recline. Workmanship on leather and wood is of high order for a semi-production car

Rolls-Royce . . .

antly "What do you think of the steering?" and we wonder what answers they expected. It is of power-assisted, recirculating-ball type. It is sensitive and lighter than previously and has constant ratio. We think it is good except in one respect, and that is the response around the neutral position. From the straight ahead point the wheel has to be moved up to a sixth of a turn before the nose of the car follows—almost leisurely. The slower you drive the less apparent is this slow response.

One result is that gusty side winds, and uneven cambers such as we met across France, take the car from side to side before the driver can check it. A second result is that corners must be taken early and held firmly or the car will run wide. Third, we did not find it easy to place the car to an inch—to six inches, yes. The more pronounced understeer, disguised in part by the power steering, no doubt arises from the forward position of the engine and the 55 per cent front, 45 per cent rear weight distribution.

In other respects the steering is good and the car's own directional stability is excellent at all speeds, the new independent rear suspension coping admirably on rolling Continental roads. With tyre noise and steering in mind we look forward to trying a Shadow on radial covers one day.

Performance, as we have said earlier, has an American (and now home) market bias towards the 0-70 m.p.h. bracket. The acceleration from a standstill is excellent for this sizeable car. By speeding the engine up in gear and holding momentarily on the foot brake—a reasonable procedure where a maximum getaway is desired—we recorded a best standing quarter-mile time of 17.5sec on more than one occasion. The acceleration continues vigorously to about 75 m.p.h., when the automatic up-change to top occurs (and if set at 3, the manual hold is also over-ridden). Thereafter it is quite brisk up to 90. Occasionally, in France, when beginning to overtake at about 70 m.p.h., we were "left hanging" with no kickdown and insufficient punch in top (4). Left-hand drive export cars, like one we sampled in Geneva, are better placed, having an American-type three-speed transmission with torque converter, which gives about 85 m.p.h. in middle ratio, so closing the big gap of the Rolls-Royce 4-speed epicyclic box.

The gear lever is light and sweet to use now that an electric selector is incorporated. Changes are very smooth — often undetected — either when selected or automatic. The single exception is the occasional jerk of a double kick-down that can occur, for example, when turning on to, and pulling away up, a steep hill.

Maximum speed these days is little more than a talking point; in favourable conditions owners may get 120 m.p.h. from the Shadow. A representative mean top speed is 115 m.p.h., but more important is the effortless all-day cruising speed of around 100 m.p.h. at which, incidentally, fuel consumption is 10.0 m.p.g. A Continental journey figure of 13 m.p.g. is normal, or in Britain up to 15 m.p.g. The 24-gallon tank thus gives a range of over 300 miles. The combination of very brisk acceleration and high cruising speeds returned unexpectedly high journey averages even when we were not trying particularly hard. We would plan on an easy 400 miles a day across Europe and expect to have a good lunch and time for a bath before dinner.

Rolls-Royce take tremendous trouble over their brakes and are using discs all round for the first time. They are excellent, with a reassuring bite, light pressure and more than average progression. One of our drivers thought there was a fraction of delay as compared with the early drum and engine-servo system of the Silver Cloud II which we re-tested at the same time (*Autocar*, 2 March).

The best stopping figure of 0.97g was difficult to record consistently, although the figure was always over 0.9g on dry surfaces. The rebound after the initial weight transfer forward would sometimes allow one wheel to lock prematurely. The tyres would also grip much better on some smooth surfaces than others. Con-

A formidable array under the bonnet can be sorted out quickly. Fillers are high and accessible and sparking plugs are now angled up for removal above the manifolds. Design and quality of parts is excellent

A large but unobtrusive tail with wrap-round bumper provides ample luggage space without upsetting the good balance of the car's appearance

sistent wet-road stops of 0·89g were usually good.

In rain we found that the first application of brakes after some miles without use, as on motorways, gave weaker and less even response. This is a well-known characteristic of cold, wet discs and pads. If our test car is representative, Rolls-Royce will have to think again about the hand parking brake. With a man's hard pull it gave a best figure of only 0·29 and barely held the car on a 1-in-6 slope.

With the 4-speed automatic transmission there is no "Park" position for the selector, but if reverse is selected and the engine switched off, a positive parking lock engages. On one occasion the reverse cone stuck in engagement with the selector back in neutral.

Intricate details of the air conditioning system and its controls are described elsewhere. A driver quickly learns to use them, and we do not know of another car which can produce so many different currents or blasts of hot, warm or icy cold outside or recirculated air on request. The numerous flap valves have actuators which buzz when operating "so you know they are working." If we say that when the next Rolls-Royce is designed we hope it will have just as good and flexible a system but with fewer (and safer) control knobs, we are not being ungrateful for this one, which does its job very well indeed. The engine warm-up to give heating was slow. With the refrigeration system goes tinted glass all round, and there is an efficient electrically heated rear window.

The refrigerator control overrides all the heater controls. The amount of cold air delivered is adjusted with the aid of one main blower motor with four strengths and two adjustable outlets for fresh air. In addition, an air-conditioning flap which looks like a drawer in the middle of the instrument panel, can be opened to allow a very large extra flow into the car.

The eight-way front seats, controlled by a tiny horizontal "joy stick" working in the natural sense, are well shaped and luxuriously padded, with just the right amount of firmness in the cushions. Leather is used generously for seats and trim. Some people thought that the rear seat cushions should be at least an inch longer fore-and-aft; otherwise they, too, are very comfortable. Small, separate, wedge-shaped footrests are provided. There are folding central armrests front and back, also ashtrays, lighters, the usual two folding tables, mirrors, a reading and a general lamp and a map pocket for each rear passenger.

A four headlamp system is used which, considering the relatively small diameter of the individual lamps, provides an unusually penetrating full beam. The dipped position is well cut off and gives an adequate spread of light. No separate auxiliary lamps are provided.

Among lighting details in the exceptionally complete system may be noted the parking lamp selector switch —left or right. This is in circuit with the starter, which cannot be energised if either parking lamp is on. If the main lighting switch is pulled out, though everything else is locked off, it lights a small lamp above the ignition keyhole.

There are four warning lamps in a group: reds for the two brake accumulators; green to warn that only three gallons of fuel remain; and amber to show both that the handbrake is on or that a stop lamp bulb has failed. To test these warning lamps, switch on the ignition and press the oil level button. In addition to indicating oil level on the petrol gauge, the four warning lamps should all show.

A switch above the rear view mirror is for hazard warning and it has its own tell-tale. This causes all four front and rear signal lamps to flash automatically—as allowed by new safety regulations. Door-open red warning lamps are provided only for the rear doors and they are situated in the backs of the armrests. They are switched on also by the front doors. In a tray under the facia is a businesslike fuse box with all its 20 circuits labelled on a separate pull-out panel.

In the carpeted boot there is an auxiliary or charging socket to save baring the battery terminals behind their covers. Here also is a fitted tool-

A trimmed and carpeted boot of generous capacity. The carefully packed tools are laid out for inspection and the battery uncovered. Just visible behind it is the accessory socket

kit and, in the floor under the carpet, a small cover which lifts to expose the spare wheel valve for checking pressure. Inertia reel harness is provided in front, and all doors have safety locks. It is a pleasure to insert and turn the key and feel the smooth action of the Rolls-Royce Yale locks after the rubbishy stuff on some other cars. Perhaps a master lock for the doors and boot lid will be introduced one day; we have found it a great convenience on certain other cars in this class.

R-R and Bentley owners of a few years ago asked for a smaller, livelier car but with more room inside; this Shadow is 6·75in. shorter (16ft 11·5in.) and 3·75in. narrower (5ft 11in.), yet it has greater internal and boot space than the Silver Cloud III. It is also 4·5in. lower but it weighs fractionally less. It feels a smaller and more stable car to handle in traffic or on the open road, and you can see out better all round. Having lived with the Rolls-Royce Silver Shadow for several weeks, we have no hesitation in describing it as a worthy successor to the Clouds and Phantoms. Nor have we any doubts that fortunate owners will enjoy tremendous satisfaction and service from their Shadows. ■

SPECIFICATION : ROLLS-ROYCE SILVER SHADOW (FRONT ENGINE, REAR-WHEEL DRIVE)

ENGINE
Cylinders .. .. 8, in vee
Cooling system .. Water; pump, fan and two 90 deg thermostats
Bore 104mm (4.10in.)
Stroke 91.4mm (3.60in.)
Displacement .. 6,230 c.c. (380.5 cu. in.)
Valve gear .. .. Overhead; hydraulic tappets, pushrods and rockers
Compression ratio 9.0-to-1
Optional 8.0-to-1
Carburettors .. 2 SU HD8
Fuel pump .. 2 SU electric
Oil filter .. .. Full flow, renewable element
Max. power .. Not disclosed by manufacturer
Max. torque .. Not disclosed by manufacturer

TRANSMISSION
Gearbox .. .. Rolls-Royce automatic epicyclic with fluid flywheel
Gear ratios .. Top 1.0; Third 1.45; Second 2.63; First 3.82; Reverse 4.30
Final drive .. Hypoid bevel, 3.08-to-1

CHASSIS AND BODY
Construction .. Integral, with steel body, aluminium alloy bonnet, boot lid and doors; sub-frames for running gear

SUSPENSION
Front .. .. Independent, coil springs, wishbones, telescopic dampers, anti-roll bar. Automatic hydraulic height control
Rear .. .. Independent, coil springs, single trailing arms, telescopic dampers. Automatic hydraulic height control

STEERING
Type .. .. Recirculating ball, with Saginaw power assistance
Wheel dia. .. 17in.

BRAKES
Make and type .. Rolls-Royce Girling, front discs each with two single piston calipers, rear discs each with one dual piston caliper, three independent foot brake systems
Servo .. .. Power assistance from two engine-driven hydraulic pumps
Dimensions .. F, 11.0in. dia. R, 11.0in. dia.
Swept area .. F, 227 sq. in. R, 286 sq. in.
Total .. .. 513 sq. in. (228 sq. in.) per ton laden

WHEELS
Type .. .. Pressed steel disc, 5-stud fixing 6.0in. wide rim
Tyres: make .. Dunlop, Avon or Firestone (Dunlop RS5 on test car)
   —type .. Low-profile cross-ply tubeless
   —size .. 8.45—15in.

EQUIPMENT
Battery .. .. 12-volt 64 amp. hr.
Alternator .. Lucas 11AC 45-amp
Headlamps .. Lucas four headlamp system 100/37.5 watt
Reversing lamp .. Standard
Electric fuses .. 20

Screen wipers .. 2-speed self-parking
Screen washer .. Standard, Lucas electric
Interior heater .. Standard, air-blending
Safety belts .. Extra
Interior trim .. Selected English hide seats, pvc headlining
Floor covering .. Wilton carpet
Starting handle .. No provision
Jack .. .. Screw pillar
Jacking points .. 2, one each side under sills
Windscreen .. Laminated
Underbody protection .. Bitumastic on all surfaces exposed to road
Other bodies .. H. J. Mulliner, Park Ward and James Young 2-door saloons

MAINTENANCE
Fuel tank .. 24 Imp gallons (107 litres) (low level warning lamp)
Cooling system .. 28 pints (including heater) (16 litres)
Engine sump .. 14.5 pints (8 litres) SAE 10W/30. Change oil every 6,000 miles; change filter element every 6,000 miles
Gearbox .. 24 pints ATF. Change oil every 12,000 miles
Final drive .. 4.5 pints SAE 90EP. Change oil every 24,000 miles
Grease .. 6 points every 12,000 miles
Tyre pressures .. F, 26; R, 26 p.s.i. (normal driving)

PERFORMANCE DATA
Top gear m.p.h. per 1,000 r.p.m.   26.2
Mean piston speed at mean max speed 2,630 ft per min
B.h.p. per ton laden.   Not disclosed

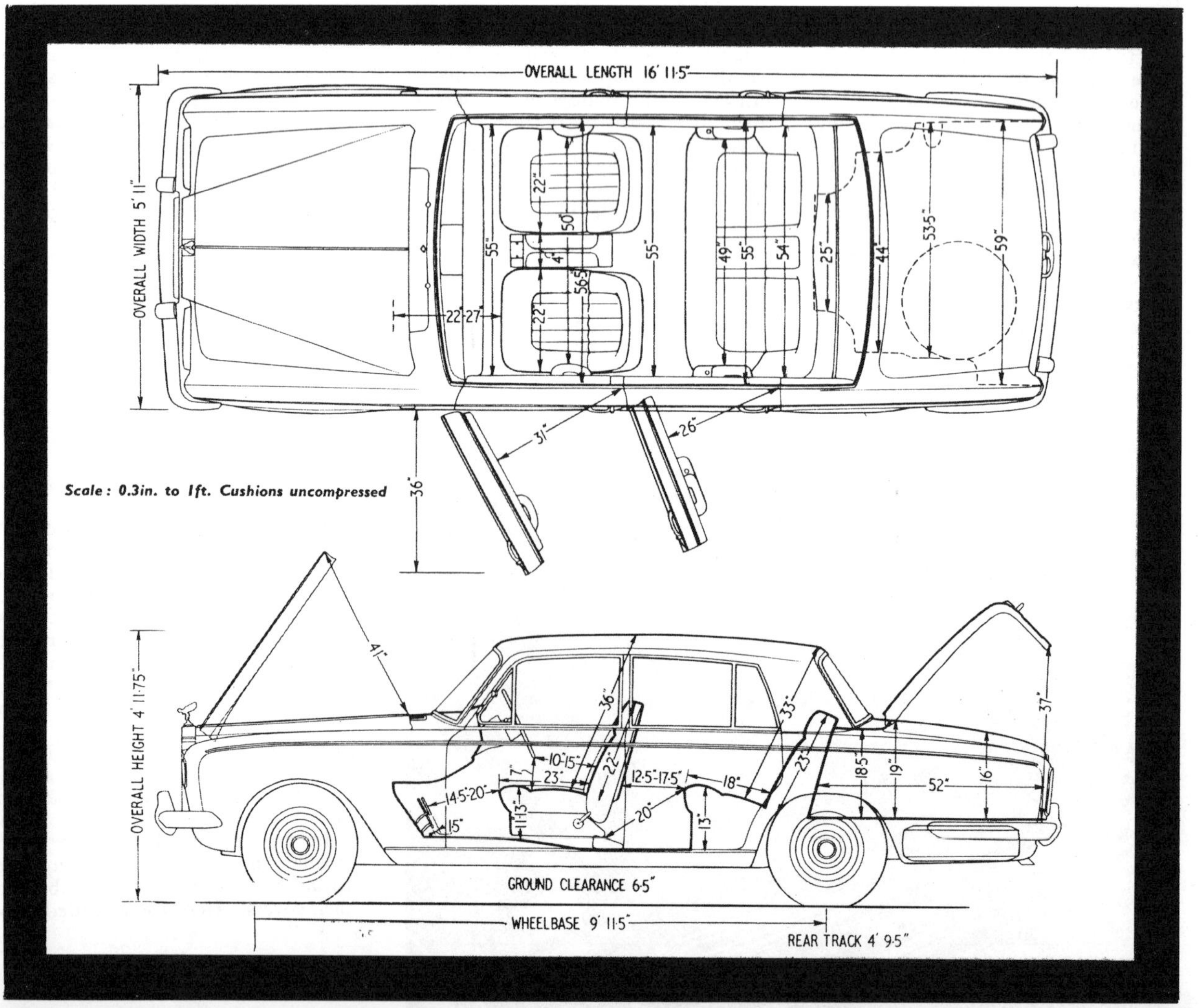

OVERALL LENGTH 16' 11·5"
OVERALL WIDTH 5' 11"
OVERALL HEIGHT 4' 11·75"
GROUND CLEARANCE 6·5"
WHEELBASE 9' 11·5"
REAR TRACK 4' 9·5"

Scale : 0.3in. to 1ft. Cushions uncompressed

Autocar road test number 2125
Make: ROLLS-ROYCE SILVER SHADOW 6,230 c.c.

TEST CONDITIONS
Weather: Cloudy, dry. Wind: 15-20 m.p.h.
Temperature: 18 deg. C. (64 deg. F.)
Barometer: 29·1in. Hg.
Humidity: 46 per cent
Surfaces: Dry concrete and asphalt

WEIGHT
Kerb weight 41·6 cwt (4,660 lb-2,067 kg) (with oil, water and half-full fuel tank)
Distribution, per cent: F, 54·6; R, 45·4
Laden as tested: 45·1 cwt (5,048 lb-2,243kg)

Figures taken at 4,100 miles by our own staff at the Motor Industry Research Association proving ground at Nuneaton.

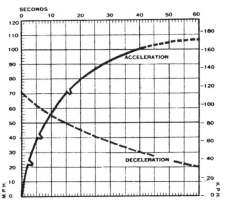

MAXIMUM SPEEDS
Gear	m.p.h.	k.p.h.	r.p.m.
Top (mean)	115	185	4,390
(best)	118	190	4,500
3rd	72	116	4,000*
2nd	43	69	4,350
1st	24	39	3,500*

*Maximum engine speed limited by automatic transmission in these gears

Standing ¼-mile 17·6 sec 76 m.p.h.
Standing Kilometre 33·0 sec 96 m.p.h.

TIME IN SECONDS	3·4	5·3	7·8	10·9	14·3	19·7	26·9	37·8
TRUE SPEED M.P.H.	30	40	50	60	70	80	90	100	110
INDICATED SPEED	31	41	52	62	72	82	92	102	112

Mileage recorder 1·2 per cent over-reading. Test distance 2,285 miles.

Speed range, gear ratios and time in seconds

m.p.h.	Top (3·08)	3rd (4·46)	2nd (8·10)	1st (11·75)
10— 30	—	—	2·9	—
20— 40	—	4·8	3·3	—
30— 50	7·2	5·1	—	—
40— 60	7·5	5·7	—	—
50— 70	8·5	6·8	—	—
60— 80	10·5	—	—	—
70— 90	13·1	—	—	—
80—100	15·8	—	—	—

FUEL CONSUMPTION

(At constant speeds—m.p.g.)
30 m.p.h.	19·6
40	17·9
50	16·5
60	15·2
70	14·4
80	13·1
90	11·6
100	10·0

Typical m.p.g. Fast Continental touring 11 (25·7 litres/100km)
Typical m.p.g. Gentle touring 15 (18·8 litres/100km)
Calculated (DIN) m.p.g. 13·1 (21·6 litres/100km)
Overall m.p.g. 12·2 (23·2 litres/100km)
Grade of fuel, Premium (96·8-98·8 RM)

OIL CONSUMPTION
Miles per pint (SAE 10W/30) .. 800

BRAKES (from 30 m.p.h. in neutral)
Load	g	Distance
25lb	0·43	70 ft
50 "	0·75	40 "
75 "	0·88	34·2 "
90 "	0·97	31·0 "

Handbrake 0·29 104 "
Max. Gradient 1 in 6

TURNING CIRCLES
Between kerbs, L, 37ft 8in.; R, 39ft 0in.
Between walls L, 40ft 0in.; R, 41ft 2in.
Steering wheel turns, lock to lock, 4·25

HOW THE CAR COMPARES.

MAXIMUM SPEED (mean) M.P.H.
Rolls-Royce Silver Shadow
Buick Riviera
Cadillac Fleetwood
Daimler Majestic Major
Mercedes 600 SE

0-60 M.P.H. (sec)
Rolls-Royce Silver Shadow
Buick Riviera
Cadillac Fleetwood
Daimler Majestic Major
Mercedes 600 SE

STANDING START ¼-MILE (sec)
Rolls-Royce Silver Shadow
Buick Riviera
Cadillac Fleetwood
Daimler Majestic Major
Mercedes 600 SE

M.P.G. OVERALL
Rolls-Royce Silver Shadow
Buick Riviera
Cadillac Fleetwood
Daimler Majestic Major
Mercedes 600 SE

PRICES
Rolls-Royce Silver Shadow	£6,670
Buick Riviera	£4,402
Cadillac Fleetwood	£5,164
Daimler Majestic Major	£2,749
Mercedes 600SE	£8,968

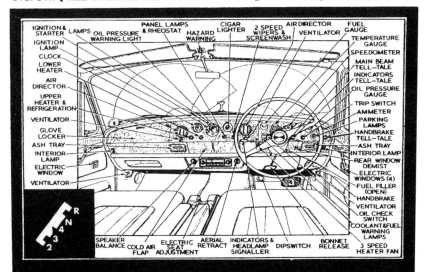

NOTES ON THE ROAD TEST

Substance of the Shadow

IT IS NOW 18 months since we described the Silver Shadow (18 October 1965) so not everyone will remember the details of this intricate and beautifully engineered car and its many mechanisms. Some inside knowledge of the first all-independent, integrally built car made by Rolls-Royce is needed for a full appreciation of its qualities. These notes are therefore added in support of the first full road test on the preceeding pages.

Engine. Aluminium 90 deg vee-8 6¼-litre introduced in 1959 and later given different combustion chamber shape and porting for the Silver Shadow, which slightly increased the (undisclosed) power output. Sparking plugs moved to above exhaust manifold (more accessible). Two hydraulic plunger pumps added (for power brakes and suspension height control) driven by extra lobes on camshaft. Hydraulic tappets. Triple silencer stainless exhaust system.

Transmission. Right-hand drive model: Four-speed automatic epicyclic gearbox with fluid flywheel. Gear range selected by steering column switch controlling electric actuator on gearbox. Ranges are:—

"R"=reverse with engine running; reverse and parking pawl engaged with engine stopped.

"N"=neutral.

"4"=fully automatic driving using 1st, 2nd, 3rd and 4th—engine braking available while 3rd and 4th are engaged —free-wheel operates in 1st and 2nd (as usual with other automatics).

"3"=1st, 2nd and 3rd for medium speed traffic—"safety change" to 4th to prevent over-revving. Engine braking as in "4".

"2"=transmission locked in 2nd regardless of engine r.p.m. except will drop to 1st if speed and load demand— no "safety change"—full engine braking.

Kick-down provided.

Emergency manual operation available. Small lever fits into socket under carpet.

One piece open propeller shaft with Detroit type ball and trunnion, constant velocity joint at gearbox end and Hooke type joint at final drive end. Half shafts have Detroit type joints inboard and Hooke joints outboard.

Construction. Steel body has aluminium alloy doors, bonnet and boot lid, mounted on front and rear running-gear sub-frames with stainless steel mesh resilient supports (Delanay Gallay "Vibra-

shock" type). Rear sub-frame compliance movements restricted by small double-acting telescopic damper.

Suspension. Basically conventional double wishbone layout at front with coil springs, telescopic dampers and anti-roll bar. Upper wishbone is fabricated pressing; lower wishbone is forged and has tilted trunnion axis providing about 68 per cent anti-drive under heavy braking. Rear suspension consists of coil-sprung single fabricated trailing arms with telescopic dampers. Trailing arms are angled to give slight negative camber of tyre during hard cornering.

Automatic Height Control System. The upper end of each suspension spring is supported by a hydraulic ram with a 3in. stroke. Each rear trailing arm has its own hydraulic height control valve linked to it and gauging its deflection. A single similar valve mounted on the anti-roll bar senses front end suspension deflection. As the car's load and therefore ride height alters, these three valves admit hydraulic oil at 1,150 to 2,500 p.s.i. to the rams and restore the ride height and attitude to normal. The front control valve also restricts roll by reducing the cross-flow of oil from one ram to the other.

While the gear selector is "in gear" the rate of height correction is slow (about 1-2 min per ½in. at front and ½-1 min per ½in. behind); on the move, only the slowly emptying fuel tank has to be compensated. At rest, with the selector in neutral or a door open, a solenoid valve in circuit with the interior lights and the selector permits a much quicker rate of height control (about ½in. per sec.). Thus loading an unloading changes are rapidly allowed for.

The levelling system is powered by the rear hydraulic pump on the engine.

Power Braking System. Front discs each have two separate calipers; each back disc has one four-cylinder caliper. The front hydraulic pump on the engine feeds pressure to one pair of front calipers and one paid of rear caliper cylinders in the ratio of braking 31/16 (a total of 47 per cent of total braking). The other pump (the one which also "drives" the levelling suspension system) provides a further 31 per cent of the total braking, feeding the other pair of front calipers. The remaining 22 per cent is provided by an ordinary unassisted master cylinder on the brake pedal connected to the other rear caliper cylinder pair; this gives the driver "feel." The Rolls-Royce Silver Shadow is thus unique in having three separate braking systems in addition to its handbrake.

A "ball-and-ramp" type of pressure limiting valve in the master cylinder circuit reduces the chances of rear wheel locking. Two spherical accumulators inflated to 1,000 p.s.i. with nitrogen are pressurized to 2,500 p.s.i by each pump, supplying a reserve of pressure and damping any sudden fluctuation in demand. A leak in either of the high pressure systems is shown by its own dashboard lamp. Main hydraulic reservoirs fitted with visual level glasses.

Another warning lamp tells when brake pads are getting thin. The discs are wound round their edges with stainless steel wire to absorb high frequency squeal.

Power Steering. Recirculatory ball type (Saginaw) with hydraulic damper to reduce road shock. A Hobourn-Eaton pump belt-driven off the crankshaft supplies pressure to a torsion-bar operated hydraulic valve concentric with the steering column in the steering box. The torsion bar varies the opening of the valve according to steering wheel effort; strong power assistance via an integral ram is given when manœuvring slowly but far less when running fast with the steering at or near straight-ahead.

Air Conditioning System. Consists of one pure ram air system and a scuttle inlet-fed air-blending heater and ventilator supplying "Upper" and "Lower" systems. "Upper" system supplies heated or fresh air to windscreen, "Lower" system supplying the same to feet and waist-level vents. The optional recirculating refrigeration works via the "Upper" system only and its controls override all other systems when switched to "cold air." Each system can be controlled individually. Trunking under the floor feeds the rear compartment from the "Lower" system.

The ram air system which, if opened, blows out of the cockpit vents is fed by inlets in the front of each wing. Two blower motors boost heater and ventilator air flow (four degrees of "blow" being available) or act as recirculating pumps for the refrigerated air, drawing it from the interior through extra outlets over the front occupants' knees.

To cool this air, refrigerant vapour is compressed and pumped through a condenser (mounted in front of the radiator), liquified and then evaporated in a scuttle-mounted heat exchanger The refrigerant pump is belt-driven (with the alternator) off the crankshaft.

Electric. Models ordered without refrigeration are normally supplied with a Lucas 35-amp dynamo. With refrigeration, a Lucas 11AC 45-amp alternator is standard. Printed circuits and multi-pin connectors are used wherever practicable.

Each passenger door carries its own electric window-raising switch; the driver's door has switches controlling all windows. Front seats are moved fore-and-aft and raised or lowered electrically by movement of ingenious 8-position switches between each seat. The aerial of the various standard radios available (according to what part of the world the car is to be used in) is electrically erected. There are front and rear speakers. The rear window has in it an electric demisting element. The fuel filler door is electrically released.

There are warning lamps or tell-tales for low level of fuel and coolant, handbrake left on (or stop-lamp bulb failure), lack of hydraulic pressure, and worn brake pads (as already mentioned), oil pressure, oil (and fuel) level, ignition.

THE ROLLS-ROYCE SILVER SHADOW

Road Test by
JOHN BOLSTER

*Who has not felt with rapture-smitten frame
The power of grace, the magic of a name?*

THE Rolls-Royce car has always been like no other. Accepted even before I was born as the best car in the world, it has never lost its pre-eminent position. Delaunay-Belleville, Hispano-Suiza and many others have challenged the great British make, *mais ou sont les neiges d'antan?*

When the new Silver Shadow was announced, the late and great Pininfarina was ecstatic about it. He pointed out that the very low radiator shell of the original Silver Ghost had classical proportions which had been lost in the later models. In returning to a low radiator shell, Rolls-Royce had once again built a beautiful car.

There are other parallels. The early Silver Ghost was not a very large car by the standards of its time. It was not particularly expensive and it was not among the most powerful machines. Though luxurious, it was small enough to be handy, and by superb chassis design it had phenomenal roadholding and was consequently very light on tyres. Anybody who drove it at once realized that it was more controllable than any other car and that it therefore made a better average than monsters with a higher ultimate maximum speed. Its silence also rendered it a most untiring car to drive.

The virtues of the Silver Shadow are very similar, when translated into a modern idiom. After the early Ghosts, the Rolls-Royce became a big car, more suitable for a chauffeur to drive. My Phantom II was splendid for a trip up to Scotland but absurd for running an errand to the village. Now, we have a real owner-driver's Rolls again, which above all is *fun* to drive. It is substantially lower, narrower, and shorter than the Silver Cloud but, as is apt to be the way with modern designs, it can hold more people and luggage.

The Shadow is the first Rolls with a combined chassis and body, the first machine from this factory to have independent rear suspension, and the first to have air-conditioning built in as part of the original structure. It is also the first model with disc brakes and at last, after more than 40 years, Sir Henry Royce's excellent gearbox-driven servo has been replaced by an engine-driven pump.

Sir Henry never feared complexity and so the Shadow would gladden his heart. A brief description must leave out many features, but the basis of the car is a steel body with light alloy bonnet, doors, and boot lid. The suspension is mounted on subframes, front and rear, which are insulated from the body by a stainless steel wire mesh system. In front, there are wishbones with an anti-roll torsion bar, the rear suspension being by trailing arms.

There are helical springs all round, each of which has a height control consisting of a hydraulic jack, that operates at high pressure from an engine-driven pump. These are controlled by valves which are connected to the suspension so as to sense the deflection. Very rapid levelling is given when the car is stationary but the effect is not in fact perceptible. When a gear is engaged the process is slowed right down so that the car cannot alter its attitude in response to temporary road conditions, but it can differentiate between a full and an empty tank. Another engine-driven pump looks after the power-assisted steering, this one having a belt drive while the other two are direct driven.

The brake discs are wire-wound to prevent noise. They each have multi-cylinder pad actuation and there are actually three separate braking systems (excluding the hand brake and the gearbox parking lock). Failure is therefore impossible. One system works from the main engine-driven brake pump and another takes its pressure from the suspension-levelling pump, both with spherical pressure accumulators containing nitrogen as reserves, and both having warning lamps on the instrument panel to indicate a drop in pressure. The cluster of warning lamps has a press button so that all the bulbs may be tested frequently. Finally, there is a conventional, non-servo circuit from the master cylinder, which allows the driver's foot to supply 22 per cent of the braking effort to avoid the "dead" sensation of 100 per cent power braking.

The existing aluminium 6230 cc V8 engine has been redesigned to incorporate the new hydraulic pumps and the sparking plugs have been repositioned for accessibility, which has, incidentally, resulted in a slight increase of the undisclosed power output. There is a fluid coupling (not torque-converter) to the 4-speed automatic gearbox and gear selection is performed electrically. The subframe-mounted hypoid unit has a ratio of 3.08 to 1.

Worthy of a full-length article of its own, the air-conditioning has two separately controlled circuits. The lower system supplies hot air to the feet and at waist level while the upper system looks after the windscreen and the breathing air. One can have cool air for the face and hot air round the feet—a favourite setting—or there are extra cold air inlets for hot-footed mortals. By turning the upper control to its cold end, the optional refrigeration is brought into action, all the complicated operations involved being carried out by little electric motors. There are also two blower motors and a refrigerator pump, which is belt-driven from the engine.

The interior is as beautifully furnished as one would expect. Naturally, the instruments are separate and round, while the front seats move backwards and forwards, upwards and downwards, or in a combination of any of these movements, all controlled by a little switch on each seat. Again, hidden electric motors do the work and normal reclining squabs give additional variety to the positions which may be adopted.

At first, I thought that the steering wheel was too close to the driver, but I eventually found a rather reclining attitude that gave me enough room to stretch my arms. The steering is so light that there is no need to hang on to the wheel, and after realizing this I achieved the most comfortable seating position that any car has yet provided for me.

The gear selector moves with the greatest ease. As the car gathers speed, one realizes that the engine develops massive torque at low and medium revolutions, and even a half-open throttle can leave the rest of the traffic lights queue standing open-mouthed. The actual change is much smoother than on the Silver Cloud, it being virtually impossible to detect many of the movements from gear to gear. It is only at around 20 mph that a slight jerk, usually on a downchange, may be experienced. The engine has so much torque that it is rarely desirable to use the kick-down.

Such a lively power unit does much to alleviate the frustrations of poor, pathetic 70 mph England. To get the full flavour, however, I pointed that shapely nose at Dover and Townsends did the rest. After lunching on the good ship *Free Enterprise*, I was soon entering Belgium at Zeebrugge, with the whole of the Continent at my beck and call.

Let us deal with performance first. This big, luxurious car accelerates very quickly up to 90 mph, giving absolutely no sign that it is hurrying. It soon reaches 100 mph, which is a convenient and silent cruising speed, and 110 mph is attained on any reasonable straight. Thereafter, the acceleration becomes gradual, and the maximum is just short of 120 mph, although that speed may be surpassed when conditions are favourable. The occupants of the car are never conscious of the machinery and the level of wind noise is very low.

The character of the engine is such that little is to be gained by manual selection of the gears, and an almost lazy driving style may achieve outstanding journey times. It is in the domain of roadholding that the new Rolls-Royce excels and most of the credit for this must go to the independent rear suspension. The angle of the trailing arms has been chosen to give a slight negative camber angle during hard cornering which, in conjunction with the big tyres, seems exactly right. There are still some very bad roads on the Continent, with sudden dips, violent changes of camber and all those unexpected hazards that trick most suspension systems. The roadholding of the Shadow under such conditions is incredibly good, and 100 mph cruising can be safely maintained when lesser cars are wagging their tails at 60 mph.

The cornering power is phenomenally high for a big car. There is some roll, but it is checked before it becomes excessive and none of my passengers criticized this movement. An understeering characteristic ensures stability on the straight and blustering side winds on the *autoroutes* caused no deviation. The behaviour during very fast cornering approaches nearer to neutrality, but the rear end never breaks away, the controllability on wet roads remaining excellent, with impressive traction away from the traffic lights.

Car Tested: Rolls-Royce Silver Shadow 4-door saloon, price £6669 19s 0d; extra—refrigeration system, £178 4s 7d, all including PT.

Engine: Eight cylinders, 104 mm x 91.4 mm (6230 cc). Pushrod-operated overhead valves. Compression ratio 9 to 1. Twin SU carburetters.

Transmission: Fluid flywheel and 4-speed automatic gearbox. Hypoid bevel final drive, ratio 3.08 to 1.

Chassis: Combined steel body and chassis. Independent front suspension by wishbones, helical springs, and anti-roll torsion bar, with hydraulic automatic height control. Recirculating ball steering with power assistance. Independent rear suspension by trailing arms, helical springs, and hydraulic automatic height control. Telescopic dampers all round. Disc brakes all round with two separate hydraulic servos and direct connection from master cylinder. Bolt-on disc wheels, fitted with 8.45-15 ins tyres.

Equipment: 12-volt lighting and starting with alternator. Speedometer. Ammeter. Combined fuel and oil level gauge. Oil pressure and coolant temperature gauges. Clock. Heating, demisting, ventilation, and refrigeration system. Rear window demister. Electric window controls. Electric seat adjustments. Windscreen wipers and washers. Flashing direction indicators with hazard warning. Radio with front and rear speakers and electrically raised aerial. Cigar lighter.

Dimensions: Wheelbase, 9 ft 11½ ins; track, 4 ft 9½ ins; overall length, 16 ft 11½ ins; width, 5 ft 11 ins; weight, 2 tons 1 cwt 2 qrs.

Performance: Maximum speed, 118 mph. Standing quarter-mile, 17.4 secs. Acceleration: 0-30 mph, 3.5 secs; 0-50 mph, 7.6 secs; 0-60 mph, 10.8 secs; 0-80 mph, 19 secs; 0-100 mph, 38.5 secs.

Fuel Consumption: 10 to 15 mpg.

SPECIFICATION AND PERFORMANCE DATA

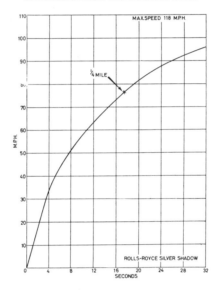

As is often the case with understeering cars, one tends initially to oversteer the wheel. This is a well-known effect, and even racing drivers are sometimes confused when trying to set up their cars. After a few miles, the driver does less work and moves the very light steering more slowly. He then finds that the car has a delightfully sensitive control at all speeds and that it is particularly easy to park, the remarkably small turning circle being a much-appreciated feature.

Though the suspension is unusually soft, there are no sick-making movements and pitching is completely absent. The engine can hardly be heard inside the car but certain road surfaces produce a good deal of tyre noise. Curiously enough, Belgian *pavé* does not have this effect, but certain comparatively smooth British roads can make the tyres protest audibly, although fast cornering does not aggravate the disturbance. The very large tyre sections which are now used on big cars do tend to pose a noise problem, but I feel that this is something which will soon be overcome, probably by collaboration between the tyre experts and the car designers. One recalls that a special silent tyre with an unusual tread design was once available for Rolls-Royce cars.

With all their elaborate design and their many built-in safety features, the brakes should certainly be good. This they definitely are, and they combine freedom from fading with great power and absolute consistency of operation. Fierce emergency braking does not cause the wheels to lock and, although the pedal is quite light in operation, it does not have that excessive lightness which can be dangerous. After the car has spent a cold night in the open, one sharp application in order to warm and dry the discs will ensure that they are ready for any emergency.

There is a parking lock, which is operated by engaging reverse gear with the engine switched off. It cannot be applied accidentally when the ignition is on. There is also a pull-out hand brake of the usual type but it has no great potency. The warning light, which tells one not to drive off with the hand brake on, also gives warning if one of the brake lights fails— a very good idea, as so many cars on the road are driven with one bulb only functioning. A switch above the screen causes all the direction indicators to flash at once in case of emergency, while parking lights on one side only may be employed, the engine then very properly being unstartable until the switch has been returned to the normal lighting position. The air-conditioning system has simple and logical controls which are easy to master.

The Rolls-Royce Silver Shadow is a car which, by great complexity and elaboration of design, makes driving easier and safer than ever before. It is a wonderfully untiring car for long journeys, and this applies to all the occupants, but above all, as I have said before, it is *fun* to drive. Having studied all its technical details and admired the superb workmanship, I am at a loss to understand how it can be made for its admittedly high price. No car offers greater value for money.

LONG WEEKEND WITH A ROLLS-ROYCE SILVER SHADOW

NOT wishing to postpone any longer the opportunity of comparing the current Rolls-Royce motor car with the earlier Silver Clouds, and finding ourselves confronted with tasks which would entail a long weekend of motoring, we borrowed from Rolls-Royce Ltd. one of their Silver Shadow demonstration cars, which had over 30,000 miles to its credit, and set off to enjoy again the unique fascination and mystique of motoring in what the majority of people willingly accept as the World's best car. It so happened that the journey commenced from Barlby Grove in London, from where, many years ago, another classic car was engineered by Georges Roesch. Soon we were proceeding along the M4 with the Frigidaire refrigerator maintaining a pleasantly cool temperature within the car. Luggage had been stowed in the boot, the lid of which is contrived for easy loading even of cabin-trunks, the comfort of the front seats, electrically-adjustable in eight ways, and upholstered in soft hide, proving acceptable at the commencement of a long journey. The Silver Shadow has been compared unfavourably with the Silver Clouds, in some quarters, as being a less imposing-looking car in which the occupants sit as they do in ordinary common all garden saloons, in full view of the populace, lacking the privacy which the high back-seat squabs and different body lines of the earlier models provided. Let me say at once that, from within, the dignity of Rolls-Royce motoring has been very little diminished in a Silver Shadow, which is actually more spacious in spite of having more compact overall dimensions than the model it replaces. The beautifully-veneered facia with its complete and unobstructive instrumentation, the thin-rimmed steering wheel, the unchanged style and layout of the manual controls, and the sheer quality and dignity of the interior appointments are just as impressive and satisfying as those in any previous Rolls-Royce model. Ahead of one the "Spirit of Ecstasy" mascot continues to ride at the head of a convincingly long bonnet, but this Silver Lady dips and curtseys far less than she once did when the pace of a Silver Cloud had to be checked. Indeed, the level ride of the Silver Shadow is a masterpiece of suspension engineering and, as we were soon to discover, the somewhat stiffened suspension of the latest cars has obviated excessive roll, so that one enjoys both an exceedingly comfortable pitch-free ride and very effective rapid cornering.

The M4 provided a rapid exit from the Metropolis, but the town of Reading represented a fearful bottle-neck, which it must have taken us some twenty minutes to negotiate, on this sunny Friday morning. At the fork where a policeman once controlled traffic from a sort of covered throne there is now a roundabout but this does not help appreciably to speed the flow of traffic. Incidentally, a public house at this junction is called "Jack of Both Sides," presumably because there are entrances to it from both roads. In contrast to congested Reading, Newbury has an effective ring-road, and soon we were on the Bath Road and making very good progress. The swans were out at Hungerford, close to which an old Austin Seven Opal two-seater was encountered. The old toll-house on this road, we noticed, was up for sale, and the one-time R.A.C. camp near Calne is no more. We were bound for Monmouth, and picking up the M4 again, crossed by the new Severn bridge, for a fee of 2s. 6d., where the Army Apprentices College is to be seen on the right-hand side; 17 miles after leaving the Motorway, we arrived in Monmouth, approximately three hours since leaving London.

While in Monmouth we took the opportunity of photographing the Silver Shadow beneath the statue erected by the townspeople in memory of Charles Stuart Rolls, M.A., F.R.C.S., A.M.I.Mech.E., who was killed in a flying accident at Bournemouth in July 1910. We also took the opportunity of visiting the impressive house, the Hendre, where the parents of the Hon. C. S. Rolls, Lord and Lady Llangattock, lived and where Rolls conducted some of his early motoring and ballooning experiments. It was pleasing to find that this impressive old house is in a good state of preservation, having recently ceased to be used as a school. It is still owned by the Rolls family, and Col. Harding-Rolls is himself an enthusiastic owner of a Silver Shadow.

After conducting some business in Monmouth we turned the Silver Shadow in the direction of the former R.-R. birthplace of Derby. This entailed a long run through the drabness of the industrial Midlands. It is under such conditions that the soothing effect of Silver Shadow motoring is appreciated. It is not an absolutely silent car, faint wind noise and the more pronounced sound of the tyres intruding, while the engine hums to quite an extent when approaching maximum speed in the third gear "hold" position of the automatic transmission. But mechanically this is an exceptionally quiet car and the other noises are noises only in the Rolls-Royce context, and would be far more pronounced in lesser cars. It is, for instance, possible to converse in perfectly normal voices at speeds of 100 m.p.h. and more, and the feeling of opulence, though it may break a commandment, is something which few humans can willingly forgo. The sense of isolation from the cares and troubles of the World without is something which I remarked on when writing of the Silver Cloud III some years ago. I experienced this in full measure in this more complex, highly sophisticated Rolls-Royce Silver Shadow.

Driving such a car at the 70 m.p.h. speed limit on motorways was akin to putting skids under the wheels of a horse-drawn dray, with the exception, of course, that the Silver Shadow is as quiet at 70 as it is when idling through traffic or devouring Continental roads at 110 m.p.h. and more. Driving is simplicity itself. The automatic transmission has a right-hand selector lever which can be set either in normal-drive position, flicked down into "hold"-third, in which case an indicated maximum of 75 m.p.h. is available before a change into top gear, or put it into "hold"-second position for motoring on slippery roads or down steep gradients, etc., when first gear starts are obviated. There is normal kick-down control, but this tends to lag appreciably and for instant acceleration it is better to flick into the third-speed position, the selector having electrical connection with the gearbox, and working impeccably. The gear changes are so smooth, 95% of the time, as to be almost imperceptible. The brakes, discs for the first time on a Rolls-Royce, are very very good indeed. They pull the car up undramatically from high speeds with the lightest imaginable pressure on the pedal. The handbrake is a rather mediocre affair which pulls out from under the facia, using the right hand; it has a very short travel, however, in contrast to that of the Silver Cloud, and a bright and large warning light signifies when it is in use. The power steering is perhaps too low-geared, at 4¼ turns, lock-to-lock, but is extremely light, more so than on the Silver Cloud III, but, having little castor return-action, it tends to be over-sensitive. Perhaps this is largely a matter of driver opinion, one experienced Rolls-Royce owner thinking it a great improvement on previous cars; but the writer thought it less effective than the power steering on, for example, a Mercedes-Benz, Citroën, or perhaps the Bristol 410. Under certain circumstances there is just the slightest trace of vibration at the steering wheel rim, and even very faint shake, and over certain surfaces there is slight vibration through the floor of the body-structure. If anything, these sensations are welcome, for they prevent the Rolls-Royce from seeming completely lifeless, and indeed, the suspension has a hardness which is far more acceptable than the flabby, completely-soft springing of most American cars. Indeed, it is almost impossible to define in words the satisfaction derived from this combination of ride, cornering power and shock absorption which personifies Rolls-Royce suspension engineering.

The interior of the Silver Shadow is typically Rolls-Royce, with full-width polished veneer facia, and thin-rimmed steering wheel, etc. The switches actuating the electrically-adjustable front seats are located on the inner walls of the central oddments well. Note that the bonnet-lid hinges from the front

I must confess that on entering the drab town of Derby that evening I was disappointed when, just after a Ford driver had given way to us to enable us to turn round in a busy road, the Silver Shadow's engine twice stalled. Admittedly it started again instantaneously, but it was embarrassing to find a Rolls-Royce caught out in this manner, and I can only assume that, the steering having been put on full-lock without much throttle being used, the hidden horses could not cope. I had been disappointed, too, to find that the speedometer needle, once so rock-steady on these cars, fluctuated by a matter of about one mile per hour, that the little scratch-proof rubber on the ring of the little ignition key bore the letter " B " and not the magic linked " Rs," and, as you will read later, there were a few more small disillusionments. But when all is said and done, these constitute comment on, rather than criticism of, the Silver Shadow. . . .

There are very, very few cars which perform as effortlessly and willingly as this modern Rolls-Royce, and none which behave in quite the same impeccable, characteristic manner. One covers the miles very rapidly indeed with the very minimum of effort, whether this is applied to steering the car, occasionally changing gear, if it is deemed desirable to do so to obtain maximum acceleration, or applying the brakes. These major services, like the minor controls, work not only with precision but are so light that even a slightly-built driver just recovering from influenza would feel no fatigue at all in putting up high average speeds throughout a long day's motoring in this latest version of the V8 6¼-litre Rolls-Royce.

After a night spent in Derby, we went the next morning to the Arboretum (which is simply the name to a now rather dreary park, presented to the town by Joseph Strutt in 1840) in order to pay homage at the statue of the late Sir Frederick Henry Royce. The news that morning, splashed in the headlines even of the *Daily Mirror*, that Rolls-Royce Ltd. had landed an American order for RB211 aeroplane engines worth £1,000,000,000 to Britain, was as pleasing to read as the contemplation of spending a few more days in the company of the Silver Shadow. Incidentally, in Derby we noticed an Austin Seven Swallow saloon parked at the kerbside, and a Y-model Ford Eight still in use. Soon we were speeding quietly along the M1, on which road we were passed for the only time during this long weekend, that by a Rover 2000 Automatic, which we still think may have been a disguise for an experimental car from the Rover Company. The Silver Shadow had no difficulty in redressing the balance, but top speed, especially in this country, restricted as it is by frustrated and ill-advised politicians, is not the sole purpose of a car. Indeed, there are faster cars than a Silver Shadow, more accelerative cars, cars which can be flung through bends more quickly and there may even be quieter cars. But none has the collective qualities of light controls, mechanical silence, high-speed cruising and impeccable finish and interior decor, coupled with the sort of performance and safety factors which the Rolls-Royce Silver Shadow provides. It is indeed a splendid car in which to cover the ground with much enjoyment and a complete lack of anxiety. Although we had deliberately kept the speed down we found ourselves some miles north of Boroughbridge, where we had another appointment, 1¾ hours after leaving Derby. Incidentally it was on the M1 that we encountered another Silver Shadow going south, the only one we had seen in motion since leaving London.

Near Bedale we came upon a stretch of public road, some 4½ miles from Catterick Camp, which is given over to tank instruction, the width and surface having obviously been specially prepared and a big turn-round bay provided, while all along both sides are special kerbs which, from the graunching marks they carry, suggest that the tanks are frequently, if not sideways-on, certainly considerably out of control !

On the Saturday evening we came south as far as Doncaster, noting that in this town one of the old A1 road signs remains, although the new A1 road no longer goes through the town. It was here that there was another mild disappointment, when the driver of an ancient Ford Zephyr tapped on the driver's window while we were stationary at some traffic-lights, to tell us that one of our brake stop-lights was inoperative. This is a thing which can happen to any car and it is only the very high standards set by Rolls-Royce that cause the owner discontent when this commonplace failure occurs on a car of their manufacture. We had not covered a sensational mileage in these first two days, but a good deal of business, as well as driving, had been accomplished and the Silver Shadow had certainly contributed to our lack of fatigue. The eight-way electric adjustment of the two front seats has already been commented upon. This works quietly and not only enables the very best driving position to be achieved, but enables minute changes of seat-position to be made while driving, which is an additional contribution to preventing aching or stiff muscles. The window lifts are also electrically-operated and the glasses rise and fall both quietly and quickly. These are amenities which one expects in a car of this calibre and cost, but it should be emphasised that, from within, the Silver Shadow seems virtually as impressive a car as a Silver Cloud and any diminution of dimensions has certainly not had the effect of conveying a sense of cramp or inferiority to the occupants. The faster this modern Rolls-Royce is driven the more you discover that it can be thrown about with considerable abandon, and it still corners accurately and feels like quite a small car. By making good use of the gear lever and those very powerful brakes average speeds which would be commented upon in many other cars pass unnoticed, and seem like normal unflurried motoring to the more " press-on " owners of Silver Shadows. Yet the car's whole demeanour is one of dignity and comfort. In these respects the Silver Shadow is as much a logical advance over the Clouds as a new Phantom was an improvement on the Silver Ghost, and the Phantom II, in its turn, was better than the Phantom I.

Saturday night having been spent in Doncaster, on the Sunday the passenger brought me to Essex by a cross-country route through the deserted fenlands of Lincolnshire and down into Cambridgeshire, a pause being made at Duxford in the hope of seeing some preparations for the filming of " The Battle of Britain " and, perhaps, photographing the Silver Shadow in the company of other famous Rolls-Royce-engined machines, which would have been appropriate on the 50th anniversary of the Royal Air Force. But all we saw were barred gates and closed hangars.

So it was home through the East End of London, in the unexpected congestion of the sunny last-Sunday in March. Under such frustrating traffic conditions, a Rolls-Royce, as I have said, placates the temper and proves restful whether crawling at a snail's pace or not moving at all, an attribute I have experienced to a lesser degree in another good car, the Lancia Flavia.

The Monday was occupied in taking the Silver Shadow over a familiar test-route, involving more than 300 miles in an easy day's motoring. Another Silver Shadow was encountered during this journey. When it was finally returned to Conduit Street the car had done a total of 1,335 miles. Fuel consumption varied from 11.9 m.p.g. to 13.2 m.p.g., with an overall figure of 12.5 m.p.g. of 100-octane petrol. The oil-level button remained well off the minimum mark and three pints of oil, according to a very flexible and vague dip-stick, would have restored the level in the 14½-pint sump. It is hardly necessary to add that the car had given no mechanical trouble. Coming into Brentwood a demented tramp had run to the kerbside to spit on it. But as he meted out the same treatment to a Daimler which happened to be following, this cannot be taken as any reflection on the excellence of the Rolls-Royce. And that concludes the story of a long motoring week-end.

Some Aspects of the Silver Shadow

Those who have not yet driven one of these cars may be interested in some details about it. The light-alloy 90 deg. vee-eight 104 × 91.4 mm. (6,230 c.c.) push-rod o.h.v. power unit introduced for the Silver Cloud II in 1959 is used, but with improved combustion chambers and porting and the sparking plugs rendered more accessible by locating them above the exhaust manifolding. Hydraulic pumps for the power brakes and suspension height-control of the Silver Shadow are driven by additional lobes on the camshaft. This engine gives, as Rolls-Royce would say, sufficient power. Sufficient, that is, for a maximum speed of over 115 m.p.h. and a s.s. ¼-mile acceleration time of 17.5 seconds. It also provides the kind of torque which gives extremely effective pick-up for silently and very quickly overtaking slower vehicles. It is inaudible when idling, after the initial clatter of the hydraulic tappets has died down from a cold start. It pours out suave power and the starter, actuated by the Yale ignition-key, brings it to life so quietly that on lesser cars a nearly " flat " battery would be suspected. The 28-pint coolant system is inhibited and treated with Rolls-Royce anti-freeze and the cylinder banks exhaust through a stainless-steel exhaust system with triple silencers.

Transmission on r.h.d. cars is by a 4-speed fully-automatic gearbox with fluid flywheel but l.h.d. export Silver Shadows have a 3-speed transmission with torque-converter which gives a maximum in middle-speed of some 85 m.p.h. This 4-speed transmission functions very smoothly, but there is still a noticeable jerk when second speed engages hurriedly to cope with abnormal road conditions. The ratios are 3.82, 2.63, 1.45 and 1.0 to 1, with an axle ratio in the hypoid-bevel final drive of 3.08 to 1. There is normal engine braking when in third and top gears but no sprag or hill-hold. When the R-position is in use with the ignition off a parking lock is engaged. A tiny lever fits into a floor socket for emergency gear selection.

Left: *Status symbols.* Right: *A picture showing the subtle differences in appearance between Silver Shadow and Silver Cloud.*

The Silver Shadow, which has been in production for 2½ years, represents an epoch, as the first Rolls-Royce to have unitary construction, all-independent suspension, and disc brakes. A steel body with aluminium doors, bonnet and boot-lid is mounted on two sub-frames which are supported on Delaney Gallay "Vibra-shock" stainless-steel-mesh resilient supports, movement of the back sub-frame being damped by a small double-acting hydraulic shock-absorber. The suspension has automatic height control by rams operating at hydraulic pressures of 1,150 to 2,500 lb./sq. in. This gives notably pitch-free travel and has eliminated the nose-dip under braking and most of the nose-up surge under fierce acceleration of the Silver Cloud cars, although there is still a jerk as it settles as it comes to rest. A double-wishbone layout with coil springs and anti-roll bar is used in front and the new i.r.s. is by trailing arms, angled to give slight negative tyre camber during fast cornering, in conjunction with coil springs and telescopic dampers. The drive shafts have Detroit inboard and Hooke outboard universal joints. There is an occasional metallic sound from the o/s back of the car, apparently characteristic of this model, and hump-bridges to some extent harshen the ride, but emphasise the very good damping of the suspension.

This suspension system endows the Rolls-Royce with an excellent ride and enables it to be cornered rapidly without excessive or untidy roll. Through fast bends the car is beautifully balanced and only on the tighter turns does eventual oversteer develop, quickly stifled by the finger-light Saginaw power steering. This steering is ultra-light, so that a Silver Shadow can be controlled, and parked, with finger-and-thumb; it is just that much too insensitive about the straight-ahead position for entirely accurate placing of the car. This is, perhaps, the penalty paid for the ultra-light parking action, and is emphasised because there is very little castor return. That a good deal of road noise intrudes and that there is some roll on corners is a reminder that even Rolls-Royce have to compromise to some extent when it comes to making a 2½-ton motor car stay on the road at speeds close to two-miles-a-minute, ride level, and yet go round corners like a good sporting car. The test car was shod with Firestone de luxe 8.45-15 low-profile tubeless cross-ply tyres, which emitted mild squeal on the faster corners. After some experience the Silver Shadow could be cornered very quickly with much satisfaction to the driver. Since its introduction stiffer damping has been introduced.

The braking system is very special, with three separate systems in addition to the handbrake, and these power-hydraulically applied, from two spherical accumulators containing nitrogen at 1,000 lb./sq. in. and pressurised to 2,500 lb./sq. in. by the engine-driven hydraulic pumps. The 11 in. Girling discs are edged with stainless steel wire and give a total braking surface of 513 sq. in. The result is impeccable braking, the Silver Shadow pulling up from the highest speeds under mere feathering of the foot on the pedal, which is, however, rather too close to the slender treadle accelerator. Moreover, this kind of braking is accomplished with no noise, no fade, and no time-lag and is delightfully progressive. It is superior even to the fine mechanical-servo system used by R.-R. from Silver Ghost days.

Coming to the appointments of a Silver Shadow, these are in the true Rolls-Royce tradition and very reminiscent of those found in Silver Clouds and earlier models. The facia is of polished-veneer, and the dials have black faces with white figures. There is a deep illuminated cubby-hole on the left, with quietly-shutting, magnetically-closed lid, having a Yale lock. Surprisingly, the polished-wood screen rail had a nasty uneven join at the top o/s and it is secured by Phillips screws. In the centre of the facia is that traditional circular R.-R./Bentley switch-panel, containing generator and oil-pressure warning lamps. It is no longer possible to lock the lights-switch, to prevent tampering, but the switch still pulls out, to light up the panel from a tiny lamp under the facia sill.

Rolls-Royce now fit a Kienzle clock. The big 130 m.p.h. speedometer is supplemented by smaller dials for FUEL (E, ½, F, with a minimum oil-level marking in red, to which the pointer records if a small button is pressed), COOLANT (Hot-Cold), OIL (Low-High) and AMPS (60—, 0, 60+). The instrumentation is by Smiths, except for a Lucas ammeter. Four different-colour grouped warning lamps, controlled by a button, light up for BRAKES (fluid-level low or pads worn), FUEL (normally indicates when only about 3 gallons remain in the 24-gallon tank, but comes on whatever the level, if the button is pressed, as a check that the light is working properly), and COOLANT (low level). A larger warning lamp acts as a reminder that the handbrake is on and indicates that a stop-light has failed, although this latter "fail-safe" had failed on the test car. These days only the speedometer carries the R.-R. insignia.

Other facia services are a parking-lights control, cigar lighter, the two heater/refrigerator controls, together with the big swivelling inlets for refrigerated air with their own cut-off buttons and volume control from a slide on the facia lower edge, 2-speed wipers-cum-washers button, 2-setting panel lighting, two discreet concealed ash-trays, a big air outlet, and exceedingly neat sunken finger-switches for the roof lamps and rear-window de-misting. A Smiths Radiomobile radio is standard, mounted centrally just below facia level, with a switch for extending or retracting the very tall aerial and a knob for front/rear speaker selection. When retracted the aerial is tamper-proof, without the need for a little key to lock it, as on lesser breeds of automobile.

There is a remarkable combination of heat/cool-air settings, which alter to the accompaniment of the slight whine of electric servos opening and shutting hidden doors, especially in refrigerated cars, and screen de-misting is prompt and 100% efficient. Scuttle foot level air venting is also provided, controlled by little plated knobs on the scuttle walls. The wipers park compactly to the left after a final sweep.

The eight-way, up/down, forward/back, tilt, electric seat adjustment aforementioned now works quietly and not only enables the most effective driving position to be selected but, because minute changes of position can be effected so simply while the car is in motion, cramp and stiffness can be eased. In addition, reclining squabs are provided. Normal electric window-lift buttons are fitted; all the minor controls and buttons are plated. Upholstery is in selected English hide; the floor is Wilton carpeted, the pile fluffing, like that of a good but newly installed domestic quality carpet, on the test car.

The boot is spacious, easy to load because of the unrestricted opening when the lid is up, and it is carpeted (and illuminated), and contains a full tool-kit, and an electric charging-socket. Pressure in the tyre on the spare wheel stowed under the boot floor can be checked through a small access door. For additional safety in night motoring red lamps are fitted to the armrests of the rear doors, and above the windscreen there is a hazard-warning with tell-tale in case it should be inadvertently operated during the day time. The doors have Yale locks but these are operated by using the ignition-key upside down, there being no separate key, which is desirable when letting hotel staff garage the car. The lock in the boot-lid moves with the button but the others are properly mounted. Sill interior-locks enable the car to be locked from outside without using the key in conjunction with the outside buttons. No air-extraction is provided for the body, and this must therefore be slow, especially as the sealing is such that the doors will not shut easily unless a window is opened; even so, they shut noisily. The two-clicks precision action of coachbuilt doors is sadly missed, especially on a Rolls-Royce. VW achieve this, however, so why not Pressed Steel? This was emphasised when the button of the driver's door stuck in, and the driver couldn't gain entry!

There are folding arm-rests for the front seats as well as a central one for the back seat, pockets in the front doors, a stowage-well between the front seats which can accommodate an extra passenger when its base is moved upwards and forwards and arm-rests on the doors, the driver's adjustable by thumb catch (but it fell off when we tried this). The interior door handles are simple plated pull-up levers, with matching fixed grab-handles set at about 45 deg. behind them—simple, but nothing could be more effective. In the back compartment additional roof or reading lamps, with their own switches, movable foot-rests, two folding tables, inbuilt mirrors, leather roof-grabs, ash-trays, and duplicated lighters and stowage pockets in the backs of the front-seat squabs are strategically placed for maximum convenience.

There is a vanity mirror in the n/s anti-dazzle vizor, the ¼-lights open at the front but are fixed for the back compartment, and although the rear-view mirror is not of anti-dazzle type it swivels easily. The headlining is of washable p.v.c., the rain guttering of the body is well contrived, and small but appreciated refinements include a completely non-dazzle full-beam warning lamp, illumination of the R, N, 4, 3, 2 indicator for the r.h. gear-control stalk lever which can be switched off with the instrument lighting, and turn-indicator warning lamps, also in the speedometer, so unobtrusive I did not at first notice them. A slender l.h. stalk controls the turn indicators, which have side repeaters; they cancelled rather too quickly. Tinted glass is provided on refrigerator-equipped cars likely to be used in hot sunny climates, so that we were able to look into the March sun-light on the first day of the test without eye-strain.

The 17-in. two-spoke steering wheel is discreetly thin-rimmed, with just a plated horn-push in its hub, deletion of the late-early ignition control many decades ago and of the hard/soft suspension settings with the advent of the Silver Shadow's sophisticated height-control all-independent suspension having tidied up this aspect of the car. When you pull up at a service station to buy some 20 gallons of fuel, the flap over the screw-thread filler is opened electrically as you press a button by the steering column. The spring-loaded bonnet-lid hinges from the front, after a substantial under-facia release has been operated. The V8 engine is largely concealed by the auxiliary services; the engineering is naturally of the highest standard, the fuse-box contains 20 fuses and labelled circuits, and the main hydraulic reservoirs have visual level-glasses. Electricity is normally generated by a 35-amp. Lucas dynamo but with refrigeration a Lucas 11AC 45-amp. alternator is used.

Lighting is by a Lucas sealed-beam four-headlamp set, with a tiny R.-R. badge set between each pair of lamps. Whether this implies specially-prepared lamps I do not know but certainly the driving light is good without being glaring and when dipped with the floor button the beams are still effective. A reversing lamp is fitted. When a parking lamp is selected the starter is rendered inoperative to prevent one driving off improperly illuminated.

The Silver Lady graces a dummy radiator of classic shape, as do

The youngest passenger ever to travel in a Silver Shadow? This newly-born lamb, whose mother had deserted it, was driven in the Rolls-Royce to shelter and sustenance.

R.-R. badges, but nowhere does the name Silver Shadow appear on the body. From the back the car's appearance is less distinguished, but far more modern than that of a Silver Cloud. But the silhouette from the driving seat is as impressive as ever, particularly after dark, although the curving lines of the bonnet made the radiator stand proud to a lesser extent, perhaps, than on any previous Rolls-Royce. Six chassis points require greasing every 12,000 miles.

The foregoing notes are additional to the full description of the Silver Shadow published on its introduction, in MOTOR SPORT for November 1965, and might be read in conjunction with our road-test report on the Silver Cloud III in the issue dated September 1965.

* * *

To conclude, a Rolls-Royce is still widely accepted as a status symbol but I prefer to covet it for the enhanced ego which derives from driving such a fine car. I headed the Silver Cloud III report " Not So Much a Motor Car, More a Way of Life," the title of a then-current David Frost programme. This remains true of the Silver Shadow. These latest cars from Crewe are still comparatively rare in this country, because the bulk of the limited output is exported. So, if you are contemplating purchasing one and it becomes available, this is an opportunity to be grabbed with open cheque book. Because, although it may not be a car for the wide open spaces, like a Ford GT40 or an A.C. Cobra, or for hurling up and down Alpine passes quite as you would a Lamborghini or a Porsche, or for enjoying excessively winding roads as you do in a Lotus Elan, nor perhaps has it quite the technical *rapport* of a Jensen FF or N.S.U. Ro80, the Rolls-Royce Silver Shadow is a great devourer of distances, as well as being an impeccably made and beautifully appointed motor car. As tested, it costs £7,895 7s. 11d.

W. B.

ROAD TEST
by John Bolster
Rolls-Royce Silver Shadow

A magic carpet with a famous name

NOBODY ever starts a Rolls-Royce article without harking back to the Silver Ghost, and why should I try to be different? Henry Royce designed the Best Car in the World in 1906, the year of the first Grand Prix, and so it is not surprising that it had a high-efficiency engine which followed current Grand Prix practice. The machine was light and compact for a 7-litre car, and it appealed because it offered an unequalled combination of performance, flexibility, and silence.

The very first Silver Ghosts were almost sports cars, but later the Rolls-Royce was habitually bought for many years by wealthy people with chauffeurs, which completely changed the image of the car. Nowadays, barely 10 per cent of Royce owners keep a "shuvver," and so it was logical to return half a century later to the conception of the first short-chassis Ghost, which was fundamentally an owner-driver car. The Silver Shadow is narrower, shorter, and much lower than its immediate predecessors, with all the controls arranged for feather-light operation. It also has a powerful 6.3-litre light-alloy engine, which ensures that the Shadow is a high-performance car.

When this model was introduced, I carried out a full road test for AUTOSPORT. Since then, I have tried the car from time to time as development has progressed, but for the purpose of this article I propose to compare the latest model with the first series, as Rolls-Royces embody improvements as soon as they become available.

So, I remark that this car has higher-geared (lower numerical ratio) steering than its predecessor and a new anti-roll bar at the rear, with a stiffer one in front. The main reason for carrying out this test, however, is the new transmission, which has been fitted for some time to export cars but has only just been standardised on right-hand-drive models.

The new gearbox has three speeds instead of four. Its second gear has about the same ratio as the third gear of the previous box and first is actually a little faster than the old second. This is made possible by the use of a torque converter in place of the former fluid coupling. Let us examine the difference in performance which this gearbox has made, before considering the car as a whole.

It is often argued that the torque converter, for all its advantages, introduces a constant power loss into the transmission, and that is why some manufacturers cling to the fluid coupling, which simply acts as a clutch and gives no torque multiplication. It is therefore astonishing that the new transmission definitely introduces no losses at all. Indeed, the maximum speed may even be a little better and the same applies to the fuel consumption; the acceleration is considerably faster all the way up the range, without any possibility of argument. In case anybody doubts the wisdom of the high bottom gear, let me emphasise that the ratio is doubled with the converter at stall, while there is sufficient torque to spin the wheels on dry concrete.

By far the greatest improvement is in the smoothness of operation, and there are literally no jerks at all. Getting rid of the low bottom gear has eliminated a source of jerks, and all the changes, both up and down, are

The traditional interior finish is unique to Rolls-Royce.

A powerful 6.3-litre light alloy engine provides the drive.

now absolutely smooth. Even the kick-down is completely without the usual hiccup and to the passengers all the changes are virtually imperceptible. I used to make manual changes on the earlier box to get smoother operation, but I hardly ever touched the selector of the new box once I was on the move. It is, of course, just as easy as it always was to select a lower gear manually if desired, but although higher engine revs can be attained by delaying the up-changes, this gives no improvement in acceleration times.

The big V8 is tuned to have immense torque and one would imagine that it peaks around 4000 rpm, a speed which is not exceeded on the lower gears, though the car will reach about 4500 rpm in top, equivalent to 118 mph. It is curious that the speedometer of the test car was spot on at 60 mph but read 8 per cent fast at maximum speed. The acceleration is extraordinarily vivid for a big luxury car, right up the range, and a genuine 112 mph comes up on any reasonable straight. The full 118 mph takes a good deal longer, but once achieved it can be held with no increase of sound.

The silence of the engine and transmission is remarkable and the complete absence of wind noise, even at maximum speed, sets entirely new standards. By making such a phenomenally quiet car, the manufacturers have fashioned a rod for their own backs, as road noise is all the more noticeable. The problem of obtaining really silent running with very wide section tyres has yet to be solved, and though this car is quiet on some road surfaces, there are those which succeed in transmitting their unwelcome message into the well-insulated interior. No doubt development in this sphere is going on all the time, and probably new tyres will be required to complement the work of the chassis designers. This is the last hurdle before the real magic carpet is achieved.

A suspension compromise has been chosen which gives a soft, floating sensation, but there is no diving, lifting, or squatting under fierce braking or acceleration. Though there is now more roll resistance, the car still rolls visibly to an appreciable angle when flung through a corner. While this is noticeable to an outside spectator, it is not felt by the occupants of the car so it is of no real importance, especially as the average owner will not attempt this sort of driving. The tyres remain glued to the road at all times, and bumps never succeed in unsticking them.

Perhaps the most impressive handling feature is the way the car goes through really fast bends. Quite appreciable corners can be taken at a genuine 100 mph, while there is no tendency for either end to take charge, the balance remaining impeccable. On sharper corners, there is sufficient power to hang the tail well out if desired, an amusing driving style which is not conducive to long tyre life. However, it does underline the sports-car controllability of the big machine.

The steering is extremely light, which may be disconcerting at first. In spite of its lightness, there is in fact some feel and there is also some castor return action. The further I drove, the better I liked this steering, and nobody should attempt to judge it after only a few miles; to park a two-ton car with one finger is surely one of the ultimate luxuries of motoring! The stability at high speeds is first class, and side winds are simply ignored. When travelling at three-figure speeds, the car can be placed with absolute accuracy, both on the straight and in curves. The brakes are immensely potent with light pedal pressure, never tending to lock the wheels in an emergency.

The handling characteristics of the car remain constant, irrespective of load. This is because a self-levelling system has been applied to the front and rear suspension. It works very slowly on the move, to compensate for such things as the emptying petrol tank, but when the gear selector is in neutral or a door is opened, the levelling becomes rapid to cope with extra passengers or luggage.

The heating and ventilation system, though of great complexity, is particularly easy for the driver or passenger to control. For heating, you turn the knobs for temperature or pull them out for quantity—it's as simple as that. The refrigeration system was not required on this occasion, but in the summer I have previously found it a sheer delight in hot weather. When it is in use and the gear is in neutral, the engine is automatically set to a fast idle, so that the pump will be driven at a suitable speed.

Any driver can find a comfortable seating position because the seats are moved in all directions by little electric motors; it goes without saying that the windows are raised electrically. The back seats are just as comfortable as the front, with plenty of legroom, and their occupants are looked after by the heating system. Though the Shadow is 5 ins shorter and 4 ins narrower than the Cloud, it gives much more space to the people inside. As a design exercise, it illustrates how much more space there is for passengers and luggage when there is no back axle leaping up and down.

It is curious that one still meets people who think that the Silver Shadow is too low and unimpressive in appearance, for to them a Rolls-Royce has a chauffeur and a footman with Royalty in the back. The answer, of course, is that the more compact car is far easier to drive and quicker, too. There is no need to dwell on the superb finish and the interior finishing, for these are traditional. All the usual luxury features, such as picnic tables, are there, but I judge a car by its tool kit, and the Shadow gets full marks.

SPECIFICATION AND PERFORMANCE DATA

Car tested: Rolls-Royce Silver Shadow four-door saloon, price £7959 13s 1d, refrigeration £189 6s 1d, both including PT.

Engine: Eight cylinders, 104.1 mm x 91.4 mm (6230 cc). Pushrod-operated overhead valves. Compression ratio 9:1. 7 in twin SU HD8 carburetters. Power output not disclosed.

Transmission: Torque converter driving 3-speed automatic gearbox, ratios 1.00, 1.48 and 2.48:1. Hypoid final drive, ratio 3.08:1.

Chassis: Combined steel body and chassis. Independent front suspension by wishbones, helical springs with automatic height control, and anti-roll torsion bar. Recirculating-ball power-assisted steering. Independent rear suspension by semi-trailing arms, helical springs with automatic height control, and anti-roll bar. Telescopic dampers all round. Servo-assisted disc brakes on all four wheels with triple hydraulic circuits. Bolt-on disc wheels fitted Dunlop 8.15-H-15 RS5 tyres.

Equipment: 12-volt lighting and starting with alternator. Fuel gauge with oil level indicator. Ammeter. Oil pressure and coolant temperature gauges. Clock. Heating, demisting and ventilation system with refrigerated air conditioning and heated rear window. Flashing direction indicators with hazard warning. Windscreen wipers and washers. Unilateral parking lights. Reversing lights. Radio with electrically raised aerial. Electric seat adjustment. Electric window winders.

Dimensions: Wheelbase, 9 ft 11½ ins; track, 4 ft 9½ ins; overall length, 16 ft 11½ ins; overall width, 5 ft 11 ins; weight, 2 tons 1 cwt 2 qrs.

Performance: Maximum speed, 118 mph. Standing quarter-mile, 17.2 s. Acceleration: 0-30 mph, 3.4 s; 0-50 mph, 7.2 s; 0-60 mph, 10.1 s; 0-80 mph, 17.6 s; 0-100 mph, 30 s.

Fuel consumption: 12 to 15 mpg.

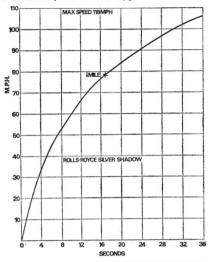

The lower, more compact lines of the Silver Shadow cater for current trends in motoring.

ROLLS-ROYCE SILVER SHADOW

NEARLY FOUR YEARS have elapsed since the Rolls-Royce Silver Shadow was introduced; for those four years we've regularly attempted to get one from the R-R distributor for a road test, but none was forthcoming—apparently because the importers are sensitive about comparison being drawn between theirs and other luxury cars. Once again Donna Crean (owner of last month's Lamborghini Espada) has graciously loaned us a $20,000 car from her stable, so that now we can bring you the first American road test of the controversial Silver Shadow.

Controversial, we say, because the Silver Shadow was (with the exception of its aluminum V-8 engine) a total break with Rolls-Royces of the past. It was smaller, lighter, simpler in appearance, and it was the first car from Crewe to use independent rear suspension, disc brakes and a unitized

ROLLS-ROYCE SILVER SHADOW
AT A GLANCE

Price as tested.........................$19,600
Engine..................ohv V-8, 6230 cc, 300 bhp
Curb weight, lb..............................4690
Top speed, mph...............................114
Acceleration, 0-¼ mi, sec....................17.4
Average fuel consumption, mpg.................9.7
Summary: operationally similar to large American sedan ... traditional R-R workmanship, comfort & prestige in a modern-size package ... excellent A/C, good brakes, dead power steering.

ROLLS-ROYCE

body. There have been cries of anguish from traditionalists that it doesn't "look like a Rolls," but surely there were similar cries when its predecessor, the Silver Cloud, was introduced in 1955. Suffice it to say that the Shadow is a far more logical car than the Cloud, being many inches smaller on the outside and offering not only more space inside but a better ride, better handling and quieter operation. As for the lack of pomp and circumstance, we feel that the passage of time will "give" the Shadow that, in a sense relative to other cars on the road; certainly the Cloud wasn't an intrinsically outstanding design.

A brief review of the Silver Shadow's engineering is in order because of the lapse since its introduction (see December 1965 R&T) and because it's such an engineering exercise. Built on a 119.5-in. wheelbase and being 203.5 in. long, it occupies about the same road area as a Chevelle or Fairlane but is considerably taller. Its unit steel body, which has aluminum doors, hood and deck, carries large subframes which are isolated from the main structure by unusual steel gauze doughnuts. The front suspension is by conventional unequal A-arms; at the rear are semi- (or diagonal) trailing arms of the type that are becoming so popular in European sedans. Coil springs are used front and rear and the latest Shadows have anti-roll bars at both ends of the car.

A hydraulic plunger pump driven by a camshaft lobe supplies pressure for the automatic suspension-leveling system, which acts on hydraulic rams atop each coil spring. The leveling system does its work very slowly when the car is being driven but if the transmission selector is in neutral or a door is open, the leveling takes place much more quickly.

Another hydraulic pump supplies the power to operate one segment of the unique 3-part brake system—one of the two pairs of front disc calipers and two of the four rear caliper cylinders. The same pump that runs the suspension works also operates the second brake circuit, the second pair of front calipers. A third, unassisted master-cylinder circuit of the usual type works the remaining two rear caliper cylinders and provides natural "feel" for the brake pedal. Finally, a mechanical handbrake operates on a set of lever-actuated calipers on the rear discs. Obviously there's little chance of running completely out of brakes.

The V-8 engine of 6.2 liters is a pushrod overhead-valve unit of thoroughly conventional design. Rolls-Royce stopped giving power and torque figures years ago because they felt the Americans were "cheating" on theirs, but we estimate the output as 300 bhp @ 4000 rpm, SAE power. Only the aluminum construction and the big twin SU carburetors set it apart from a typical American V-8.

Our test car was not entirely up-to-date, being an early 1968 model. Present Shadows have an instrument panel and collapsible steering column revised to meet U.S. safety laws (less wood and more padding on the former), head restraints, no picnic tables in the rear (pity!) and clearance lights on the body sides; the engine now has manifold air injection to control exhaust emissions. Also, a long-wheelbase version (4 in. more) with a central partition will be available shortly.

Driving the Silver Shadow

ONE OF THE R-R legends of years past has been that a nickel could be balanced on edge atop the idling engine. We've never tried this and suspect it to be apocryphal, but nevertheless the present V-8 is the most silent at idle and moderate loads of any engine we can recall. It's almost impossible to tell whether or not the engine is running and only the slight bump of the automatic gearbox gives a clue that a gear has been shifted. There is some clatter from the hydraulic valve lifters when it's first started, and on hard acceleration it takes on a familiar V-8 throb. With nearly 4700 lb of curb weight the 380-cu-in. size provides only modest performance by today's standards; using automatic shifts that occur at 39 mph (3700 rpm) and 73 mph (4100 rpm) the Shadow gets through the standing ¼ mile in 17.4 sec, just chirping its rear tires as it comes off the line.

ROLLS-ROYCE

The GM 3-speed Turbo Hydra-Matic transmission, the best of its kind in our opinion, is only barely noticeable as it shifts at light throttle openings and is marvellously smooth at wide-open throttle. Running with air conditioning we got only 9.7 miles per gallon overall, which means a limited range even on the 28-gal fuel tank. Strictly highway driving would give over 300 miles between stops, however.

Rolls tradition lives on in the high seating position and step-up interior. One feels high and mighty behind the long hood and Flying Lady if not so much so as in earlier cars, and threading through traffic is no problem with the car's clearly defined corners. The seats are bulky—R-R engineers would prefer thinner seats but the customers wouldn't—and soft, but not too soft, and are supremely comfortable for both front and rear occupants. The individual front seats each have an 8-way power adjuster in addition to separately adjustable backrests that take center armrests along with them! The leather smells fabulous, but that's probably the only justification left for using it.

Traditional readable, white-on-black instrumentation is complete and is set into burl walnut; pushing a dash button switches the fuel gauge to an oil level reading. A really comprehensive heating-ventilation-A/C unit works independently at upper and lower levels and has a very quiet 4-speed blower, though the little servo motors that do the work of flipping flaps and such make more fuss about it, and take longer to do it, than seems necessary. Curiously the A/C has no provision for the intake of fresh air into its evaporator. An incidental note is that a servo motor (electric) also does the work of changing transmission range, making the selector lever a fingertip proposition.

The Silver Shadow's ride is quite good over large or gentle undulations but rather disappointing on sharp, small road irregularities—on the whole, not as refined a ride as in a large American sedan. The car bottoms easily on dips also, and the large cross-bias tires cause it to jerk somewhat when crossing longitudinal ridges. Road noise is prominent, partly because of the extreme silence of the powertrain but perhaps partly because of the unit body construction. Wind noise too is not as low as we expected.

One doesn't expect a Rolls to take kindly to sporty driving and the Shadow holds no surprises. It rolls a lot (current ones with larger front anti-roll bar and a rear one added would roll less), and this fact coupled with the very dead power steering—GM, by the way—really discourage one from proceeding with anything but dignity. The handling is predictable, however, with no tendency at all for the tail to slide, and we found the independent rear to help keep the rear wheels on the road when negotiating tight turns up our favorite hill.

Braking is very good indeed, as it should be with so elaborate and expensive a system. The pedal feels absolutely like that of an unassisted system except that the effort required is moderate, and a proportioning valve for the rear brakes leaves no tendency for the rears to lock up in panic braking on dry pavement. Fade resistance is also satisfactory. The handbrake is less happy, however, getting nowhere near holding on our 30 percent test hill. A trace of squeal is occasionally heard from the discs despite R-R's clever use of a spring steel band set into the disc's periphery.

W**HAT, THEN,** does the Silver Shadow offer when compared to other luxury, prestige sedans such as, for instance, a Lincoln or Cadillac at less than half the price or a Mercedes 6.3 at two-thirds? Well, there's the traditional R-R finish and materials, every bit as good now as they were in the past though a bit anachronistic now. Pin striping, leather and burl walnut do have their appeal. In performance and handling characteristics it is so close to the Lincoln and Cadillac (at least under American conditions) that the difference is negligible, although it's certainly more maneuverable in traffic than these cars which are as much as two feet longer. It does stop better than the Americans, too. It cannot be driven in the sporting manner the Mercedes can, nor is its ride that much better to compensate for the fact.

The conclusion, then, is that the Rolls has to justify itself on psychological grounds. It is most likely a durable thing—the whole car is warranted for 50,000 miles—and will satisfy the buyer who is attracted to it in the first place. But it also demonstrates that a small manufacturer is hard-pressed these days to match the standards set forth by the giant automakers—American or otherwise—and that Rolls-Royce can no longer be considered the completely magical motoring experience it had the reputation for in the past.

ROAD TEST: ROLLS-ROYCE SILVER SHADOW

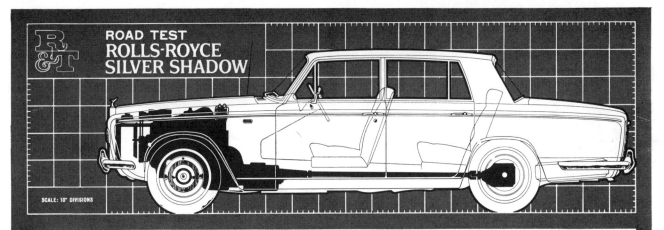

SCALE: 10" DIVISIONS

PRICE
- Basic list..............$19,600
- As tested.............$19,600

ENGINE
- Type...................V-8, ohv
- Bore x stroke, mm...104.1 x 91.4
- Equivalent in........4.10 x 3.60
- Displacement, cc/cu in..6230/380
- Compression ratio.........9.0:1
- Bhp @ rpm......est. 300 @ 4000
- Equivalent mph............105
- Torque @ rpm, lb-ft est. 400 @ 2500
- Equivalent mph............65
- Carburetion.........two SU HD8
- Type fuel required.....premium
- Emission control......air injection

DRIVE TRAIN
- Transmission type: automatic, torque converter with 3-speed planetary gearbox
- Gear ratios: 3rd (1.00).........3.08:1
- 2nd (2.48)................4.56:1
- 1st (1.48).................7.62:1
- 1st (1.48 x 2.22).......10.13:1
- Final drive ratio..........3.08:1

CHASSIS & BODY
- Body/frame: unit steel with some aluminum panels
- Brake type: 11.0-in. disc front & rear; triple system with power assist
- Swept area, sq in...........513
- Wheels........steel disc, 15 x 6 JK
- Tires.........Dunlop RS5 8.45-15
- Steering type: recirculating ball, power assisted
- Overall ratio............19.3:1
- Turns, lock-to-lock.........4.0
- Turning circle, ft..........38.0
- Front suspension: unequal-length A-arms, coil springs, tube shocks, anti-roll bar; self leveling
- Rear suspension: semi-trailing arms, coil springs, tube shocks; self leveling

INSTRUMENTATION
- Instruments: 130-mph speedometer, 99,999 odo, 999.9 trip odo, oil press, water temp, ammeter, fuel & oil level, clock
- Warning lights: oil press, alternator, water temp, brake system, fuel level, handbrake on, leveling system

ACCOMMODATION
- Seating capacity, persons.......5
- Seat width, front/rear 2 x 22.0/51.0
- Head room, front/rear...38.0/37.0
- Seat back adjustment, deg......30
- Driver comfort rating (scale of 100):
- Driver 69 in. tall............95
- Driver 72 in. tall............90
- Driver 75 in. tall............75

MAINTENANCE
- Engine oil capacity, qt..........8.8
- Every 6000 mi: chg eng oil & filter, cln plugs & points
- Every 12,000 mi: lube steering fittings, chg trans oil, cln air filter, chg plugs & points
- Every 24,000 mi: chg trans & power steering filters, chg diff oil
- Every 48,000 mi: cln flame traps, chg fuel filter
- Tire pressures, f/r, psi....26/26
- Warranty period, mi........50,000

EQUIPMENT
- Standard: air conditioning, power steering, electric windows & seats, etc

GENERAL
- Curb weight, lb.............4690
- Test weight................4985
- Weight distribution (with driver), front/rear, %......53/47
- Wheelbase, in..............119.5
- Track, front/rear........57.5/57.5
- Overall length.............203.5
- Width......................71.0
- Height.....................59.8
- Ground clearance, in..........6.5
- Overhang, front/rear...34.5/49.5
- Usable trunk space, cu ft.....18.2
- Fuel tank capacity, gal.......28.0

CALCULATED DATA
- Lb/hp (test wt)..............16.6
- Mph/1000 rpm (3rd gear)....26.2
- Engine revs/mi (60 mph).....2290
- Engine speed @ 70 mph......2660
- Piston travel, ft/mi..........1600
- Cu ft/ton mi.................101
- R&T wear index...............37
- R&T steering index..........1.52
- Brake swept area sq in/ton....206

ROAD TEST RESULTS

ACCELERATION
Time to distance, sec:
- 0–100 ft...................3.4
- 0–250 ft...................6.0
- 0–500 ft...................9.3
- 0–750 ft..................12.1
- 0–1000 ft.................14.5
- 0–1320 ft (¼ mi)..........17.4
- Speed at end of ¼ mi, mph...75

Time to speed, sec:
- 0–30 mph..................4.2
- 0–40 mph..................5.9
- 0–50 mph..................8.1
- 0–60 mph.................11.0
- 0–70 mph.................14.9
- 0–80 mph.................20.6
- 0–100 mph................36.5

Passing exposure time, sec:
- To pass car going 50 mph...5.5

FUEL CONSUMPTION
- Normal driving, mpg..........9.7
- Cruising range, mi..........271

SPEED IN GEARS
- 3rd gear (4350 rpm), mph.....114
- 2nd (4100)...................73
- 1st (3700)...................39

BRAKES
Panic stop from 80 mph:
- Deceleration, % g...........74
- Control................excellent
- Fade test: percent of increase in pedal effort required to maintain 50%-g deceleration rate in six stops from 60 mph........19
- Parking: hold 30% grade......no
- Overall brake rating....very good

SPEEDOMETER ERROR
- 30 mph indicated.....actual 31.3
- 40 mph.....................42.0
- 60 mph.....................63.3
- 80 mph.....................84.5
- Odometer, 10.0 mi...actual 10.2

ACCELERATION & COASTING

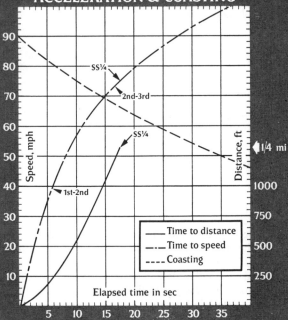

Road Test

A WCG 1st
Rolls Royce Silver Shadow

by Joseph Lowrey

CAR AT A GLANCE: Costly credit card on wheels... Fine detail applied to unexciting design... Quarter-mile in 17.5 seconds, 115 mph maximum... Not utterly silent, just quiet comfort.

Test Rolls is suitably backstopped by the gate to Stowe, once the Duke of Buckingham's home but now a boys' school. Rolls stands almost exactly five feet tall, second fiddle to author Lowrey.

Buying quality clothes is sensible. The design, craftsmanship and materials provide style, comfort and life advantages over cheap stuff. At tremendously high prices hand-made shoes, Savile Row tailored suits and made-to-measure silk shirts may be less practical than good factory gear, but wearing them does make you feel like a million dollars! Whether you aim to set up a big and legitimate deal or to sell some sucker the Brooklyn Bridge, it's no handicap to look like you might own Brooklyn Bridge.

And that's the way it is with a Rolls Royce. Taking a test car back to their London offices, I was amused at the enthusiasm of one of their top men for the cornering of Jaguar's latest XJ6 sedan. So Jaguar offers fine value in quantity-produced fine cars. Rolls Royce still offers the most universally recognized credit card you can drive!

Yes, if you need to watch value, three times a Jaguar's price for a car which is inferior in some respects although superior in others is too much, but a "Royce" does the most for your ego and for your prestige. Also to my horror, it's habit forming. I've always

40

said that for so much gold I'd prefer a Jaguar plus a few world tours, but six days with the Silver Shadow weakened my sales resistance.

Writing a factual road test on an ego-builder seems almost silly, for at a Rolls Royce price you obviously get good transportation — not a sports job but a gentleman's carriage, that's quiet, elegant, extremely comfortable and inconspicuously brisk. So far as they can, Rolls Royce Ltd. insulate their clients from the world, inviting them to look down upon it rather than to fight it. With a Rolls Royce you go about your affairs in fine surroundings, quite smoothly and quietly, always comfortably and with fingertip control over every feature of your car.

Frankly, I wish Rolls Royce Ltd. built cars of more inspired (and more inspiring) design. Their fame grew during the lifetime of Sir Henry Royce, a perfectionist engineer who saw a car as one entity. Today, a number of fine engineers contribute excellent detail

The Flying Lady (it's no longer made of silver) was created in 1910 by the late Charles Sykes, a famous sculptor of his day. It's now cast in the "lost wax" process, a technique used when intricate detail is required.

". . . Sir Henry Royce was a perfectionist who saw a car as one entity. Today, a number of fine engineers contribute excellent detail work to a design lacking the coherence which one dominant individual — an Alec Issigonis or an Enzo Ferrari or a Bill Lyons — can sometimes achieve."

Descendents of Royce's Rolls gather for a rare family portrait. Shadow owner-driven saloon (left) costs $19,600; Mulliner coupe (right), $29,200; Mulliner convertible, $31,600; and the new long wheelbase formal sedan (foreground), $24,900 with chauffeur's partition.

It's a shame that some neater way cannot be found to duct air to the carburetors (twin SU HD8's) but the front opening hood at least leaves the $300 grille (cost over Bentley's simpler version) and its mascot undisturbed.

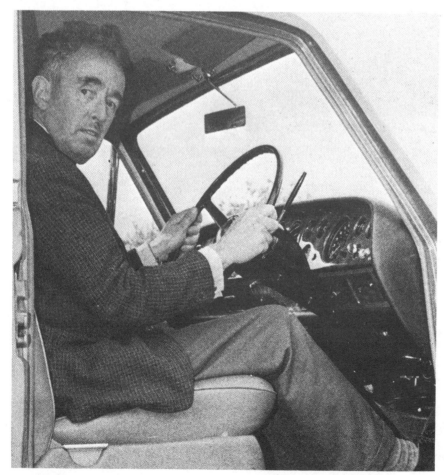

Joe Lowrey's six foot, six inch frame fits comfortably into the Rolls cockpit which, of course, has right-hand drive in its homeland. That front seat, incidentally, requires four full cowhides to upholster.

work to a design lacking the coherence which one dominant individual — an Alec Issigonis or an Enzo Ferrari or a Bill Lyons — can sometimes achieve.

What you get in a Silver Shadow is a four-door sedan of integral steel construction, dimensioned to make five people really comfortable. Neither length nor height have been stinted. There's a good trunk, its carpeted floor being absolutely flat. Air conditioning is an integral part of the car, as of course are power windows and powered all-directions front seat adjustment.

There're no flashy decorations, just such aids to physical and mental ease as well-designed leather armchairs to sit in, pile carpets laid over foam which in turn rests upon felt and fine polished walnut wood bordered with compulsory crash padding.

To propel this 6450-pound sedan of 119.5-inch wheelbase, Rolls Royce build their own V-8 motor of 380.5 cubic inch (6230 cc) displacement. Its power output is secret but the design emphasis is strongly upon torque at low and medium speeds. Aluminum block and head castings are the norm for an airplane engine factory, not conscious weight-savers. For years Rolls Royce built their own version of Hydra-Matic under licence, but now they buy 3-speed torque converter transmissions from General Motors and make a superb job of matching them to British V-8 engines.

Beneath the integral steel hull, Rolls Royce put front and rear suspension sub-frames as mounts for four-wheel independent suspension linkages. Rubber mounts could not provide satisfactory noise insulation without impairing steering precision; if the damping was right the flexibility was wrong or vice-versa. So, from the aerospace industry Vibrashock mounts were

Capacious trunk is fully carpeted, thus closing out a major source of road noise in lesser cars. Spare lurks under the floor along with a 28-gallon gas tank.

obtained, mounts which are different for each location but which damp coil springs by embedding them in "pot scourer" blocks of crinkled steel wire.

Front suspension is by unusually long transverse links, rear suspension by independent trailing arms, and four coil springs carry the weight. Hydraulic pressure from an accumulator and engine-driven pump goes to four jacks which regulate ride height, slowly when you are driving, or with almost violent promptness when the transmission is in Park or Neutral and a door is open for loading the car.

Power braking works on four discs, and just about anything which isn't duplicated is triplicated instead. There are double operating calipers on each wheel, double power assists, and a mechanical follow up in case both hydraulic systems fail. A fault tell-tale lamp on the instrument panel is itself checked out whenever you press a button to convert the fuel contents gauge into an engine oil-pan contents gauge. Power steering which Rolls Royce used to make themselves, with proportional feel, is now bought from Saginaw and is lighter but virtually devoid of feel.

One walks into the Silver Shadow rather than getting down into it and then you savor a fine view forward over a long hood to the famed Silver Lady mascot above that Grecian temple of a radiator. The starter is quiet and the automatic strangler effective without giving so fast an idle as to cause much transmission creep. Selecting Drive needs only one finger's pressure, as there is an electric link between lever and transmission to eliminate a possible route for mechanical noise to enter the car. However, a get-you-home control can be plugged in if the electrics should fail.

ROLLS ROYCE SILVER SHADOW

SPECIFICATIONS FROM THE MANUFACTURER

Engine:

Type: Front-mounted, overhead-valve V-8, water-cooled with aluminum block and heads
Bore and stroke: 4.1 inch x 3.6 inch (104.1 mm. x 91.4 mm.)
Displacement: 380.5 cu. ins. (6230 c.c.)
Horsepower and torque: Not disclosed
Compression ratio: 9 to 7

Transmission:

Type: GM 400 torque converter and 3-speed automatic
Gear ratios: 1st—2.48, 2nd—1.48, 3rd—1.00
Rear axle ratio: 3.08

Suspension:

Hydraulic ride height control on four coil springs, operating slowly with car in motion, fast when loading stationary car. Telescopic Girling dampers. Front linkage transverse independent; rear linkage trailing independent

Steering:

Saginaw power-assisted recirculating ball gear, with variable ratio. 3.6 turns from lock to lock, curb to curb—35 ft

Wheels:

Bolt-on steel disc wheels with 6-inch rims.
8.15-H-15 tires

Brakes:

Fully duplicated Girling hydraulic operation of four power assisted disc brakes. Discs 11 inch diameter

Fuel capacity:

28 gallons with low level warning lamp

Oil capacity

Engine: 8.75 quarts, with gauge on instrument panel
Transmission: 11.8 quarts

Cooling system:

Water capacity—17 quarts

Body and frame:

Integral steel body. Front and rear sub-frames isolated from main structure by Vibrashock all-metal damped-flexibility mountings

Dimensions

Wheelbase: 119.5 inches
Track: Front and rear, 57.5 inches
Overall: Length 203.5 inches, width—71 inches, height—59.75 inches
Ground clearance—6.5 inches

Curb weight: 4650 pounds

Silky is the word for every single control, minor as well as major. Touch the gas pedal and you can ooze gently away, or move off pretty fast with not much more than a purr from under the hood. This is rather an irregular purr on wet days, as traction on slippery surfaces could be better. In the dry, you can reach 78-79 mph at the end of a quarter mile in about 17.5 seconds elapsed time, and whether you use

ROLLS ROYCE

Continued

manual or automatic shifting matters very little. Top speed on the level is about 115 mph. Anywhere in the speed range, either a kick-down or fingertip selection of Intermediate or Low ratio will produce good overtaking response.

Smooth and very light steering needs 3.6 turns from lock to lock, its variable ratio quickening on acute turns. It will center itself when you are cornering normally, just, but you don't put a Silver Shadow through turns the way you would a race-bred Jaguar, and must tolerate some body roll if you try. Confidence comes with familiarity, but you use appreciable lock to put this understeering car into a turn, then gradually take off lock if the curve is a long one. Perhaps the ride height control is by then jacking extra weight onto the outside rear wheel? Not bad, but it's not Jaguar or Mercedes Benz.

Silkily smooth also is the ride over most road surfaces, for a back seat passenger as well as for the driver. It's a quiet ride but not I think so silent as in past Rolls Royce models as some sorts of road noise do filter through all the insulation. At a reasonable speed, really bad dirt surfaces are ironed flat but if you hit a big undulation fast, suddenly the back of the car is up skywards and down to the rubber buffers as if springs and dampers were too soft. Really one notices the big bump more because one has been so unaware of the small bumps, but this is a car for mature gentlemen, not for Rallye drivers.

Do you enjoy playing pin tables? If so you'll love the Rolls Royce's air conditioning controls, which operate electrically. Two neat switches, one for upward and the other for downward-going air, are pulled out separately to regulate air flow and twisted left or right to set the temperature of each air stream. Move either switch and strange whirring sounds come from behind the instrument panel, going on for varied numbers of seconds as motor-driven fluid and air valves automatically take up new settings. There does not feel to be a great range of temperature available, but apparently this is because thermostats prevent your roasting or freezing yourself. They allow you to call only for temperatures within a sensible range.

Always this car should be driven with closed windows to avoid noisy wind buffeting. At its slowest setting, a four-speed fan provides plenty of ventilation, unfortunately creating a gentle hum which is just audible in so quiet a car. The clock, incidentally, is now a quiet one. If at first the Silver Shadow doesn't seem so utterly silent as you'd expected at a legal 70 mph, look again at the accurate speedometer which may have crept round to the 90 mark without your noticing any extra mechanical or wind noise – a wailing noise will be from the car behind with the flashing blue lamp!

If paying $20,000 for a car would be a strain, forget about this Rolls Royce. Gas consumption is no worse than you'd expect as you may quite often cover as many as 10 miles on a gallon, meaning upwards of 200 miles before a light flashes to suggest refilling the 28-gallon tank. Essential services are lube level checks at 6000 miles and greasing of six joints at 12,000 miles, but this is a complex machine and if you want to enjoy it for life, you are advised rather than instructed to spend time and money on quite a long list of "optional preventive maintenance" jobs every 6000 miles.

A friend whose everyday transportation is an immaculate old Rolls Royce with 208,000 miles on the odometer tells me they could never afford a car which wears and depreciates. I'd doubt whether the complex Silver Shadow will be as reliable 35 years hence as my friend's car is today, but the details of all its gadgetry are expensively well executed. This remains true deep beneath the visible excellence of layer upon layer of smoothly rubbed-down paint, representing the opposite philosophy to Detroit's skill in assessing which parts must be precise and which can be crudely cheap.

It was the cynical enthusiast for Vintage cars, Bunny Tubbs, who once suggested as the Rolls Royce motto "even our mistakes are beautifully made." A pervading awareness of deep-rooted refinement must be what makes a sound but uninspired design into a tremendously potent ego-tonic. ●

NEW ROLLS

(Continued from page 17.)

has been how to dampen them and stop noise coming up either from the suspension or the roadway.

Rolls answer seems to have come through the use of two sub-frames (following some other more specialist makers), which are insulated from the body by rubber-metal mountings.

The fully-independent suspension has double triangle lever coil springs at the front with telescopic hydraulic shock dampers (not absorbers as such) and automatic height control.

\All are attached to the front subframe and located by a Panhard rod and mounted by what Rolls call "resilient metal mountings." These comprise metal like steel wool, located in rubber and wrapped between the dampers and the coil

In this way noise is apparently damped out (I'll take Rolls' word for it: they claim it is quieter than the S3–so apparently they've had a soft-ticking clock made).

The rear suspension is by coil spring with single trailing arm, telescopic hydraulic dampers. All are attached to the body sub-frame, located by a torque arm link and mounted again by resilient metal.

Brakes, Steering

The disc brakes are magnificent affairs—like everything else – to Rolls-Royce perfection. The front discs have two single callipers to each disc—that is, a combined pair of normal callipers for each wheel. And wait for it: yes, each pair—one on each wheel—of callipers has its own hydraulic pipeline, so that if one fails there's always another to take over. And if both **THOSE** fail, there's a third hydraulic system!

The rear brakes (specially developed by Girling to Rolls specification, and quite different from anything else Girling make) have dual callipers on each wheel.

Each "dual" has exactly double the power of any normal pair of disc brakes. There are the three independent footbrake systems plus a handbrake system. The rear wheels have a brake-pressure control valve, which prevents rear wheels locking.

The discs are 11in. in diameter; the reason for the big callipers at the rear is so the handbrake will work.

The handbrake is mechanical with an equaliser link to operate the rear dual discs.

Overall steering ratio is 19.32 to one with four turns lock-to-lock. Steering is recirculating ball, but power-assisted.

There are three silencers to make sure the majestic Rolls does really purr along—they are all in stainless steel and flexibly mounted on the body. The front one is lagged with asbestos and encased in aluminium; the middle one is just a silencer by Rolls-Royce standards (it would probably silence most trucks); the rear is a special "high frequency" job to make sure all noise is filtered out.

The hypoid bevel gears of the final drive are enclosed in an aluminium casing mounted to a cross member, which is attached to the body by those resilient metal mountings. (Ratio is 3.08 to one and top gear speed is 26.2 m.p.h. for every 1000 r.p.m.)

Other Details

Minor detail in the electrics section is that gear selection, front seat adjustments, windows on all doors, the rear window demist, the wing aerial (there's a choice of three radios as standard equipment), cigar lighters front and rear, windscreen washer, petrol-filler flap, horn and heater - demister controls are all electrically-driven.

~~Warning lights on the facia tel~~/ou when the hydraulic pressure fails, when the engine oil pressure is getting low and when the engine coolant level is getting too high. There's a warning light for too little petrol and too little charge to the large-capacity 11-plate battery.

And there's a handbrake light to warn that it is still on, and one to tell you when either or both of the stoplights aren't working.

Four headlamps to shine your way --though, as yet, Rolls don't recognise iodine vapor lamps.

The cooling system holds 28 pints, the engine holds 14½ pints of oil and the gearbox 24 pints.

Greasing is required only at the steering and height control ball joints and then only every 12,000 miles.

Availability? Production has started and deliveries to the home market will take place almost immediately; first exports will be sometime in the new year.

Price hadn't been settled at time of writing. But, if you must be so crassly commercial as to discuss these things, you could put the cost of all the extras (electric window winders, etc.) on to the price of the replaced S3 (about £9000 in Australia), add a few more quid—and not be far out.

THERE AND BACK

The Rolls-Royce Silver Shadow on the Continent with Maurice Smith, and in England with Martin Lewis

DON'T let the "100 LG" trick you; this registration number is switched to each new Rolls-Royce demonstrator. The car was red last time, now it is a silver-blue one, different in several externally invisible respects. Although the Silver Shadows, as they were introduced, suited the great majority of traditional R-R purchasers very well indeed, a few critical drivers, including some of the press, variously found fault with the American-market-orientated suspension, the dated automatic transmission, some aspects of the performance, road noise and the unassuming appearance. That was in March 1967, and since then development has gone methodically on at Crewe under Harry Gryles, technical director for the past year, John Hollings, who took over from him as chief engineer and the fresh 'atomic' outlook of Geoffrey Fawn, managing director.

We find as an unpublicized result, Silver Shadows (and Bentley Ts) with altered—improved we would say—suspension; new, more responsive automatic transmission, largely attributable to General Motors; slightly higher steering ratio and some detail improvements in anticipation of extensive internal re-styling to meet the latest export and home secondary safety requirements.

This time we took two good bites at the Shadow, first bringing it back from Switzerland and then testing it in England. While in Switzerland we made one rather revealing subjective test by driving up to Villar on the sinuous hill climb course and then repeating the dose with an earlier T Bentley, a twin in colour and being an export car, also having the American type torque convertor three-speed auto transmission. The difference in performance was marked, the later Shadow being more stable, easier to guide accurately round the bends with much less roll on the hairpins, and apparently more acceleration between

the corners. The engine and transmission were also appreciably quieter, particularly at high rpm.

The main reasons for the improvements in handling, we are told, are the fitting of a stiffer front anti-roll bar and a new one at the back to keep the same front-to-rear balance as before. As well as this, the steering ratio is a fraction higher, 3.6 compared with 3.9, and in fact it may be increased again later on. The change by stages is intended to give Rolls-Royce clientele time to condition their thinking to quicker response.

We do not drive these cars frequently enough to be really familiar with them and when we have driven them, a high proportion of the mileage has been covered abroad. But the conclusions always come out the same; first and foremost, you cover very long distances in considerable comfort and, even more important, you complete a 400-500 mile journey quicker than you expect and with the minimum of fatigue.

On anything like a main road, the steering, handling and ride are very good indeed—very safe too. The steering needs time for getting used to; to start with, the inclination is to turn too late, pull a bit harder to prevent the car running wide and then find you have overdone it and must pay-off again. You get round well enough but not what you might call a Rolls-Royce line. After a few hundred miles it all becomes right and natural and if you are on radial tyres the steering response is that bit better.

On rough roads, or those with a coarse top dressing, there is more noise than we have come to expect from Crewe cars. You hear it most when relaxing in the rear seats —a distant wheel rumble creeping up the body sides and bouncing between the solid rear quarter panels.

Over the winding mountain roads such as those around the frontier, north of Lake Geneva, the Shadow is best driven sedately otherwise the driver's elbows and the car's roll attitudes become progressively less dignified—though the car remains reassuringly safe—as you increase speed.

At least two of our staff, most of whom are fortunate enough to drive several cars each year that cost as much as the Rolls, find the Shadow well worth its price. As much as anything it is the intangible qualities which count; the mature way in which the cars perform, the quality of fits, finish and materials and the satisfaction they bring to a driver. M.A.S.

NO matter how many different cars I drive, I always have that slight feeling of apprehension when I sit down for the first time in a stranger.

Thoughts vary from "Who on earth designed that?" to "How on earth did he manage to park this monstrous heap

Abroad, the Silver Shadow in the road to Glutieres, above the Ollon-Villars hill-climb in Switzerland, with the Dents des Midi behind. In Britain, motorways and racing circuits make it seem equally at home.

46

THERE AND BACK...

just there," and the next five minutes are spent trying to extract a vast American sedan from between two Escorts.

With 100 LG it was different. I got in, did up the seat belts, started and drove away, very conscious of having £8,000 under the seat of my pants, but not at all worried about any gimmicks dreamed up by some Mittel European emigrée engineer. To be fair, I must admit to getting the air conditioning almost tied in knots, but before I blew the whole thing to pieces, I did stop and read pages 47 to 49 in that wonderful handbook, to find out how to make it work.

Once the worst of London traffic was over, I set about tentativly exploring how 100 LG would behave in these hands unaccustomed to such gracious living. Silence is all-pervading; at about 30 mph, you are faintly aware of the outside world, and as the clock is now electric, even that does not shatter the peace with its ticking. Once you have got over the novelty of driving yourself up and down the seat runners on little electric motors, and have remembered that the *rear* button operates the driving door window, the Rolls-Royce Silver Shadow grows on you.

Initially, the power steering seems a little over-sensitive, making fast lines through corners look rather like a high-speed slalom. Perhaps it is the thought of that female on the bonnet turning round and asking you just what you think you are doing with *her* car, but after a few miles you soon learn the art.

Performance comes in the sweeping progress class, and for some reason, one never seems to be caught behind queues of lorries or carved up by emergent London buses with that imposing bonnet in view. It is not a car that takes kindly to being driven crudely, and one's natural tendency is to treat this superb piece of motor car engineering with the respect it deserves.

At speed, there is a good deal of roar from the front tyres, while at the rear, as mentioned before, sound seems to be pumped up the rear quarters. These noises are relative, however, and after a drive of 200 miles one can emerge unruffled, calm and in an "instant" frame of mind. The only thing which worried me was the rather alarming rate at which five-star petrol disappeared from the 22-gallon tank. But when you have paid £8,000 this is a minor point...

Dress is another important Rolls-Royce requisite. I was wearing a sweater and flannel trousers when I took the car out, and people tended to look at me as either a car thief or as a young whizz-kid power-house executive. If I were either of these I would not be writing this...

Rally jackets and scruffy casuals are not recommended... unless you want the Spirit of Ecstasy to look remarkably like a frightening great aunt who has just had her Pekingese kicked.

I fell for 100 LG.

When I parked her outside my flat, between the go-faster Ford Cortina GT and a humble Viva HA, it looked as if Buckingham Palace had been moved to a housing estate. And as she is valued at rather more than my flat, I was tempted to spend the night curled up on the back seat.

It may have been luck or just imagination, but when I took her back to the office the next morning, all traffic lights seemed to be green, and all coppers seem to have a sixth sense about a Rolls, always stopping the car *behind* you. **M.L.**

Above: 100LG on M6, cruising at a silent 70 mph. Right. Leaving the Chateau d'Oron, high above Lausanne. The Chateau is open to visitors daily, but is closed in January

West in Luxury

By Peter Garnier

Sampling the H. J. Mulliner, Park Ward Silver Shadow Convertible

The ageless styling of the Silver Shadow convertible combines something of the stateliness of the 'thirties with an up-to-date, sporting look—and ensures that the man who spends £12,078 13s 8d will not have something "dated" on his hands for a great many years. Some of the craftsmanship in wood and leather can be seen in the cockpit picture

HOWEVER supercilious a motoring journalist may become through a professional familiarity with cars far beyond his means, there remains always a reverence for the name Rolls-Royce. It is partly because there aren't many of them; and partly because the driver of a Rolls is at once judged to be extremely rich and probably famous too. There is the instinctive association of reliability, comfort and engineering perfection with the name—even to the extent that, seen through a window of an airliner, the Rolls-Royce badge on an engine cowling gives a profound reassurance of security. The pleasure of driving one is always tempered with slight misgivings as to the responsibility one is accepting in respect of the car itself, and one's conduct at the wheel—for one is very much in the public eye.

It was with these inner thoughts that I set off in an H.J. Mulliner, Park Ward Silver Shadow Convertible on the very familiar run down to

WEST IN LUXURY...

west Cornwall one Friday night after work—a journey I have made three or four times a year for the past 35 years in (or on) every possible sort of vehicle. I had studied the handbook (£2 10s, and produced like a limited edition of some rare book—which I suppose, in effect, is what it is); and I had recalled the traditionally accepted edict of the Rolls-Royce School of Instruction—if you're involved in an accident when you're moving, you are certainly to blame; if you have one when you're parked, you very probably are. It didn't leave a lot of room for doubt. Of all the company's products, this is essentially the one for the owner-driver, and although it is a fairly large car, there is never any question of wondering where the extremities are—you very quickly learn to judge the length and width without any bother.

One does not drive these beautiful cars often enough to grow accustomed to them—so that first impressions are of greater value perhaps. There is the feeling that one gets on entering one's bedroom in a *very* expensive hotel—that no single requirement in the way of comfort or convenience has been overlooked; and there is the quality of fits, finishes and the working of controls—smooth and precise—that bring satisfaction to the driver. And, as we threaded the long bonnet—blue like a jay's wings—through the rush-hour traffic we found ourselves talking quietly, as in a church, because everything else was so quiet. Even the ticking of the clock—traditionally the only sound in a Rolls-Royce—was absent, the clock being electric. There was the novelty of being able, at the touch of a switch, to raise or lower the front seats, or tilt or move them bodily backwards or forwards, which beguiled us in traffic jams.

Because of the silence even at high speeds—a great achievement with a convertible—it is difficult at first to judge speeds, and one finds that the speedometer reading is always much higher than one expects. One sweeps along in an atmosphere of good living, gobbling up the smaller fry—even some highly sporting ones—with no effort whatever, and thinking what fun it would be on a Continental *autoroute*, cruising proudly at the speeds for which the car was designed.

I had been warned not to be alarmed at the rate at which the 22-gallon tankful of 5-star fuel tends to disappear—but at our first top-up, which had included getting out of London, the consumption worked out at 12.8 mpg. On the return journey, a shade over 100 miles covered at an average of 48-odd mph gave a consumption figure of 14.1 mph, which isn't bad for a 6.3-litre vee-8 pulling roughly 2 tons of luxury.

It takes a few miles to "learn" the steering. This particular car was on radial tyres, which give better response, but even so the car runs wide on fast bends—so that instinctively one puts on more helm, and overdoes it, making an untidy line through the corner. After a while, however, one grows accustomed to it completely.

It is perhaps not until the end of a long drive that one most appreciates Rolls-Royce motoring, when one realises suddenly that the journey's over; that it has been as near effortless as motoring can be, and that one is neither tired nor stiff. We had purposely postponed our open-air motoring for pottering round the Cornish lanes, but unfortunately the weather broke up. The hood-lowering operation is extremely simple and quick—with the car stationary, handbrake on, gearlever in neutral and ignition switched on, all one has to do is release the two big over-centre catches that secure the hood to the windscreen top rail, and operate a switch. It folds itself away in no time, stowing itself in the big recess behind the rear seats that serves as useful in-board luggage space when the hood is raised. Due to the height of the rear suspension units, it projects somewhat above the body, but packs away very neatly under its press-stud cover.

The return journey was in heavy, continuous rain and in memory has become a sort of leap-frog from heavy lorry to heavy lorry along the narrow roads of the West Country. Worse almost than the dust clouds that follow one along the route of the East African Safari, these were accompanied by vast clouds of spray, making overtaking very tricky. Once or twice I eased out amid the spray looking for an opportunity to overtake, only to find something coming from the opposite direction. It is here, perhaps, that one notices the steering characteristics most; it requires a much greater than usual wheel movement to pull the car back again.

By the time we eased off finally at the outskirts of London we had covered 700-odd miles, becoming thoroughly familiar with the car. I don't think we had brought the name of Rolls-Royce to shame, and we had grown to like the car and its way of eating up the distances so much that it was a very sad moment when we returned it to Conduit Street.

Ideas above our station, maybe; but not until you have travelled far in a Rolls-Royce do you understand where—in the case of this convertible—the £12,000-odd goes. It is motoring in a class of its own, with intangible qualities playing their part.

Rolls-Royce (and Bentley) interpretation of the "Coke-bottle" kink over the rear wheel arch is unique to the Mulliner, Park Ward bodywork. For photographic purposes, the hood was hurriedly lowered on a rainy day—we apologise to the creators of this elegant car that we did not fit the cover, which tidies up the hood in its stowed position

---

ROAD TEST / John Bolster

Rolls-Royce Silver Shadow

The exterior is unchanged, but within is a 6.7-litre engine.

Still a car apart

Quidvis Recte Factum Quamvis Humile Praeclarum

"Whatever is rightly done, however humble, is noble." That Latin hexameter was carved on the mantelpiece in the sitting room of Sir Henry Royce's house at West Wittering. The great designer always claimed that he was only a mechanic, but if genius is an infinite capacity for taking pains, there is no doubt that Frederick Henry Royce was the greatest genius of them all.

Though Royce died in 1933, his principles are followed implicitly in every car that comes out of the factory at Crewe. He never accepted anything as perfect or incapable of further development, and that is why it is always so exciting to try the latest Rolls-Royce car. The reason for my recent test of the Silver Shadow was simply that the size of the engine had been increased, but I knew that I would find many subtle differences, for development goes on all the time.

The design of the Silver Shadow is now well known. The big light-alloy V8 has a new crankshaft, bringing the dimensions up to 104.1 mm × 99 mm (6745 cc), but the power output is still not revealed, though as the compression ratio is 9 to 1 and peak revs appear to be about 4500, one can make a shrewd guess. The increase in capacity has certainly made the low-speed torque even greater than before.

The last major change was the substitution of a torque converter and three-speed automatic transmission for the former fluid coupling and four-speed automatic box, which gave a slight performance increase and a vast improvement in smoothness of operation. The gear selector lever is an electric switch, the actual work of selection being performed by an electric motor. It is now impossible to withdraw the ignition key unless the lever is in the Park position, and the ignition cannot be switched on if the unilateral parking lights are still lit.

The Shadow is, of course, the first Rolls-Royce with an integral chassis and body. There are three sub-frames, however, mounted on stainless steel mesh insulation, the two at the back carrying the hypoid unit and the pivots for the independent rear suspension respectively. The front sub-frame locates the suspension pivots, but it also supports the engine and gearbox unit and the power steering.

Two camshaft-driven pumps, backed up by two nitrogen-filled hydraulic accumulators, supply the high-pressure fluid for the self-levelling suspension and the brakes. Recently the self-levelling of the front suspension has been deleted because it was found to do virtually no work, but the rear self-levelling is constantly employed in compensating for the varying weight of passengers, luggage and fuel. The appearance of a good-looking car can be ruined if the back rides too high or too low, and the angularity of the driveshafts is reduced if the car is kept level. There are two entirely separate power-braking systems, and a third circuit is operated directly from a master cylinder connected to the pedal, which provides only a small proportion of the braking but does give some "feel."

When I took over the Silver Shadow, I moved my seat in any direction I wished by touching a switch; electric window operation, of course, one now takes for granted on a car of high quality. A new device, which is so useful that I cannot imagine how I have lived without it, is a switch that instantly locks or unlocks all four doors—no fumbling to get them open while your friends shiver outside.

The bigger engine justifies itself because it is even more effortless than the 6.3-litre unit. The difference in acceleration is scarcely perceptible at low speeds, but a fifth of a second is saved on a standing quarter-mile and over a second from a standstill to 100 mph; there is barely 2 mph difference in the maximum speed. The car surges forward so quietly that the acceleration never seems as brilliant as it is and, although I am very very familiar with the Shadow, I sometimes tended to enter corners excessively fast on this latest version until I had become fully accustomed to its deceptive speed.

The engine never seems in a hurry, and it is literally no noisier at its maximum speed than at quite moderate velocities. The car accelerates remarkably quickly up to 118 mph, and will just encompass 120 mph if it is kept flat out on a very long straight. The gearchanges go through so smoothly that it is quite difficult to know which gear is in use. I used to employ the manual selector a good deal on the previous transmission, but with the latest box there is seldom any temptation to take over oneself; I make exceptions on such occasions as when corners follow one another in quick succession and it may be advantageous to hold the intermediate gear, or when awaiting an opportunity to overtake on a hill. Generally, though, I do it the lazy way.

In addition to the very low level of mechanical sound, the absence of wind noise is remarkable, presumably due to such things as the perfect fit of the windows and doors. With the earlier Shadows road noise was something of a problem, but each one I have driven has shown an improvement, the latest car being quite a step forward in this direction. There are some road surfaces which are still just audible, but this is only because the car is otherwise so quiet.

The restrained interior is flawlessly finished.

During the lifetime of this model higher-geared steering has been fitted and stiffer roll bars have been incorporated in the front and rear suspension. The springing is still very soft but the car never wallows, feeling taut and responsive. The power-assisted steering is extremely light, but it has quite rapid caster-return action and sufficient feel. An inexperienced driver might tend to oversteer the car at first, but after a little experience the effect disappears. The balance is in fact very good, with just enough understeer for stability, which can be converted to oversteer on corners if desired by the use of that massive torque. This is a big, heavy car, but it is very much at home on the faster bends.

The complexity of the braking system is justified by the results and, though the pedal pressure is moderate, one does not tend to overbrake. The brakes are progressive and their effectiveness does not seem to vary whether they are hot or cold. The discs are wound with iron wire to prevent whistling noises, and this is completely effective.

Compared with smaller cars, the Rolls-Royce runs with a lack of effort and a feeling that there is a vast power reserve. Even at a silent 100 mph cruising speed the accelerator pedal is scarcely depressed and the passengers, lolling in their leather club armchairs, casually estimate the speed at 60 mph. Curves in the road are taken in the car's stride, with no swaying or screaming of tyres. If hard braking is called for, the built-in antidive geometry of the suspension prevents the occupants from being disturbed.

There are separate heating systems for the upper and lower levels inside the car, so cool breathing air and warmth for the toes can be dispensed simultaneously. The elaborate heating and ventilation system is allied with a remarkably compact refrigerator installation for cooling the air, with a compressor driven by twin belts from the engine. Full air-conditioning is now standard on all Shadows, as is a radio with an electrically raised antenna. An optional extra, which was installed on the test car, is the Radiomobile Stereo 8 tape player with loudspeakers concealed in the doors. The quality of reproduction is superb and the driver can "post" another tape while travelling at full speed without sparing a glance from the road.

All these things add to the pleasure of the journey and ensure that the passengers arrive feeling fresh after a long day's motoring at high average speed. There is also a wonderful lack of anxiety, for the Silver Shadow is not only extremely stable, but incorporates more safety features than could possibly be catalogued here. To some of us the price may seem very high, but the Rolls-Royce is still a car apart for which there is no substitute. In proof of this, more of them are being sold than ever before in the company's history, and the production now exceeds 2,000 cars a year, while provision is being made for expansion up to 2,500.

The complex auxiliary equipment helps the big V8 to fill the engine bay completely.

SPECIFICATION AND PERFORMANCE DATA
Car Tested: Rolls-Royce Silver Shadow four-door saloon, price £9,271, extra Radiomobile Stereo 8 tape player, price £150, both including tax.
Engine: Eight cylinders, 104.1 mm x 99 mm, 6745 cc. Pushrod-operated overhead valves. Compression ratio 9 to 1. Twin SU HD8 carburetters. Power output not disclosed.
Transmission: Torque converter driving 3-speed automatic gearbox, ratios 1.00, 1.48 and 2.48 to 1. Hypoid final drive, ratio 3.08 to 1.
Chassis: Combined steel body and chassis. Independent front suspension by wishbones, coil springs and anti-roll torsion bar. Recirculating ball power-assisted steering. Independent rear suspension by trailing arms, coil springs with automatic height control, and anti-roll bar. Telescopic dampers all round. Servo-assisted disc brakes on all four wheels with eight pads per disc and triple hydraulic circuits. Bolt-on disc wheels, fitted Dunlop 8.15-15 low-profile cross-ply tyres.
Equipment: 12-volt lighting and starting with alternator. Fuel gauge with oil level indicator. Ammeter. Oil pressure and coolant temperature gauges. Clock. Heating, demisting and ventilation system with refrigerated air conditioning and heated rear window. Flashing direction indicators with hazard warning. Windscreen wipers and washers. Unilateral parking lights. Reversing lights. Electric seat adjustment. Electric window lifts. Radio with electrically raised aerial.
Extra: Radiomobile Stereo 8 tape player.
Dimensions: Wheelbase 9ft 11½ins. Track 4ft 9½ins. overall length 16ft 11½ins. Width 5ft 11ins. Weight 41½cwt.
Performance: Maximum speed 120 mph. Standing quarter-mile 17 s. Acceleration: 0-30 mph 3.2 se, 0-50 mph 7.2 s, 0-60 mph 9.8 s, 0-80 mph 16.5 s, 0-100 mph 28.4 s.
Fuel Consumption: 12 to 15 mpg.

The sober, restrained lines of the Silver Shadow are perfectly in character with its traditional heritage.

New Rolls-Royce CORNICHE

By Geoffrey Howard

More power, better interior and revised trim details for two-door H. J. Mulliner, Park Ward models

It was singularly appropriate that just when the overall fortunes of the company had suffered a devastating setback, Rolls-Royce should launch a new car. Of course there had been virtually no warning in the car division that a financial crisis was brewing, and being still a very solvent and profitable offshoot of the main aero-engine division, Rolls-Royce cars had specific instructions from the official receiver to continue as normal.

At first glance the idea of taking a small party of British journalists to the South of France to try the new model seemed totally incongruous with the immediate state of the company. Judged objectively the setting was absolutely right for sampling this new model, which is specifically named after the three roads which link Nice with Monte Carlo at different levels up the hillsides that rise so dramatically from the deep blue of the Mediterranean in this region. And after all, as David Plastow (managing director of the car division) pointed out, the whole operation cost substantially less than half one of the new cars.

Out-of-season Nice is still a busy place and an impending film festival brought many visitors to the town. The sight of no fewer than nine brand new Rolls-Royces lined up in front of the airport created quite a stir. When they moved off in a ragged sort of convoy along the *Promenade des Anglais*, the crowds out walking in the warm winter sunshine stared in wonderment. We felt like the liberation troops entering Paris and the whole scene began for me to take on a dream-like quality that never faded until we splashed down on to the wet tarmac at Gatwick the next day.

Above: Formidable line-up outside Nice airport. On the left is David Plastow, managing director of the car division

Below: The two-door saloon is particularly elegant in this view. Coachwork is by H. J. Mulliner, Park Ward

As well as introducing a new car, Rolls have instigated a new policy whereby the special-bodied models are to be engineering and styling leaders for the standard production cars of the future. The two-door saloon and convertible built by H. J. Mulliner, Park Ward have therefore been made more different from the standard Silver Shadow and will in future be known as the Corniche, in either Rolls or Bentley guise. They have a new, more powerful engine, a revised facia incorporating a rev counter, a wood-rimmed 15in. dia. steering wheel, new wheel trims and a deeper (fore and aft) radiator shell.

The engine is the latest 6,750 c.c. version of the Silver Shadow vee-8 unit with 10 per cent more power and better breathing in the upper rev ranges. Retiming the camshaft, removing the Federal anti-smog equipment, increasing the bore of the exhaust system from 2 to $2\frac{1}{4}$in. and fitting a more efficient air cleaner have all played their part in improving the engine characteristics. Because it does not comply with the US regulations on exhaust emissions this engine cannot be fitted to cars destined for the North America market; they will have the normal 6,750 c.c. Silver Shadow engine.

To improve engine cooling in heavy traffic the drive ratio to the fan has been stepped up, and the new wheel trims allow more air to get to the brakes.

The name Corniche was already registered by Rolls-Royce before the war when it was a prototype Mark V Bentley with a body incorporating some unusual aerodynamic features, built by van Vooren of Paris. It had completed 15,000 miles of endurance testing

Left: The radiator shell is slightly deeper from front to back and is actually raked forward by 3 deg

on the Continent when war broke out and it was blown up by a bomb while waiting to be shipped home on the quayside at Dieppe. *Corniche* itself can be translated as cornice, which in mountaineering terminology means an overhanging outcrop, hence the three coast roads along the Mediterranean cliffs.

By comparison with the last Silver Shadow we tested, which had the original 6,230 c.c. engine, the Corniche is a much more lively car. It really feels now as if the engine is large enough for the huge size and weight of the body, and its step off the mark from rest is now most impressive. At the top end it will run quite easily into the red sector on the rev counter (from 4,500 to 6,000 rpm) even in top gear (4,500 rpm is equivalent to 118 mph). The claim that the car will exceed 120 mph is therefore undoubtedly true, the example I drove reaching this speed on several occasions without any difficulty.

Smooth automatic transmission

Despite the extra torque, gearshifts in the three-speed General Motors automatic transmission are as silky smooth as ever and barely detectable most of the time. When using the selector to get engine braking on a descent it is very easy to stay in intermediate by mistake, so near silent is the engine.

No claims are made for suspension improvements, but there has been a progressive programme of development in this department and the latest car is much better than previous ones. Most of our criticisms of poor directional stability and slow steering response have been answered and the new small wheel helps a lot in giving the driver more feel of what is going on. There is still very strong understeer, largely disguised by the power steering, and excellent straight line running. Rolls now have their tyres made specially by Dunlop, Firestone and Avon, the car I drove being on Firestone F100 radials. There were some out-of-balance tremors at about 100 mph, but much less bump thump than before and no harshness.

One Rolls-Royce passing through the Franco-Italian border at Menton might cause little comment, two in close proximity take longer to clear, and with nine altogether we were in trouble on the Italian side. They were

Below: Rolls-Royce Corniche in convertible form, still a very popular version

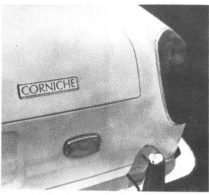

Above: Script for the Corniche nameplate is in traditional Rolls style

Above: New hub caps are plain with space around for ventilated wheels to cool the brakes

Left: The new wood-rimmed steering wheel is only 15in. dia. and there is a rev counter for the first time

suspicious of £115,000 of rolling bullion naturally enough and had not Rolls-Royce (France) thought to have a smooth-talking solicitor on hand our driving might have been confined to France alone. As it was we were able to make our way slowly to Genoa and then rush quite a way up the Turin *autostrada* before retracing our steps to Monaco and a night's stop at the Vistaero hotel, appropriately situated on the *Grande Corniche* overlooking the Cap Martin.

The price of the Corniche is about 10 per cent more than that of the previous two-door model, the saloon costing now just over £12,829 as a Rolls-Royce and £12,758 as a Bentley. The convertible is about £600 more. We look forward to carrying out a full test of the Corniche saloon later this year. □

ROAD TEST / John Bolster

Rolls-Royce introduce the Corniche

In Edwardian times a fair number of Rolls-Royce owners kept villas on the South Coast of France, and it was largely for commuting to and from this area that the Silver Ghost was introduced. This high-performance model was built at the suggestion of the Hon. C. S. Rolls, and many photographs still exist of the Honourable Charles driving his exquisite car along the Grand Corniche. It was therefore natural, when a new grand touring Bentley was on the stocks in 1939, to call it the Corniche. The car was destroyed by enemy action at the start of the second war, and the model never went into production. However the name has at last been used, for it has now been applied to the fastest Rolls-Royce and Bentleys that have yet been catalogued.

It was a splendid idea for Rolls-Royce to invite a few of us down to this delectable area for a preview, and to let us loose on the roads of France and Italy with their new creations. These beautiful machines caused something of a sensation and the local press broke the announcement embargo in depicting our appearance on the streets of the Principality! From the crawling traffic in the sunshine of Nice and Cannes to the fast curves and gradients of the Autostrada, this was an ideal playground for testing the great cars in every mood.

The Corniche is powered by the 6750 cc engine which was recently introduced for the Silver Shadow, but in this case the secret bhp figure is increased by 10 per cent. A special camshaft, allied with an exhaust system of larger diameter and an air silencer giving a free flow, are the principal ingredients, while the breathing is not restricted by American anti-pollution equipment. The gearbox is the three-speed General Motors automatic which replaced the former four-speed transmission, with performance and smoothness advantages.

The independent four-wheel suspension is that of the Silver Shadow; it is self-levelling at the rear according to load. As befits a more sporting car, the well-equipped instrument panel contains a rev-counter, with a red sector starting at 4500 rpm. A smaller steering wheel is in line with current fashion, and the new stainless steel trim of the road wheels gives better brake ventilation.

A gentleman can no longer instruct his own coachbuilder in these days of monocoque construction. However Rolls-Royce own the only major company which continues the tradition of English coachbuilding—H. J. Mulliner, Park Ward. Two styles, a two-door saloon and a convertible, are built on the floor pan of the four-door shadow. The origins of Mulliners go back into the 18th century, and it is fitting that they employ the finest craftsmen in the country. On the other hand Park Ward were pioneers of steel construction, and they have evolved the most complete series of processes ever employed for protecting a car against corrosion. The door bonnet and boot lid are of aluminium for light weight.

H. J. Mulliner, Park Ward employ 770 men and eight Corniche models are under construction at once, each car taking an average of four months to complete. A week of testing and tuning by Rolls-Royce follows, and every vehicle undergoes a 20-minute monsoon test for complete exclusion of water under the most extreme conditions. Refrigerated air conditioning and a radio are standard, with much optional equipment and various extras

The interior is very plush and the dashboard extremely comprehensive as would be expected for £13,000.

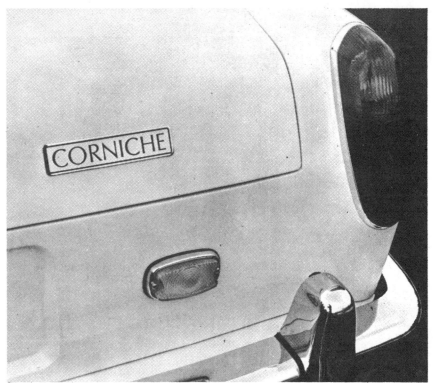

The new wheels give a better flow of cooling air to the disc brakes (left). The name plate is the external giveaway of the Corniche (right).

to make each car an individual construction. The styling is based on the previous two-door models with detail changes: the low radiator grille of the Silver Shadow is retained, but for the Corniche it is a little deeper from front to rear, which emphasises the famous Corinthian shape.

One feels a certain responsibility in handling a £13,000 car. Nevertheless I risked this masterpiece among the slightly crazy Italians and the somewhat casual Monégasques without incurring an unforgivable scratch. It has perfect manners in town, as one would expect, while the extra power has been found without marring the silence or smoothness. The car cruises with restful ease at 100 to 110 mph, and will reach a genuine 120 mph very quickly. I did not try beyond 120 mph as the rev counter was then just entering the red section, but the sound level did not differ from that at 30 mph. The rev-counter is an essential piece of equipment as the more powerful engine can achieve injudicious speeds so easily; an ignition cut-out would be a worthwhile addition, which shows how racing practice can be usefully employed on the most luxurious of cars.

In spite of the great weight of such a car, it can be handled in a truly sporting fashion. It hangs on splendidly on fast curves, only the sound of the tyres betraying what is afoot. On sharper corners the degree of understeer never becomes excessive, and the independent rear suspension *almost* prevents wheelspin at the traffic-lights grand prix. The steering is so light that it might seem disconcerting at first to one unaccustomed to the car, but this is a matter of practice. As would be expected, the big machine is rock steady in side winds.

The brakes are really silent at all times, controlling the weight with ease. Useful engine braking can be employed to steady the car by slipping the gear selector into the Intermediate position on descending a steep hill. The suspension is firmer than that of the earlier Shadows and the anti-roll bars are stronger, which gives one greater confidence for fast-driving, but the ride is still supremely comfortable. The transmission of road noise to the interior, always a problem with an extremely quiet car, has at last been completely overcome, the monocoque Corniche reaching the high standards set by earlier Rolls-Royces with separate chassis. However,

it is the surge of torque from those eight big cylinders that is the most impressive thing of all.

It is necessary to mention the problems that are currently besetting Rolls-Royce, and to emphasise that these do not directly concern the Car Division as such. It is widely believed that the design of the cars owes much to the aircraft side, but in fact no technical assistance has been forthcoming, except that which has been paid for at full price. The Car Division is a successful, profit-making concern in its own right and the production of " The Best Car in the World" is greater than it has ever been before. Heroic measures will have to be adopted to salvage the Turbine Division, but this surely cannot be allowed to interfere with car production. The order books are full and the Company now has a most competitive range of models—if, indeed, there is anything to compete with them. I have faith in the great name, and I am confident that the best days of the Rolls-Royce car still lie ahead.

SPECIFICATION AND PERFORMANCE DATA
Car described: Rolls-Royce Corniche two-door saloon, price £12,829.36 (including tax, convertible £13,410.34. Bentley T-Series £12,757.56, convertible £13,332.01.
Engine: Eight cylinders, 104.1 mm x 99.1 mm, 6750 c.c. Light alloy V8 with pushrod-operated overhead-valves. Compression ratio 9.0 to 1. Power output not stated. Twin SU carburetters.
Transmission: Automatic 3-speed transmission with torque converter and electrical gear selector, ratios 1.00, 1.48 and 2.48 to 1. Torque converter multiplication at stall 2.04 to 1. Hypoid final drive, ratio 3.08 to 1.
Chassis: Combined steel body and chassis. Independent front suspension by wishbones, coil springs and telescopic dampers on sub-frame with resilient metal mountings. Recirculating ball power-assisted steering. Independent rear suspension by trailing arms with coil springs and telescopic dampers, with automatic ride control and anti-lift geometry, on sub-frame with resilient metal mountings. Anti-roll bars both ends. Power-assisted disc brakes on all four wheels with three independent circuits. Bolt-on disc wheels fitted 205 x 15 radial ply tyres.
Equipment: 12-volt lighting and starting with alternator. Speedometer. Rev-counter. Fuel, oil level, oil pressure and coolant temperature gauges. Ammeter. Voltmeter. Clock. Full refrigerated air-conditioning with heating, demisting and ventilation. Heated rear window. Electrical window operation, seat adjustment and centralised door locking. Electrically raised radio aerial (radio standard). Electrically released petrol filler flap. 2-speed windscreen wipers and washers. Flashing direction indicators. Reversing lights. Cigar lighters front and rear. Extra: stereo tape player.
Dimensions: Wheelbase 9 ft 11.5 ins. Track 4 ft 9½ ins. Overall length 16 ft 11.5 ins. Width 6 ft.
Performance: Maximum speed 120 mph. 26.2 mph per 1000 rpm on top gear.

The H. J. Mulliner, Park Ward coachwork is largely hand built and each car takes an average of four months to build.

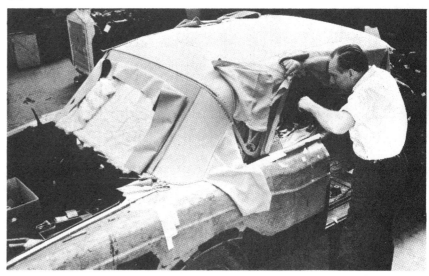

55

Above: For safety reasons, "Spirit of Ecstasy" is now spring loaded to prevent injury to unfortunate pedestrians.

IT WAS with trepidation that we approached the latest Silver Shadow Rolls-Royce.

But it was the legend, rather than the car which produced the feeling. The car itself simply proved to be the easiest luxury car to conduct in crowded conditions that we've tried.

As Kellow Falkiner's managing director backed his deep red Silver Shadow from its spot in the company car park, it seemed he was about to wipe off at least a mudguard — so close was he going to the pillars.

But this is all part of the Rolls-Royce "tight manoeuvring" syndrome. When backing and filling, you can be sure the car's extremities stop at the side window line.

Sallying into peak hour traffic proved the immediate benefit of a parthenon shaped stainless steel radiator topped by "The Spirit of Ecstasy."

Despite an introverted approach to traffic produced by the Rolls' aura and the knowledge that it costs $27,000, the car carried all before it.

At intersections, room was left for the Rolls. At car parks, attendants leapt . . . and so on.

Guiding a Rolls for a couple of days produces sufficient anecdotes to fill this column without telling you anything about the car.

The first thing to note is that this is not the same car as was released some five years ago. At least 100 changes have been made, many with United States' safety regulations in view.

The 6.2 litre V8 car has all independent suspension and more modest dimensions and styling than previous models.

The less ostentatious look of the car was one of early critic's favorite harps. But it has advantages. There have been some 83 of the current body style Bentley and Rolls-Royce cars sold (5:1 in Rolls favor, incidentally) but it wouldn't appear so looking about the roads.

The design of the Silver Shadow followed two basic tenets — do what the others do a little better and keep the final car smaller outside and bigger inside to please customers.

Rolling along in the air-conditioned, green windowed world of the Rolls, we were able to confirm the attainment of the objectives.

Sound deadening snuffed the sound of surrounding traffic and replaced it with the sough of air, temperature regulated for maximum comfort.

The physical business of driving the car was ludicrously simple. The gear selector was placed in "D" and then everything was power assisted with the exception of the accelerator. In fact even the gear selector, really an electric switch, was power assisted.

for the man who has everything...

$27,000 will buy you just about any worthwhile car available in the world today — but there are not too many that are able to challenge the "Best car in the world" claim made by Rolls Royce. We tested the latest Silver Shadow model to find out why.

At first the power assisted steering needed care because it was very light. But the smaller wheel of the current car helped here and the variable ratio steering hadn't lost its feel in the search for effort — free parking.

The brakes were the car's piece de resistance. It had three separate braking systems to operate the four disc brakes.

One looked after the front, another the back, another shared some of the back with power assisting the automatic suspension levelling device and there was the handbrake of course.

If there should be a failure of one system, at least 60 per cent. of the brakes would always be there. Helping guard against the phantom failure which might go undetected were a pair of warning lights which watched pad wear, circuit condition and fluid level.

And to check these, you just pushed a dashboard button to set them alight. This button also checked the fuel warning and oil pressure warning system globes and gave a reading on the petrol gauge of the oil level in the sump.

Although the Rolls-Royce with its backup systems sounds rather complicated, it is quite a straightforward car to drive.

Mastering the controls for the interior's climate proved the hardest part. There were two basic controls but these had two scales of lateral adjustment, left and right, and then vertical adjustment for fan boosting and volume increasing.

The flow through ventilation system worked well. The electric window mechanics no longer have standby handles because with the comfort gear built into the car, there is practically no need to open them.

Power output of the motor is still not a subject Rolls cares to talk about. But there is sufficient there to run all the comfort systems, including the heavy drain of an air-conditioning pump and still propel the car to 50 mph in 7 sec.

The standing quarter mile took 16.8 sec., definitely sports sedan class.

The three speed transmission, a GM 400 automatic, worked impeccably giving 43 mph, 73 mph and 116 mph.

The fuel consumption of 13 mpg. was good.

But figures are rather too angular a measure for the Rolls.

It should be judged against comfort, serenity of travel, and by these parameters, it scores heavily.

Sitting at 90 mph, the Rolls is incredibly relaxed. There is plenty of power on tap for overtaking and there is ample braking for emergencies.

It is able to handle bends with a sort of dignified aplomb which should not be confused with the ponderous handling of many traditional luxury cars.

Suspension alterations have produced a tendency to transfer weight to the rear under hard cornering, enabling the driver to hurry the Rolls along in a sporting fashion. There is plenty of feel in the steering and the Avon cross ply tyres gave a very secure grip — at the cost of a short life we suspect at Australian road temperatures.

The suspension was by coil springs all round, the rear suspension being independent with semi trailing arms. All the suspension bits were isolated by wire wool type pot links on their body-mounting subframes to muffle noise.

The automatic levelling even worked quicker at rest then on the move to prevent road undulations being read as ride height changes by the mechanism.

Inside, the Rolls seats were excellently shaped, high at the back and electrically adjustable, of course.

The glove box has been reduced in size, unfortunately, to meet United States safety demands.

But Government regulations haven't been able to net the Rolls-Royce spirit. Not even the Spirit of Ecstacy on the grille, which now is spring loaded for safety.

Summing Up: Rolls-Royces still meet their legendary claim of taking you hither and yon a little better than the others can.

data sheet

SPECIFICATIONS

CAR FROM:
Kellow Falkiner's, Russell St., Melbourne.

PRICE AS TESTED:
$27,000 inc tax.

OPTIONS FITTED:
None.

ENGINE:
Type....V8, push rod, alloy construction
Bore and Stroke....104 mm x 91.4 mm
Capacity ....6230 cc
Compression ratio 9:1
Power (gross)........................NA
Torque ....NA

TRANSMISSION:
Three speed, automatic, steering column selector.

CHASSIS:
Wheelbase......................119.5 inches
Length ....203.5 inches
Track F............................57½ inches
Track R............................57½ inches
Width ....71 inches
Clearance (Minimum)............6½ inches
Test weight ....4600 lbs
Fuel capacity ....24 gallons

SUSPENSION:
Front: Independent, coil springs, wishbones, stabiliser bar.
Rear: Independent, coil springs, semi trailing arms, stabiliser bar, automatic height control.

BRAKES:
Power assisted two of three systems.
Front: disc 11 in.
Rear: disc 11 in.

STEERING:
Type: Power assisted, recirculating.
Turning circle: 38 ft.

WHEELS/TYRES:
8.45 x 15 in., Avon nylon cross plies.

PERFORMANCE

Zero to
30 mph 3.2
40 mph 4.6
50 mph 6.9
60 mph 9.5
70 mph 13.0
80 mph ....NA
90 mph ....NA
100 mph ....NA
Standing quarter mile 16.9 seconds
Fuel consumption on test 13.1 mpg on super fuel.
Fuel consumption (expected) 13.5 mpg.
Cruising range 300 miles.

Speedometer error:
Indicated   30 40 50 60 70 80 90 100
Actual   29 NA 49 NA 69 NA NA NA

MAXIMUM SPEEDS IN GEARS:
1st ....43 mph
2nd ....73 mph
3rd ....118.1 mph

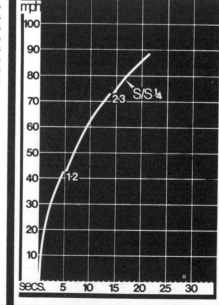

COMMENTS

ENGINE:
Response ....Good
Vibration ....Low
Noise ....Low

STEERING:
Effort ....Very Low
Road feel................................Good
Kickback..................................Nil

SUSPENSION:
Ride comfort ....Good
Roll resistance ....Good
Pitch control ....Good

HANDLING:
Directional control ....Very Good
Predictability ....Very Good

BRAKES:
Pedal pressure ....Very Low
Response ....Very Good
Fade resistance......................Good
Directional stability ....Excellent

CONTROLS:
Wheel position......................Good
Pedal position ....Good
Gearshift position ....Good

INTERIOR:
Front seat comfort ....Good
Front leg room......................Very Good
Front head room ....Very Good
Rear seat comfort ....Good
Rear leg room ....Good
Rear head room....................Good
Instrument legibility..............Very Good

VISION:
Forward ....Good
Front quarter ....Fair
Rear quarter ....Good
Rear ....Good

CONSTRUCTION QUALITY:
Paint ....Very Good
Chrome ....Good
Trim ....Excellent

GENERAL:
Headlights — highbeam ....Good
Headlights — lowbeam..........Good
Wiper coverage ....Fair
Wipers at speed ....Good
Maintenance accessibility......Poor

Rolls-Royce

The normal sedan body for the Silver Shadow is built by Rolls in its own plant, obtaining some stampings outside. While the radiator shell is no longer nickle silver, it is painstakingly fashioned.

Prices ranging from $19,600 to $31,600 for a car that looks from a side-view like a Rambler of a few years back and which is not much bigger don't stop some 600 Americans each year from buying either a Rolls or its companion Bentley T-Series which is the same car except for the radiator grille. Once you penetrate the Rambler-like exterior facade, though, you'll see that all resemblance between the two ends. The Rolls is truly luxurious, especially the coupes and convertibles which are fashioned by H. J. Mulliner, Park Ward, Ltd., the last of the great coach-builders.

The present Silver Shadow series was introduced in 1966 and of course, Rolls is not one for annual model changes. The normal sedan body for the Shadow and its post-WWII predecessors is built in the Rolls plant and is integral with the chassis. This enabled lowering the "new" model by a full five inches but it still towers above current U.S. traffic with a height of 59.8 inches. Other new features of the Shadow were a V-8 engine with an aluminum block and detachable steel liners, standardization on a version of GM's Turbo-Hydramatic transmission and fitting disc brakes to all wheels. Rolls has never published torque and horsepower ratings but it is reliably estimated that the new engine produces about 330 horsepower. Still, this is sufficient to propel the 4,650-pound car to 115 mph sedately and without fuss.

While it would be ridiculous to consider the Rolls in the same light as a sports car, one of these sedans could undoubtedly keep up with most of them on winding roads. Suspension is by coil springs front and rear with provision for automatic leveling. Brakes are disc at each wheel with no less than three independent systems, one being the Girling mechanical type should both hydraulic circuits fail. The handbrake even operates on all four discs and forms a fourth method of emergency stopping.

Air-conditioning and every form of power assist is standard equipment. Silky is the word for every major and minor control. The re-circulating ball steering needs only 3.6 turns from lock to lock but the car, of course, understeers. The technique is to yank it into turns and then straighten the wheel gradually but most Rolls drivers won't try this and besides, the tires are worth about $100 each.

Data in Brief

Silver Shadow and Bentley T

ENGINE: V-8 cylinder ohv, 380.5 cu. ins., 6230 cc, hp and torque not disclosed, Dual S.U. HD 8 carburetion.

DRIVELINE: 3-speed Turbo-Hydramatic, rear drive.

SUSPENSION: Coil front and rear, automatic leveling.

BRAKES: Girling discs front and rear, power assisted with auxiliary mechanical linkage.

STEERING: Re-circulating ball, power assisted.

DIMENSIONS: Wheelbase 119.5 ins., overall length 203.5 ins., width 71.0 ins., height 59.8 ins., weight 4700 lbs.

Rolls-Royce Silver Shadow

The Silver Shadow doesn't do anything other luxury cars don't do; it just does everything a little better and with more grace, precision and manners/By John Christy

Symbols of Empire are rarely, if ever, attainable by mere mortals, however wealthy or powerful. Lacking the proper blood-lines and Family background one can hardly aspire to the Throne. Not even an heir-apparent or occupant of that exalted place could possibly conjure dreams of ownership of the Bank of England. Nor of the Royal Navy. Command, possibly, with diligence, background and luck, yes; ownership, never. There is, however, one such symbol upon which the sun never sets (and probably never in our lifetime shall) that with diligent application to the work ethic, a modicum of luck, a great deal of ambition and the good taste to desire it, *is* attainable—a *new* Rolls-Royce.

A mere motorcar a Symbol of Empire? In the face of a shrinking Navy, devaluation of the pound and an honored though all-but-powerless Throne the Rolls-Royce is the sole remaining cachet of a British presence that once girdled the globe. Until recently, before the Federal Trade Commission began demanding absolute computer proof of all claims the men charged with advertising the Rolls-Royce had one slogan for it: The Best Car In The World. Not the biggest, not the fastest, not even the most technologically advanced. Simply the *best*. But how do you show proof of a sensation, a *gestalt* of sensations? How do you prove to a bureaucrat that this or that particular example will, if given proper care, be running 40 years hence? If the consumerists wish to consider this last claim as apocryphal they can also consider the fact that of approximately 55,000 Rolls-Royce cars built since the beginning, roughly half that number are known to be still running, many of them in regular use.

The phenomenon of longevity is no accident. It is the result of a form of engineering overkill that began with the first car that Frederick Henry Royce ever built and which continues to this day. Another reason can be discerned in the fact that employees charged with the mechanical function of constructing the cars, or any function other than administration or sales, are not referred to as "laborers" or "workers." The term is "workman" and it is significant. Sir Henry Royce, even after his elevation to knighthood, was known to list his occupation the same way. It is an example in action of one of Royce's favorite aphorisms which, translated from the Latin, reads: Whatever is rightly done, however humble, is noble. In building the Rolls-Royce car, now as in the beginning, each detail, however humble is treated as an exercise in nobility by those who perform them.

The resulting product of this form of thinking is more than a mere mechanical device but rather a melding of art, engineering, technology and artisanship. Consider, if you will, the wood used on the dash panel and door trim. Whether it is the standard Circassian walnut, or some such rare wood as coromandel, myrtle, paldao, Persian burr, birdseye maple, tola or sycamore, all available on order, each piece in a given Rolls-Royce has come from the same tree. You will be further informed that each tree was at least 100 years old at time of cutting and that a portion of the log in question has been placed in permanent storage and coded to that particular car. Should you be so unfortunate or careless as to mar a panel it can be replaced to match exactly. Even were all that unknown, the finish alone is enough to impress all but the most unaware. Unless a matte finish has been specified, the surface has the consistency and glister of fine plate glass, the result of a week of lacquering and rubbing in numerous layers by a cabinet-maker capable of impressing Louis Quatorze. Colleague Jim Brokaw was treated to a lesson in just how good the finish really is on a trip through the premises. At the end of a demonstration on finishing, the master cabinet-maker with seeming indifference ground—not stubbed, but *ground*—his cigaret out on a finished dash panel. He casually took his handkerchief out and whisked the mess away. There wasn't a mark on the flawless surface.

While still on the subject of such gantlets thrown in the face of a world devoted to production quotas, cost efficiency and proliferating plastic, we might also consider the upholstery, paint and trim. There are, in the upholstery of the average long wheelbase Silver Shadow, 10 hides, each the survivor of selection by a man who has worked at Rolls-Royce for 45 years. He chooses a mere thirty from some 10,000 hides offered for sale and has been known to express a preference for those whose previous owners grew up behind electrified fences rather than barbed wire. One barb scratch or insect bite scar is sufficient cause for rejection and the rejects from Rolls-Royce are used for the most expensive ladies' purses.

The exterior of the Rolls-Royce is protected by no less than 14 coats of very special lacquer, each coat sealed, rubbed and air dried in stages over a period of several weeks. Before the process is even started the entire monocoque body/chassis is totally immersed in a vat of corrosion-inhibiting chemicals. If, during the painting process, at any stage, a flaw such as a run or unevenly applied coat occurs, all the lacquer is removed and the process is started over again. It doesn't happen often, obviously, but if it does the *whole* thing is totally redone. Whatever is rightly done . . .

And then there is the trim—the brightwork that on lesser cars freckles with rust in a couple of years and turns cancerous in a few more. On a Rolls-Royce much of it is highly polished stainless steel of the highest Sheffield quality. That which can't be made from stainless due to forming difficulties or other technical reasons wears a triple coat of chromium over heavy layers of copper and nickel. Most prominent, of course, is the handmade, flat-planed radiator shell that, since the beginning has proclaimed the car to be a Rolls-Royce. There is an interesting point about the planes of the shell. The human eye sees a truly flat surface as slightly concave so the planes on the Rolls-Royce shell are almost imperceptibly convex so that the eye perceives them as flat. You can tell by looking at a close-up photograph or by laying a straight-edge along them but if you didn't *know* you would swear before the Queen's Bench that they were truly plane.

Stainless steel, triple chrome plate, rustproof dipping, attention to the minutest detail, silence as deep as technology and material can make it, longevity beyond the imagination of other carmakers, all are there because Henry Royce decreed it so in the beginning and the decree is followed to this day as though Sir Henry were still there to make sure of it. To Sir Henry, the best was only a starting point and he was an absolute believer in testing to see that he got it. Every possible part had a test tab formed on it which could be snapped off and sent out for metallurgical testing. If it failed the part and possibly even the whole lot would be sent to the discard pile. He was quite possibly the first carmaker to use what he called a "bump machine," the forerunner of today's dynamometer road simulaters on which any road surface can be duplicated with the car *in situ.* Royce's was much less sophisticated than that; it was simply set up to produce the most drastic pounding a car could ever be expected to take. The machine was capable of breaking up some of the world's best automobiles of the time in an hour or less but Rolls-Royces were expected to endure as long as the engineers wished to continue the torment. It was this sort of engineering overkill that allowed T. E. Lawrence to use them in his WWI Arabian campaign without any significant failure. They just didn't break axles or springs in spite of being equipped with armor plate and equipment that far exceeded normal loading.

This devotion to testing to a point far beyond normal use continues to the present. Engines are totally run in on a dynamometer and then run in some more on the road *before* the car is finally painted (so that any required adjustments won't subject the finish to damage.) The result is that there is no break-in period in a Rolls-Royce—you can run one flatout from the dealer's back door if you feel you can get away with it. Proof of the product is sometimes carried to a point that others might consider extreme. The late Ken Purdy once told me the story of the hood ornament that takes the place of the Flying Lady mascot on the royal Rolls-Royces. H.M. Queen Elizabeth's preference was a silver statue of St. George slaying the dragon. When the statuette arrived on the premises, R-R engineers agreed that it was quite in order as an ornament for a royal Rolls-Royce—as art at any rate. They firmly attached the piece to a small version of the early "bump machine" and proceeded to give it the equivalent of a year's worth of runs up and down the Baja California peninsula. After this they surveyed the piece carefully. When the car was delivered back to the Master of Horse, St. George's lance was firmly imbedded in the dragon's flank and the dragon's tongue curled up under the horse's belly. The statuette was now quite in order as R-R engineering as well as it was in order as art and a Symbol of Realm. This devotion to quality and the work ethic permeates the personnel of Rolls-Royce from the lowliest floorsweeper to the most senior administrator. "We couldn't build a bad car," Royce once said, "the doorman wouldn't let it out."

Possession, even temporarily, of the current Rolls-Royce, the Silver Shadow, is an exercise in discovery as well as an experience in emotions entirely different from those engendered by other forms of automotive exotica. A Ferrari, for instance, hammers at your senses and sets the blood racing almost before you turn the key—an instant turn-on. Not so the Rolls-Royce. If you have driven almost any large luxury car, there is nothing off-putting about your entry into the driver's seat of a Silver Shadow, nothing overbearing other than a faint sense of the *rightness* of things.

The feeling is reinforced when you insert the key in the ignition and twist it to start. Each movement is as precisely designated as is the cocking of the bolt on a custom Mannlicher carbine. It is only when you start to move the shift lever into one of the drive positions that the message begins to filter through that someone you've never met *cares* for you. There is no resistance, just a defined click into the proper detent, felt not heard. The reason is that *you* do not do the shifting; your movement of the lever has sent a message to a servo which does the shifting for you—instantly. As the car moves smoothly, almost eerily, away, the shifts are made imperceptibly and without a sound. Until you are used to it, it is very nearly impossible to tell whether the car is in Intermediate or Drive. The car is quiet but not silent. Contrary to legend the loudest noise you hear is *not* the clock (it's electric and therefore silent) but, at moderate speeds, the swish and rumble of the tires on the pavement. The quiet is of an order that it is no wonder that it is occasionally exaggerated. There is a faint mechanical hum from the V-8 engine. At speeds above 60 mph it does become rather audible but not obtrusively so because the loudest noise still comes from the tires.

As the car prowls through traffic another sensation makes itself known—a sense of precision that totally belies its 4800 lbs. laden weight. Move the narrow rimmed steering wheel just so much and the car turns in direct proportion. Except in really hard turns there is very little feeling of lateral force, either to the driver or the passenger. It is one effect that never ceases to surprise, especially when one is traveling in company with some other such luxury device as a Cadillac Fleetwood or Lincoln Town Car. Watching the movements of the other car as it negotiates its way through traffic, if there is a sufficient rate of knots being developed, leads you to expect the same water-borne action. When it doesn't happen there is a sense of mild astonishment that is only one of the pleasances rendered by the car. Another is the ride. To one used to the pliancy of the domestic luxury product the ride seems just a bit firm at first, not harsh, mind, just a little less pliant than expected. The impression lasts only until the first visible bump or dip which just simply is not felt. Rather than being soft or pliant it is, rather, compliant, ironing out every irregularity, regardless of speed. The feeling is one of awesome suspension control that begets a confidence unknown and unexpected in a conveyance of this genre and approached only by one of the more distinguished products of Daimler-Benz. Nothing less approximates or draws nigh.

Rolls-Royce never reveal, and never have revealed, the horsepower of the engines that have motivated their equipage much less trumpet power figures from the housetops. When pressed with questions about "how much" the answer is either "adequate" or "sufficient." Either way it means that what's under the hood is competent to the purpose of propelling the Rolls-Royce at a rate equal to, or better than all but the most muscular carriages of the commonalty.

That which propels the latest Silver Shadows and Corniches is a 411.9 cubic-inch V-8 of silken smoothness begot of total balance. Since we are eternally curious we put a Silver Shadow on the rollers of a chassis dynamometer and by a process of calculation deduced that the flywheel horsepower of the engine was approximately 300 as installed. In any event it is, as the people at R-R claim, adequate. Its adequacy is of an order that, if one is clod enough to punch out from a standing start, it will propel the car over a quarter mile in just over 17 seconds, not quite as quick as a Cadillac Coupe de Ville but quicker than a Fleetwood by a significant amount. It is also adequate to pull away from most traffic without being actually forced. And it is adequate to take the car to an actual 120 mph at which speed it is no more unsettling than it is at 80, which is to say not at all.

A car, especially one as heavy as a Rolls-Royce, should stop as well as, if not better than, it accelerates. The men of Rolls-Royce from the days of its founding have been devoted to braking to an extent beyond all but builders of pure racing cars. The first post World War II car, the Silver Dawn had not one but two hydraulic braking systems and one mechanical system. The current Rolls-Royces have three hydraulic systems, separate from each other, two of them servo assisted. They operate two separate calipers on each front wheel and a dual caliper on each rear wheel disc. They will stop 5000 lbs. of Rolls-Royce from 30 mph in from 28 to 30 feet which means the rate of retardation is on the order of 1.0 g or higher. From 60 mph they will bring the car to a standstill in 128 feet if you do the braking properly. They will also, so progressive is their rate of appliance when properly applied, bring the car to a normal stop with absolutely no final negative g-force apparent to driver or passengers, a trick only chauffeurs to the gentry usually bother to learn.

Driving a Rolls-Royce to its full capability of comfort is nearly as much an achievement as being able to drive a Ferrari to its limit of performance safely. It takes practice and skill to do either. The Silver Shadow, if driven as it is intended to be driven—with dignity and smoothness, whatever the speed or lack of it—will do things with a gentleness and precision that borders on the uncanny. There are other luxury motor carriages that are faster, others that are quicker and still others that are openly more technologically advanced, but none at all that will do it with the silkiness and quiet dignity of the Rolls-Royce. There never have been in any era.

It is inevitable that such a device as a Rolls-Royce would engender a certain amount of cultism. Among such, to contract the name to "Rolls" is to commit a gaucherie akin to bathing with a lady while wearing a hat. To a 32nd-degree Rolls-Royce cultist the only allowable contraction is "Royce" since it was he who actually built the first one. Charles Stewart Rolls merely bankrolled the ensuing ones. To a second-generation owner the best form is merely "my mo-

62

torcar'' as though there were no other.

It is a form that might be appreciated, but never admitted, by its makers. Competition there is but rivals, no. The Rolls-Royce is, after all, the Best Car In The World. Not the fastest, not the biggest. Simply—The Best.

Is the Best worth 28 to 36 thousands of dollars? If one has enough money that buying a Rolls-Royce is, like buying a Purdey shotgun, the same as buying a Ford or an Ithaca,. the answer is yes. If one's income and position demands the Best for business reasons, the answer is again yes. If one's income is not so high but adequate enough to pay for a Rolls-Royce on time and one is of an age that it is felt that only one more car is wanted for the rest of a lifetime, however long it lasts, there is no other choice. As for the rest of us, housing is, regrettably, more needful.

The final answer to the question might well have been given by Sir Henry Royce: ''Quality will be remembered long after price has been forgotten.'' Too, there is something to be said for a conveyance in which one rides tall enough to knock one's pipe ashes on the roof of any lesser car with the temerity to pass by. Because it is a Rolls-Royce the act could probably be gotten away with but to do so would be to betray the trust of those who made it. ■

SPECIFICATIONS: ROLLS-ROYCE SILVER SHADOW

PHOTOS:

No matter where you look in a Rolls-Royce the message comes over that somebody you never met cares for you. While such little pleasances as an ashtray and lighter in each rear door, mirrors in the quarter posts, courtesy lights and deep-pile carpeting are found in other luxury cars, the way they are thought out in the Rolls-Royce is unique. The reading lights above the passenger seat may be turned on without bothering the driver in the least. Courtesy lights in the doors light not only the floor but the ground when opened and flash a red warning to traffic approaching from the rear. Front seats are electrically adjustable five ways and have a further mechanical adjustment for rake. The trunk (excuse us, boot) is fully lined with the same pile carpet as the interior. Leak-proof battery and tool box reside in a compartment on the left, also covered with the same carpeting. The engine shows the same attention to detail as every other part of the car and was obviously assembled by people with more than a nodding acquaintance with aircraft techniques. One detail should endear the Rolls-Royce to insurance people: next to the steering column is a cubby with a removable bridge that, when taken out and pocketed, deactivates all electrical circuits.

LONG WHEELBASE SEDAN

Engine	90° V-8 OHV
Bore & Stroke—ins.	4.1 x 3.9
Displacement—cu. in.	411.9 (6750cc)
HP @ RPM	Adequate
Torque: lbs.-ft. @ rpm	Sufficient
Compression Ratio	9.0:1
Carburetion	2 SU HD-8
Transmission	Turbo-hydramatic
Final Drive Ratio	3.08:1
Steering Type	Recirculating ball
Turning Diameter (curb-to-curb-ft.)	39.2
Wheel Turns (lock-to-lock)	3.5
Tire Size	205VR x 15
Brakes	Disc disc
Front Suspension	Self-leveling, independent, coil springs, concentric shocks, lower A-arm, upper control arm.
Rear Suspension	Self-leveling, independent, coil springs, trailing arms, tube shocks
Body/Frame Construction	Integral w steel body, aluminum hood, doors and lid.
Wheelbase—ins.	123.5
Overall Length—ins.	207.5
Width—ins.	71
Height—ins.	59.75
Front Track—ins.	57.5
Rear Track—ins.	57.5
Curb Weight—lbs.	4850
Fuel Capacity—gals.	27.5
Oil Capacity—qts.	8.5

PERFORMANCE

Acceleration	
0-30 mph	3.4
0-45 mph	5.6
0-60 mph	10.1
0-75 mph	16.4
Standing Start ¼-mile	
Mph	77.5
Elapsed time	17.2
Passing speeds	
40-60 mph	5.75 (5.25 in 2nd)
50-70 mph	7.27 (6.3 in 2nd)
Speeds in gears*	
1st ... mph @ rpm	NA
2nd ... mph @ rpm	NA
3rd ... mph @ rpm	119.8 @ 4600
Mph per 1000 rpm (in top gear)	26.2
Stopping distances	
From 30 mph	29.5
From 60 mph	137.4
Gas mileage range	12-14
Speedometer error	
Electric speedometer	30 40 50 60 70 80
Car speedometer	29 39 50 60 70 80

Rolls-Royce Silver Shadow

FOR: very roomy and comfortable, particularly in back; extremely quiet and relaxing; outstandingly smooth automatic gearbox; comprehensive heating and air conditioning system; good visibility; superbly made and finished

AGAINST: indifferent performance and handling for price; very expensive to run; poor side support from slippery seats; rumbly brakes and (relatively) noisy tyres

Whatever the future of Rolls-Royce Ltd may be, the car section of the company is no lame duck. Representing 55 per cent of R-R Motors division the cars made a profit of £1,456,000 in 1971 and in the first nine months of last year they made £778,000. The Silver Shadow is the company's mainstay and out of a total of 2473 cars made in 1972, 84 per cent were Shadows or their Bentley equivalent; the Corniche is rather more exclusive, only 358 being built last year, and the Phantom VI, which is more suitable for state occasions than normal motoring, accounted for 2 per cent of the total, 47 being made in all.

We last fully tested a Silver Shadow in 1968, publishing a brief supplementary test on the then new three-speed transmission, early in 1969. Five years later the Shadow's body design doesn't look too dated, its subtly discreet and conservative appearance somehow precluding any hint of anachronism in the car's character. If you must drive a status symbol, the choice is still obvious, as rivals without the classic R-R radiator and elegant Silver Lady are no more than pretenders to the throne in terms of sheer prestige. But no firm can survive indefinitely on such a reputation without continuing to satisfy its customers' demand for engineering excellence as well. The fact that the waiting list for Silver Shadows fluctuates between 15 and 18 months suggests that the customers are satisfied.

Although the car looks the same as it did five years ago, it has changed greatly in many hidden respects. Last year, Rolls-Royce switched the Shadow from cross-ply to radial-ply tyres, and made basic changes to the front suspension in the interests of improved handling, as well as increasing compliance and reducing road noise. In addition to tuning the chassis to suit radial tyres, the steering wheel has been reduced in size and the steering ratio raised, making the car much more responsive. Later last year a new electrical package was introduced and the somewhat haphazard minor control layout redesigned on more logical lines. The electrical pack included centralised boot and door locking, intermittent wiper action, better motors for the window winders, a courtesy light delayed action off-switch, improved instrument panel illumination, a very modern and comprehensive warning light cluster, and brighter tail lights for use in fog.

But the main change is under the bonnet: the engine has been enlarged from 6230cc to 6750cc by lengthening the stroke, in order to help comply with different international exhaust emission regulations. Power output figures are not given but the new engine seems to be similar to the old in this respect, if a little noisier. An interesting optional extra on our test car, not available in 1968 was the automatic cruising speed control device.

Development is a continuous process. Perhaps no car is really worth £10,403, but when you buy a Silver Shadow you get superbly engineered and durable motor car. We found the finish and fittings, both in the main body panels and in tiny details,

ROAD TEST

The Shadow's V8 engine has been modified to accord with American emissions regulations since we last tested the model in 1969. The bore remains at 104.1 mm but the stroke has been increased from 91.4 mm to 99.0 mm, thus raising the capacity from 6230 cc to 6750 cc, though the compression ratio of 9:1 is unchanged. Rolls-Royce keep any differences in specification for different markets down to a minimum and they tell us that American market engines only differ from European ones in that the former have air passed into the exhaust system to ensure complete burning of the fuel mixture.

Power and torque figures are not quoted for the old or the new engine, of course, but the differences must be negligible. Our figures taken at MIRA show that up to 80 mph the '69 and '73 cars are effectively the same, though above that speed the acceleration of the latest test car tailed off a little by comparison: but subjectively it would be hard to tell any difference. The Shadow reached 60 mph from rest in 10.3 sec and 100 mph in 38.1 sec.

Those are the straight facts. What is hard to convey is the delightful smoothness of the big V8 engine at all times, its complete lack of unpleasant boom or vibration periods, and its discreetly muffled roar under full acceleration: it suggests power and dignity in noble union. Perhaps we should simply state that it meets "Rolls-Royce standards." From a cold start or after a long high-speed run it invariably provided the same excellent qualities of smooth response without fuss.

The optional automatic speed control system enables any cruising speed between 30 mph and maximum to be selected and held without touching the pedals. Braking disconnects the system until the "Resume" button is pressed, whereupon the car accelerates gently back to the pre-set speed. The device copes with all but the steepest hills well and is smooth in operation. It costs an extra £71.50.

ECONOMY

★★ A big frontal area, over two tons unladen weight, and 6.75 litres of engine are not exactly aids to fuel economy: our overall petrol consumption figure of 10.9 mpg for the latest car is virtually identical to previous Silver Shadows we have tested, the old four-speed automatic recording 10.8 mpg back in 1968. With fairly gentle driving in the country we managed to obtain just over 13 mpg on one run and at a steady 70 mph we recorded 16.5 mpg with our equipment at MIRA. Full use of the acceleration eats up the fuel, but we feel sure that few owners will be concerned about that, unless that word "conservation" pricks their conscience.

Of greater interest is the car's range between refuelling stops: the fuel tank is an apparently vast 23.4 gallon affair, giving a maximum range of 372 miles, based on our calculated touring consumption figure of 15.9 mpg. In practice this would not be easy to achieve without the aid of motorways and the more dubious assistance of a 70 mph speed limit. Normal brisk driving on ordinary roads, and the consequent drop to 10.9 mpg, reduces the range to a less satisfactory 255 miles. At current five-star fuel prices, the Silver Shadow is strictly for the well-heeled.

TRANSMISSION

★★★★ Any improvements here would be hard to imagine. Rolls-Royce dropped the old four-speed GM Hydra-Matic-based automatic in 1968, replacing it with the GM 400. The newer design has a torque converter and three forward ratios ideal for the Silver Shadow. When the box was first used in the car it was designed to change up at full throttle at 3500 rpm and 3900 rpm, corresponding to 37 mph and 70 mph, but the latest cars change up at 3800 rpm and 4200 rpm (40 mph and 74 mph) and we found that these points were correct for maximum acceleration. Hanging on to the gears in manual hold does not make the car go any faster, and the intermediate and low ranges should only be selected for heavy traffic, occasional towing, the descent of steep hills, or to prevent unwanted upward changes under hard cornering. For normal driving the "D" range, permitting automatic use of all the ratios, does the job perfectly.

Like most automatics, the Silver Shadow jerks perceptibly when selecting Drive or Reverse from Neutral or Park, but on the move it would be extremely hard to pick out the gearchanges but for the accompanying change in engine note. In addition to making the gearchanges uncannily smooth whether on light or full throttle, the torque converter makes the most of the engine's excellent torque characteristics. Response to kickdown is quick and completely jerk free, and the torque converter transmits the maximum torque available for snap acceleration. The transmission makes the initial acceleration from rest excellent too; strong, relentless but smooth, it feels like a jet aircraft setting off along a runway. Its only fault is that it can be jerky when changing down manually.

Removal of the ignition key automatically shifts the gearbox into "Park" whatever the position of the gear selector.

HANDLING

★★★ In the past we have described the Shadow's handling as safe and predictable, if understandably lacking in sporting appeal. Though few drivers are likely to succumb to competitive urges in such a machine, Rolls-Royce quite rightly consider this to be no excuse for not making continuous improvements to the car's handling. Several modifications have been made since 1968. The principal suspension changes were intended to reduce road noise by building compliance into the wishbones and to improve handling by using harder rubber mounts for the front subframe: this new, compliant front suspension was introduced last year and described in Motor week ending September 2, 1972. Early Shadows were strong understeerers by nature though the tail could be provoked into a slide under power on tight corners, particularly in the wet. Current models still understeer under most circumstances, but not to the same excessive degree as their predecessors when driven hard: to reduce understeer a stiffer anti-roll bar was fitted to the rear when the compliant suspension was introduced though, no doubt, the higher ratio of the steering, raised from 3¾ to 3¼ turns from lock to lock since the model first appeared, helps to mask the sensation of understeer. But while the Shadow is safer and more predictable than ever, giving the driver and passengers a feeling of security at high cornering forces, it is designed for comfort not for speed. We noticed that a spell of high speed work caused the outer shoulder of the front tyres to wear very rapidly, drastically shortening their life.

BRAKES

★★★★ Most of the braking is powered by the same pressurised hydraulic reservoir as sustained the ride height adjuster. We say "most of" because there are no fewer than three separate braking circuits, one of which is a completely independent non-power assisted circuit acting on the rear discs only and contributing 16 per cent of total braking effort. Should the power system fail completely, though such an event is virtually impossible, the car can still be stopped without pumping the footbrake. Of the powered circuits, one provides 53 per cent of braking effort by acting on the front and rear wheels and the other, which only affects the front brakes, contributes 31 per cent. To accommodate this ultra-fail-safe system there are two twin-cylinder calipers on each front disc and one four-cylinder caliper on each rear one. Should anything go wrong in any of the three systems a warning light shows "Partial Brake Failure" and is complemented by another light to indicate low pressure in whichever of the three circuits the fault lies. A warning light test button will reveal any faults in the warning light circuits and also indicates sump oil level.

Although the Shadow's brakes are virtually as safe as modern automotive technology can make them—seemingly immune from fade, they were also unaffected by the water splash too—and they do emit a throbbing, almost a grating, noise when the car is slowed from high speed, even quite gently. This impression of effort is very disturbing but despite the hardworking noises the performance remains stable. We achieved a 1g stop without quite locking the wheels and the car invariably felt completely stable at these extremes.

---

delight. Though it is slower than some rivals, and perhaps makes slightly more road noise than some, the near absence of wind noise and its superb comfort may well justify its high price in the eyes of the very rich. And we cannot deny that riding in a Rolls-Royce Silver Shadow gives one a special feeling of contentment with the world. At the end of a journey you feel as if you've picked up a title on the way.

PERFORMANCE

★★★ Today, as ever, the message reads, "the performance is adequate," though it should be noted that Jaguar's V12 engine endows the XJ12 and the Daimler Double Six with a turn of speed that might embarrass competitive Silver Shadow drivers. Whereas the cars from Coventry achieve a mean top speed of about 135 mph, the Rolls lapped MIRA at 112.4 mph and achieved a best one-way speed of 113.9 mph. This information is probably of little interest to the typical R-R customer, however: the Shadow's natural cruising gait of 90-100 mph only becomes inadequate on Continental motorways when a few owners may be disappointed by the superior cruising speeds of certain Jaguar and Mercedes models.

Above: plumbing for various operations practically hides the big V8 engine. The under-bonnet finish is as high as that elsewhere. Above right: huge armchair seats are adjustable for height, rake and reach but don't provide much side support. Right: big low-lipped boot swallowed 14.5 cu ft of test cases.

The handbrake just failed to hold the car when facing down the 1-in-3 test hill.

Body roll, never a problem in the Shadow at normal cornering forces, has been reduced by the stiffer rear anti-roll bar and the latest cars are consequently less inclined to lurch through ess-bends. The car is still inclined to float badly when negotiating undulating roads at high speed, though this tendency has been curbed slightly by modifications to the dampers, making them more effective at high speeds without being harsh at town speeds.

The self-levelling suspension remains unchanged: above each coil spring is an hydraulic ram controlled by automatic sensors which make sure that the correct ride height is maintained front and rear, whatever the load. Normally, the system reacts slowly to any change in ride height to prevent suspension movements from being cancelled out but the system changes to instant correction whenever the transmission is put into neutral or a door is opened.

ACCOMMODATION

 Those of us who can still afford to support a chauffeur will not be disappointed by the back seats. Headroom is good and kneeroom is adequate even for a passenger 6ft 5in tall, so you can sit back in luxury to read the newspaper or catch up on some work while commuting. In this respect the Shadow scores over its rivals very well: in some luxury limousine-style saloons it is a relief to move from the cramped rear quarters to the relative luxury of the front seats. No so the Shadow, for wherever you sit, even with five people in the car, the armchair seats are comfortable, though lateral support provided by slippery leather upholstery is minimal.

Some drivers also complained of insufficient lumbar support. The front seats are adjusted electrically for height, reach and base angle, the possible combination of settings being so great that it takes quite a long time to master the confusing single control stick for each seat. The reclining back rest is not electrically controlled, however.

The proportions of the Shadow are such that at a first glance it does not look like a 17 ft car, a size which permits generous accommodation for the engine, passengers and luggage. The boot is enormous and very easy to load as there is no lip over which luggage must be lifted: we fitted an incredible 14.5 cu ft of luggage into the back.

Stowage space within the car is no more than fair. The glove box is rather small, so is the underfacia oddments pocket on the passenger's side. The pockets in each door are big enough to hold a couple of paperback books, though the magazine flaps behind the front seats can hold a fair amount of material. The stowage compartment between the front seats can accommodate several small articles, but when the car is fitted with a stereo tape player this compartment becomes a tape cartridge store and is fitted with a special rack for the purpose. The shelf behind the back seats cannot hold parcels.

RIDE COMFORT

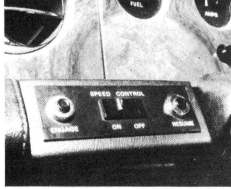

Above: generous room in the back for the long in limb to lounge in comfort. The lambswool rugs are extras. Left: the feet-off speed control panel for motorway cruising. The car maintains whatever speed it is doing when the engage button is pressed.

 The ride is very soft and resilient—everything that a Rolls-Royce owner would expect—but there have been problems with the ride of the Shadow. A trace of body shudder is still evident in the car on some poor surfaces—a trait which can hardly be claimed to affect comfort to any serious degree but which has been with the Shadow since its introduction.

Bump-thump from potholes and ridges in the road are still heard despite the compliant suspension. At most speeds the Rolls-Royce glides along very smoothly though the gentle-swell passage over long-wavelength undulations might induce car sickness. Although the range of damping has been improved the car still "floats" on bad undulations at high cruising speeds; it's only fair to add that earlier Shadows felt even more underdamped under these conditions. The rear seat ride feels as good if not better than that on the front, though perhaps the excellent seats help to give this impression.

AT THE WHEEL

 Rolls-Royce recognise that many owners do not have chauffeurs and that in this enlightened age even those who do will want to enjoy driving their Shadows from time to time. The

Motor Road Test No. 23/73 • Rolls-Royce Silver Shadow

driving position is first class, and the range of adjustment in the seat accommodates all shapes and sizes. The major controls are well laid out, the gear selector being within fingertip reach on the right while the pedals are comfortably positioned; the brake pedal is easy to hit quickly with either foot. The steering wheel is now a more manageable 16 in diameter instead of 17 in.

In the past we have criticised the layout of the Shadow's confusing minor controls. Although the basic style of the facia and controls remains the same, detail changes have greatly improved the layout. The oil pressure and water temperature gauges have been abandoned in favour of warning lights and buzzers—and the ignition switch, lights switch panel has been moved to the right of the facia from the middle: this is a definite improvement as the lights switch is much easier to reach now.

The wash/wipe control has been moved from the facia to a much more accessible position on the indicator/flasher stalk on the left of the steering wheel. A large cluster of warning lights is now situated in the centre of the facia. The action of all the switches is neat and precise, and while some people found the arrangement confusing at first it is very much better than it used to be—once it has all been learnt. We like the position of the horn button in the centre of the steering wheel, though the horn itself is not suitable for high-speed work: we have driven a Shadow with a foot-actuated airhorn, the pedal being mounted next to the dipswitch, and we found this one-off extra very useful once we had learnt which button was which. A black mark goes to the pull-out handbrake on the right where it can catch an unwary knee painfully on getting in or out of the car.

VISIBILITY

★★★★ Even in narrow London streets the Shadow does not feel too big or unwieldy because you have such a commanding view down the bonnet and through the back window, which is fitted with almost invisible hot lines as standard. The optional driver's door-mounted mirror is adjustable from the driving seat. The screen pillars do not obstruct the view badly and even the rear corners can just be seen when reversing.

The area swept by the wipers is poor, and the two-speed-plus-delay wipers were beginning to lift off the screen at about 100 mph. The headlamp range is reasonable and dipped beam well controlled.

INSTRUMENTS

★★ As we have said, there is no longer a water temperature gauge or oil pressure gauge on the Shadow. Instead you have warning lights in addition to those for the ignition, low fuel level, partial brake failure (with a

PERFORMANCE

CONDITIONS
Weather	Dry; wind 0-9 mph
Temperature	45-55° F
Barometer	30 in. Hg
Surface	Tarmacadam

MAXIMUM SPEEDS—see text
	mph	kph
Banked circuit	112.4	180.9
Best ¼ mile	113.9	183.3
Terminal speeds:		
at ¼ mile	78	126
at kilometre	96	154
at mile	105	169
Speed in gears:		
1st	40	64
2nd	74	119
(Automatic changes)		

ACCELERATION FROM REST
mph	sec	kph	sec
0-30	3.4	0-40	2.6
0-40	5.1	0-60	4.5
0-50	7.4	0-80	7.3
0-60	10.3	0-100	11.1
0-70	14.1	0-120	16.5
0-80	19.0	0-140	23.8
0-90	26.1	0-160	37.3
0-100	38.1		
Stand'g ¼	17.5	Stand'g km	32.5

ACCELERATION IN 4TH
mph	sec	kph	sec
20-40	3.2	40-60	1.9
30-50	4.0	60-80	2.8
40-60	5.2	80-100	3.8
50-70	6.7	100-120	5.4

60-80	8.7	120-140	7.3
70-90	12.0	140-160	13.5
80-100	19.1		

FUEL CONSUMPTION
Touring*	15.9 mpg / 17.8 litres/100 km
Overall	10.9 mpg / 25.9 litres/100 km
Fuel grade	100 octane (RM) 5 star rating

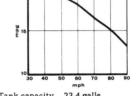

Tank capacity	23.4 galls / 106.4 litres
Max range	372 miles / 599 km
Test distance	1360 miles / 2188 km

* Consumption midway between 30 mph and maximum less 5 per cent for acceleration.

BRAKES
Pedal pressure deceleration and stopping distance from 30 mph (48 kph).

lb	kg	g	ft	m
25	11	0.56	54	16
50	23	0.91	33	10
60	27	1.00	30	9
Handbrake		0.28	107	33

FADE
20½g stops at 1 m intervals from speed midway between 40 mph (64 kph) and maximum (76 mph, 122 kph).

	lb	kg
Pedal force at start	35	16
Pedal force at 10th stop	47	21
Pedal force at 20th stop	36	16

STEERING
	ft	m
Turning circle between kerbs		
left	35.5	10.8
right	36.3	11.1
Lock to lock	3.3 turns	
50 ft diam circle	1.1 turns	

SPEEDOMETER (mph)
Speedo	30	40	50	60	70	80	90
True	30.5	40	50	60	70.5	80.5	90.5

Distance recorder: accurate.

WEIGHT
	cwt	kg
Unladen weight*	41.4	2103
Weight as tested	45.1	2291

* With fuel for approx 50 miles.

Performance tests carried out by Motor's staff at the Motor Industry Research Association proving ground, Lindley.

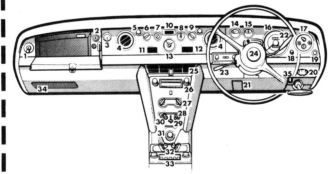

1. map light switch
2. boot unlock
3. glove box lock
4. cold air outlets
5. cold air controls
6. panel lights switch
7. speaker balance
8. aerial switch
9. wipers switch
10. partial brake failure warning light
11. warning lights: coolant, handbrake, stop lamps, fuel
12. warning lights: brake pressure and fluid level
13. clock
14. fuel/sump oil level gauge
15. ammeter
16. speedometer
17. ignition and lights switches panel
18. trip reset
19. fuel filler cover release
20. parking lamps split switch
21. bonnet release
22. gear selector
23. indicators, flasher, wash/wipe stalk
24. horn
25. air conditioning outlet
26. stereo tape player
27. heating and air conditioning controls
28. heated rear window switch
29. fan switch
30. hazard warning flasher switch
31. cigarette lighter
32. seat adjustment switches
33. radio
34. oddments compartment

COMPARISONS

	Capacity cc	Price £	Max mph	0-60 sec	30-50* sec	Overall mpg	Touring mpg	Length ft in	Width ft in	Weight cwt	Boot cu ft
Rolls-Royce Silver Shadow	6750	10,403	112.4	10.3	4.0	10.9	15.9	17 0	5 11	41.4	14.5
Daimler Double Six	5343	3794	135.7	7.4	2.6	11.5	13.5	15 10	4 11	34.8	11.8
BMW 3.0S‡	2985	4899	126.4	8.0	9.0	15.4	18.8	15 6	5 9	27.9	12.3†
Fiat 130 Saloon	3235	3941	112.5	10.7	3.8	14.4	18.7	15 8	6 0	31.8	12.7
Citroen SM‡	2670	5478	135.2	9.9	11.7	17.2	21.5	16 0.5	6 0.5	28.5	—
Cadillac Fleetwood Brougham	7735	7237	123.0	9.9	3.0	11.0	13.4	19 0	6 8	45.9	—
Lamborghini Espada‡	3929	10,785	**	7.8	10.3	11.3	16.5	15 7	6 2	34.6	4.6
Jensen SP	7212	7041	140.0	7.6	2.8	11.0	15.2	15 8	5 10	35.1	8.5

*in top/kickdown (top gear for manuals)
**Estimated at 150 mph
†measured with boxes, not cases
‡manual transmission

STAR GRADE KEY excellent good average poor bad

Test Data · World copyright reserved; no unauthorised reproduction in whole or in part.

Motor Road Test No. 23/73 ● Rolls-Royce Silver Shadow

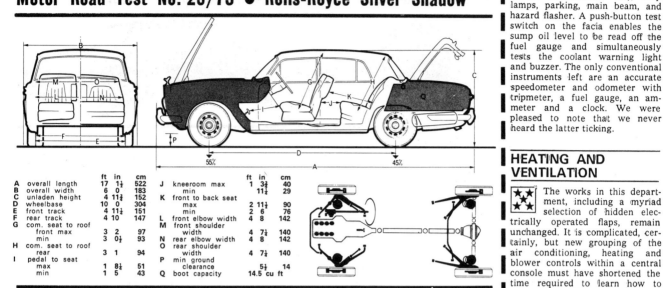

		ft	in	cm			ft	in	cm
A	overall length	17	1½	522	J	kneeroom max	1	3¾	40
B	overall width	6	0	183		min		11½	29
C	unladen height	4	11¾	152	K	front to back seat			
D	wheelbase	10	0	304		max	2	11½	90
E	front track	4	11¼	151		min	2	6	76
F	rear track	4	10	147	L	front elbow width	4	8	142
G	com. seat to roof				M	front shoulder			
	front max	3	2	97		width	4	7½	140
	min	3	0½	93	N	rear elbow width	4	8	142
H	com. seat to roof				O	rear shoulder			
	rear	3	1	94		width	4	7½	140
I	pedal to seat				P	min ground			
	max	1	8¼	51		clearance		5½	14
	min	1	5	43	Q	boot capacity	14.5 cu ft		

GENERAL SPECIFICATION

ENGINE
Cylinders — V8
Capacity — 6750 cc (411.9 cu in.)
Bore/stroke — 104.1 x 99.0 mm (4.1 x 3.9 in.)
Cooling — Water
Block — Aluminium alloy
Head — Aluminium alloy
Valves — ohv pushrod, hydraulic tappets
Valve timing
 inlet opens — 18° btdc
 inlet closes — 68° abdc
 ex opens — 60° bbdc
 ex closes — 26° atdc
Compression — 9.0 : 1
Carburetters — Twin SU HD8
Bearings — 5 main
Fuel pump — Twin SU electric
Max power — Not disclosed
Max torque — Not disclosed

TRANSMISSION
Type — Automatic 3-speed GM type 400 with torque converter
Internal ratios and mph/1000 rpm
Top — 1 : 1/26.2
2nd — 1.5 : 1/17.5
1st — 2.5 : 1/10.5
Rev — 2.1 : 1
Final drive — Hypoid spiral 3.08 : 1

BODY/CHASSIS
Construction — Unitary
Protection — Zinc-coated steel in vulnerable areas plus underseal

SUSPENSION
Front — Ind by wishbones, coils, anti-roll bar with automatic height control
Rear — Ind by semi-trailing arms, coils, anti-roll bar with automatic height control

STEERING
Type — Recirculating ball
Assistance — Yes
Toe-in — 0.060 - 0.160 in. at 17 in. dia
Camber — -¼° ± ¼°
Castor — +3° ± ¼°
King pin — 11°

Rear toe in — 0.00 - 0.060 in. at 17 in. dia
Rear camber — -¼° ± ¼°

BRAKES
Type — Rolls-Royce discs all round
Servo — Twin camshaft-driven pumps
Circuits — Two high pressure hydraulics and one master cylinder
Rear valve — "g"-sensitive valve
Adjustment — Self-adjusting pads

WHEELS
Type — Steel disc 15 in.
Tyres — Avon, Dunlop, Firestone radials
Pressures — 26 psi

ELECTRICAL
Battery — Dagenite 12V, 71 amp hr
Polarity — Negative
Generator — Alternator
Fuses — Fully fused : 20 fuses
Headlights — 4 sealed beam : rating varies according to market

STANDARD EQUIPMENT

Adjustable steering .... No
Anti-lock brakes .... No
Armrests .... Seven
Ashtrays .... Three
Breakaway mirror .... Yes
Cigar lighters .... Three
Childproof locks .... Yes
Clock .... Yes
Coat hooks .... No
Dual circuit brakes Triple circuit
Electric windows .... Yes
Energy absorb steering col ...... Yes
Fresh air ventilation .... Yes
Grab handles .... Three
Head restraints .... No
Heated rear window .... Yes
Laminated screen .... Yes
Lights
 Boot .... Yes
 Courtesy .... Yes
 Engine bay .... Yes
 Hazard warning .... Yes
 Map reading .... Yes
 Parking .... Yes
 Reversing .... Yes
 Spot/fog .... No
Locker .... Yes
Outside mirror .... No
Parcel shelf .... Yes
Petrol filler lock .... Yes
Radio .... Extra
Rev counter .... No
Seat belts
 Front .... Yes
 Rear .... No
Seat recline .... Yes
Seat height adjuster .... Yes
Sliding roof .... No
Tinted glass .... No
Combination wash/wipe .... Yes
Wipe delay .... Yes
Vanity mirror .... Yes

IN SERVICE

GUARANTEE
Duration ...... Three years or 50,000 miles on mechanical parts; one year for coachwork

MAINTENANCE
Schedule .... Every 12,000 miles
First service .... At 3000 miles

DO-IT-YOURSELF
Sump ...... 14.5 pints, SAE 20W-50
Gearbox .... 18.6 pints, Dexron
Rear axle .... 4.5 pints, 90EP
Steering gear ...... 3 pints Dexron
Coolant
 28.5 pints anti-freeze mixture
Chassis lubrication
 5 points every 12,000 miles
Distributor dwell angle ... 26°-28°
Spark plug type .... Champion N14Y
Spark plug gap .... 0.025 in.
Tappets .... Not adjustable

REPLACEMENT COSTS (incl. fitting)
Brake pads/linings (front) ... £18.75
Complete exhaust system... £200.00
Engine (new) £1050.00
Dampers (front) (pair) £55.00
Front wing £375.00
Gearbox (new) £250.00
Oil filter £6.00
Starter motor £40
Windscreen £50.00

Make: Rolls-Royce
Model: Silver Shadow saloon
Makers: Rolls-Royce Motors Ltd, Crewe, Cheshire, CW1 3PL
Price: £8730 plus £1673.25 car tax and VAT equals £10,403.25. Radiomobile Stereo 8 player £107.25 with tax, lambswool rugs £42.90 with tax, automatic speed control £71.50 with tax, internally adjustable exterior mirror £17.28 with tax: total as tested £10,642.18.

---

circuit identification too), stop lamps, parking, main beam, and hazard flasher. A push-button test switch on the facia enables the sump oil level to be read off the fuel gauge and simultaneously tests the coolant warning light and buzzer. The only conventional instruments left are an accurate speedometer and odometer with tripmeter, a fuel gauge, an ammeter and a clock. We were pleased to note that we never heard the latter ticking.

HEATING AND VENTILATION
★★★★

The works in this department, including a myriad selection of hidden electrically operated flaps, remain unchanged. It is complicated, certainly, but new grouping of air conditioning, heating and blower controls within a central console must have shortened the time required to learn how to operate everything. In essence it's simple enough, but it takes a while to get to know what to expect from the system. There are two air volume and temperature control knobs and the basic rule for both is: pull out for more air, turn clockwise for hot air. One knob controls the upper outlets (ie demist, facia, and console) the other feeds the footwells front and rear. The facia and console outlets can be opened and closed independently. The three-speed fan is only excessively noisy on full blast, but is so powerful that such a strong boost would never be needed for more than a few minutes at a time. Refrigerated air is only available through the upper outlets. Anyone who has difficulty in understanding the set-up can use a special instruction card which slips over the control knobs and explains everything at a glance.

Whenever these knobs are turned or pulled, the distant whirring of a multitude of motors can be heard as the doors within the depths of the system are opened and/or closed. When switching to refrigeration, the system should be given a few seconds to adjust: an unpleasant preliminary blast of hot, stuffy air is sometimes initially blown from the facia outlets but the air conditioning soon settles down to keep the occupants cool and comfortable in hot weather. We even found the refrigeration useful in London once or twice in March, and it was not long before we could adjust the whole works correctly in one attempt.

NOISE
★★★★

Few cars can even approach the Shadow in terms of wind noise suppression. Even at very high cruising speeds it is superb in this respect. At steady speeds engine noise is negligible, sometimes inaudible, thought the (admittedly dignified) roar under full acceleration was noisier on our test car than we recall from earlier models. Transmission noise was virtually

grounds of safety, so R-R have dropped them from the specification in their bid to keep the car as near the same as possible in all markets. For determined in-car picnickers they are still available as an optional extra. Rear seat passengers are cosseted in luxury with door-mounted armrests, a central fold-down armrest, an ashtray and cigarette lighter in both doors, vanity mirrors on each side, a choice of lighting for reading or general purpose, and grab straps.

The electric windows are individually controlled though the driver can control all of them from switches in his door. A centralised door locking system, with switches in the front door, locks all the doors and the boot: the boot can be unlocked by pushing a button in the glove box. The petrol filler flap is also released by means of a button on the facia and the interior light stays on for seven seconds after the door has been closed, giving the driver time to insert the key in the ignition switch. Each front seat has a fold-down armrest and there are adjustable rests in the doors.

Customers are given a choice of radios and stereo equipment, the latter being particularly enjoyable within the quiet confines of the Shadow's interior. The aerial is electrically operated though it does not rise automatically when the radio is turned on. The radio is mounted at the base of the new central console, and just in front of it is the voluminous ashtray. A cigarette lighter is mounted in the console.

The Silver Shadow is opulently finished without being gaudy. Beautiful carpeting, woodwork, and leather create a pleasing and very impressive interior: even the roof lining is perfectly made from fine leather. As with the bodywork, the standard of finish inside the Shadow is a delight to inspect.

IN SERVICE

After the first service at 3000 miles, the Silver Shadow requires a major service every 12,000 miles —about once a year on average— with an oil change and check on brake pad wear every 6000 miles. Routine service points such as oil, coolant and washer bottle filling points and the dipstick are easy to get at, though a plug change would not be so simple. The distributor is accessible, if at arm's reach in the rear centre of the crowded engine bay. The hydraulic reservoir levels can be checked instantly through sight glasses. The fuse box is inside the car under the facia.

The battery is in the boot and reasonably accessible. Also in the boot is the jack and a well-stocked toolkit containing a selection of spare light bulbs. The spare wheel is under the boot floor and can be released without removing any luggage. The pressure in the spare tyre can be adjusted with the wheel in situ through a hole in the boot floor.

The prices given in the specification tables include a labour charge: VAT of 10 per cent must be added to these approximate figures. Rolls-Royce were anxious to point out that several of these items would not need to be replaced within the car's normal life span.

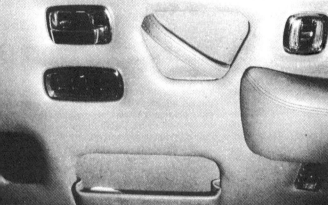
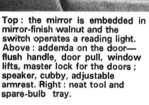

Top: the mirror is embedded in mirror-finish walnut and the switch operates a reading light. **Above**: addenda on the door— flush handle, door pull, window lifts, master lock for the doors; speaker, cubby, adjustable armrest. **Right**: neat tool and spare-bulb tray.

Above: fingertip pressure on a small joystick adjusts the seat. On the timber panel above are the heater controls. **Left**: handy letdown fuse box under the driver's shins. The pull-out handbrake on the right can be a painful knee-stop when getting out.

unnoticeable. Road noise, however, is not so well suppressed, and there are some cars that beat the Shadow on this count quite comfortably.

Some bump-thump is evident from time to time, but tyre roar can be heard most of the time. But we don't think anyone would find the level of road noise intolerable: in general, the thick glass of the windows, and carefully applied insulation materials works well. Few passengers fail to comment on how quiet the car is at 70 mph, and we are bound to agree that in all respects other than road noise, the Shadow sets the standards. This is particularly true of the rear seat area: we cannot think of any car on the market that is as quiet in the back as the Rolls-Royce.

FINISH AND FURNITURE

★★★ ★★ It may surprise you to hear that the Shadow is no longer supplied with the traditional picnic tables in the backrests of the front seats. The American regulations prevent them from being fitted on the

TAKING STOCK

What it means to own a Rolls-Royce Silver Shadow

By Edward Eves

Introduced	October 1965
Autotest:	30th March 1967
Price:	£9861
Delivery Charge:	£17.50
Plates:	£7
Year's Tax:	£25
	£9910.50

To many prospective Rolls-Royce owners the ownership of a Silver Shadow represents a career achievement and neither its technical merit nor its running costs are of paramount importance. To others, more technically minded, it represents the most perfect car they could hope to own. There is even a small body of people who buy them because they stand up to hard use year in and year out. None of these buyers is likely to be fobbed off with anything else. Although the price seems high it is not so high as that asked for some less worthy cars. And it retains its value over its life span better than almost any other make.

What it Costs

The delivered price of the car is shown in the heading table. It would be illogical if the manufacturers of the 'Best car in the World' were to charge extra for any essential item of equipment. Seat belts, heated rear window, radio, electric aerial, air conditioning and rear seat footrests are all standard equipment. Cloth trim, which is a personal choice, can be had for an extra charge of £37.50. Stereo tape equipment may not be to everyone's taste and is rightly regarded as an option. It costs £148 installed at the works.

Insurance is more of a variable than with most cars. The majority of cars sold in the UK go to the top executives of companies and are covered under 'umbrella' policies. Those that go to individuals are likely to be driven by Rolls-Royce trained chauffeurs and despite their value would be lightly 'loaded'. Just how pop stars fare I do not know. A BIA company has provided us with a typical figure of £167. Just for once we have raised the age of the 'Taking Stock Owner' from 30 to 40. But it is still assumed that he does not possess a no claims bonus and does not wish to pay an excess.

Living with a Rolls-Royce Silver Shadow

For me, no other car has the aura that surrounds a Rolls-Royce. Possibly because, having lived with cars for something like 35 years, I have learnt to appreciate what goes into making one. Rolls-Royce have always endeavoured to produce the best possible quality car in the context of the times. In the early, Derby, years there was a surplus of craftsmen. The best of them were picked by Henry Royce to build his cars. They took immense pride in doing so. Nowadays machine tools are improved and the materials they have to work on are less liable to writhe and contort during and after machining. So there is less call for hand finishing, and a new breed of toolmakers, machinists and assemblers has grown. Hand fitting is now at a minimum.

"Royces" have always endeavoured to use the best equipment available and get the best out of it. All this does not automatically cause it to be the best designed car in the world. But the old adage that the Rolls-Royce is a triumph of workmanship over design is facetious. It needs more than workmanship to produce a car as complex as the Silver Shadow. Let alone to make it work as smoothly and efficiently as it does. There are faster cars than the Shadow and cars that handle better and go round

GOOD POINTS

Miniature joysticks in the console adjust the seat fore-and-aft, up-and-down and alter the rake of the cushions

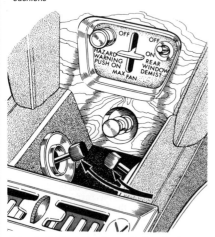

Nerve centre of the complex electrical system is a well engineered, let-down fuse-box under the panel which is properly placarded and carries a full range of spare fuses and a bobbin of fuse wire

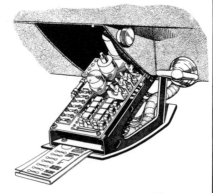

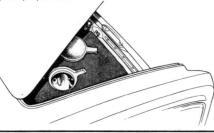

A trap in the boot floor is provided to check spare tyre pressure

The driver has full control of all the electric windows by means of this three-position master-switch with emergency position for use when the ignition is switched off

BAD POINTS

When changing from a British made second car confusion can be caused because the gearshift lever is just where the trafficator lever ought to be

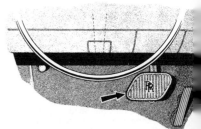

The Brake pedal is located over to the right of the steering column precluding two-foot driving. US market cars have a wider pedal

corners faster. But there are few cars which provide the combination of speed, handling and carrying capacity, combined with refinement of operation. If one merely costs the time and patience which has gone into making the Shadow so good to the touch, it is remarkable value.

Being the owner of a Rolls-Royce saddles one with certain responsibilities. On the road everyone's eyes are on you and in this socialist world it is as easy to attract the finger of scorn as the glance of admiration. Your conduct on the road and off it must be impeccable. It is not difficult because the car is so responsive and pleasant to the touch. From the finger-tip precise power steering to the heater control switches everything operates smoothly and exactly. Now that the proper brand of Avon radial tyres is available the Shadow corners very quickly — deceptively so — and it is possible to place it very precisely on the road at any speed.

Although the controls may be precise and free from backlash, it doesn't follow that they might not be better arranged. I am not alone among journalists coming from other cars who have had moments with the right-hand gear shift located just where one usually finds the trafficator switch, and it is not unlike one to look at. It is quite natural to flick it upwards or downwards to turn left or right. The former motion can land one in reverse at the worst or neutral at the best. If the engine should stall as a result, the steering, bereft of assistance becomes very heavy indeed — almost beyond the strength of a woman. On the British market models the footbrake is positioned to the right of the steering column and virtually precludes two foot driving, American style. Rolls-Royce in justification point out that a survey of owners showed that very few of them used left foot braking. American market cars are provided with a wider brake pedal. I am sure any British owner who wanted one could have one too. Until very recently one could have criticized the Shadow for its windscreen wiper arrangement, which had a plain two-speed switch on the panel, with press-to-wash action. The position of the switch called for quite a stretch to operate it. It is good to learn that on the latest cars the wiper switch has been moved to a lever on the steering column and there is an intermittent wipe action. On earlier cars the heavy wiper blade springs needed to keep the blades on the screen at speed caused bad blade judder the moment the screen became nearly dry.

Naturally it is an easy car to get in and out of gracefully. The one proviso is that the doors are heavy if the car is leaning over on a steep camber. As expected they close with a pleasant click but the electric locks which secure them go home with a frightening clunk reminiscent of the closing of a bank vault. However, fear not, you are not trapped in should the electricity fail. One simply lifts the locking knob on the door to disengage the catch.

Rolls-Royce seats prove that soft leather, the best Connolly hide, is still unbeaten as a seat trim material. It is warm in winter and does not sweat in summer. The upholstery is firm but with a soft overlay and the seat spring rate is carefully designed not to coincide with that of the car suspension. Electric power assisted adjustment forward, backwards and upwards, with a tilt thrown in for good measure go to prove that Rolls-Royce look upon the Shadow as an owner-driver car. All these movements are controlled by delightful miniature joysticks mounted in the centre control. However, seat back adjustment is not electrical; it could well be because the adjusters are difficult to reach with the doors closed. they provide a fine adjustment in a near upright position and allow the seat backs to pull right back. The rear seats are equally comfortable — like the front seats they have centre armrests. There is a touch of comfortable Victoriana in the little companion sets in the rear quarters complete with mirrors and cigar lighters. They are a nice indulgence but are more ornamental than useful. More practical are the little wedge shaped padded loose footrests on the rear floor.

The Shadow is a good chauffeur-driven car, especially with the optional division. However despite the self-levelling suspension, the chauffeur inevitably has a fractionally better ride than his master.

I like the way that controls have been devised to take human failings into account. A master control for the electric windows located on the driver's door is a good example. In the normal position the electric windows can all be operated by individual switches on each door and by the four master switches on the driver's door. A second position cuts out all the individual switches and only the driver can open and close windows. It is useful when naughty children in the back play with the window switches. A third, emergency, position allows all the windows to be operated when the ignition is switched off. To avoid the possibility of it being left on accidentally in this position and running down the battery, it is spring loaded towards the second position.

It takes a little while to learn the ins and outs of the complex air conditioning, heating and ventilating system. The whole operation is controlled by miniature switches — little kidney shaped affairs — and a small quadrant lever. All together they occupy an area on the dash scarcely bigger than a playing card and a couple of old fashioned pennies. Servo motors behind the dash do the manual work. Briefly there are two ducting systems, upper air and lower air. Upper air is the air directed to the screen or one circular and one rectangular outlet on the facia panel. Air delivered to outlets in the footwells and the rear compartment counts as lower air. Each is controlled by one of the two little switches. They are turned to control the temperature of the air and pulled out to control the volume. The fan quadrant lever simply controls the amount of air delivered and circulated when the air conditioning is in operation. One soon learns to get the best out of the system. It is easy to operate by touch, like the controls of an expensive camera. It will demist the screen in seconds with a blast of hot, dry air and not once did I have to "call up" the rear window heater. There is only one snag. It takes longer than usual to warm-up.

Radio is standard equipment. The set provided is the top-of-the-range Radiomobile, complete with electric aerial, controlled by a switch in the middle of the facia. I have yet to come upon a bad Radiomobile set. That fitted to the Silver Shadow is excellent. In "my" car it was backed up by an eight-track stereo player,

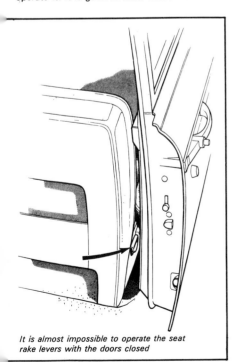

It is almost impossible to operate the seat rake levers with the doors closed

Service Interval	6000 Miles	1200 Miles	2400 Miles
Time allowed			
Cost @ £2 per hour			
Oil			
Oil filter			
Breather filter			
Air filter			
Contact breaker points			
Sparking plugs			
Total Cost:	£29.00	£44.50	£53.00
Materials included except those subject to wear and tear e.g. brake pads.			

Routine Replacements	Time	Cost	Spares	Total
	Total is for		New front 13.20	£20.00
Brake pads	supplying and fitting.		rear 6.60	£10.00
Brake shoes NA				
Exhaust systems	Stainless Steel replacement % very low.			
Clutch NA				
Shock absorbers — front	Total for supplying and fitting.		New One £11.65	£20.25
Shock absorbers — front	Total for supplying and fitting.		New One £11.65	£20.25
Replace drive-shaft, each NA				
Alternator	,, ,,		£23.10	£29.90
Generator	,, ,,		£23.10	£29.90
Starter	,, ,,		£27.00	£33.80

TAKING STOCK...

which constituted the one classed as extra and optional. Royce install a speaker in each door during manufacture complete with wiring to the central console in case you need stereo. And they have ready for you a special little leather-covered pigeon hole box to hold your tapes. It slots into the glove receptacle between the seats.

For one's own safety one tends to look after the tyres of the fast, heavy Shadow and the incidence of puncture should be low. If one does occur you are well provided with a braced screw jack and a plated, forged, box spanner and tommy bar for the wheel nuts. The end of the tommy bar is shaped to act as a wheel trim remover. You will also have to remove some luggage if the boot is really full, to get at the nut on the boot floor which operates the spare wheel cradle located below the boot. The spare is protected only by the various chassis members in the boot area and tends to come out dusty but not mud encrusted. A pair of gardening gloves would be a useful addition to the tool kit and could go in the well-made bag provided for the wheel changing kit. Small tools, a full set of them including feeler gauges and an adjustable spanner — I'd hate to use one on any nut in my Rolls-Royce — are to be found in a fitted case on top of the battery box.

The Shadow is a rewarding car to keep clean. Clean cold water only is recommended, detergents may stain lighter coloured paint. The makers recommend polishing and sealing the paint once a quarter with their Formula 2 and Formula 3 polishes. A good tip is the recommendation of ammonia for removing the tarnish from chrome, not polish as one would expect. Care of the coachwork is fully explained in the very comprehensive instruction book.

Finish and Fits

With memories of rust problems experienced with their early pressed-steel bodies Rolls-Royce leave nothing to chance with body protection. The whole of the underframe is pressed from zinc-coated steel and the body is bonderized before being dipped in primer. The doors, bonnet top and boot lid are corrosion-resisting light alloy. Thereafter no fewer than 12 — it can be 14 — coats of cellulose finish with rubbing down in between are applied to get that deep shine that we all expect. There is in all .012in. thickness of paint on the body at the end of the finishing process. Afterwards about 70 lb of bitumastic paint are applied to the underside of the shell.

Chromium-plating is in excess of the British Standard requirements and the radiator shell is hand made from stainless-steel sheet. Should you damage it you, or your insurance company will be faced by a bill for £204 for a new one. Incidentally the Spirit of Ecstasy mascot is no longer silver plated. She is now cast, along with and at the same time as the compressor blades of Rolls-Royce aero turbines, from nimonic alloy and should be as near indestructible as makes no matter. To comply with common sense safety requirements she is hinged at the base on a spring-loaded mounting and bends back on impact.

Door, bonnet and boot fits leave nothing to be desired and demonstrate that a pressed body can be just as good as a traditional coachbuilt body, in fact better, in respect of maintaining even shut lines. Probably the most perfect panel on the whole body is the bonnet top which reflects without secondary distortions the mascot and oncoming scenery in its flawless surface.

Inside the car all is traditional. The Connolly hide trim is top grain and available in eight standard colours. Cloth can be had should you want it. The carpets are deep pile and readily detachable for brushing and cleaning. They are held in place by big, serviceable press studs which do not carry away at the first sign of use. All the veneered wood panelling is produced by Rolls in their own cabinet shop and, a nice point, spare veneers are kept marked with the body number, so that it can be matched at a future date in case of damage.

On the forecourt

Normally one pulls up for fuel only. Then the only chore is to press the button located on the right-hand side of the instrument panel which opens the fuel filler flap electrically.

If oil or water needed to be replenished I would want to stand by to make sure that the attendant put it in the right hole. There are rather a lot of filler caps, two for main hydraulics and one for steering, water and engine oil. It might be damaging to get them confused. They are all quite accessible.

A green light warns you when there are three gallons left in the tank. It is properly very bright because you have just 30 miles safely to go if you do not tread too heavily on the loud pedal. The tank holds 28 gallons and would take the flow from the biggest pumps available in the UK without blowing back.

A good point is that a tiny trap is provided on the boot floor for checking the spare tyre pressure. For it to be effective remember to turn the wheel so that the valve coincides with the trap.

Doing it yourself

Doing it yourself implies shortage of cash. This hardly applies to R-R owners. However many of them prefer to do the essentials themselves. Some of them even attend the chauffeurs' courses organized by Rolls-Royce. These take a couple of weeks and used to cost about £20. A comprehensive, properly bound, instruction book fully describes all the routine jobs and warns owners off the difficult ones. For example, you are not encouraged to tamper with the power hydraulic system. To do so could be physically dangerous with pressures of around 2,000 psi stored in the hydraulic accumulators. Nor should you fiddle with the air conditioning.

Changing brake pads could be an essential on a long, fast, Continental tour with much mountain driving. Pads could wear down in as little as 3,000 miles. The job is fully described in the instruction book! All you would need would be a brake piston compressing tool — R-R don't like the idea of you using a screwdriver — and a little graphite lubricant.

If you are at all mechanically minded the Shadow is a lovely car to work on. Sharp edges are few and far between, many components are stove enamelled and come clean with a wipe of a cloth, there are no tight threads and every nut on the car is machined and plated.

It is interesting to find that sealed-for-life steering joints are avoided. This is because Rolls-Royce last longer than life and the makers like you to lubricate the six steering joints at the 12,000-mile intervals with the correct lubricant. At the same time the ball joints in the automatic levelling system have to be disconnected and lubricated.

Oil changes are straightforward, 12.75 pints every 3,000 miles if you are doing only local motoring. If you change the filter you need an extra 1.25 pints, a new filter element and a long arm to reach under the cylinder banks to get at the filter housing. The sump plug is quite accessible from below and a special spanner for it is included in the tool kit. Changing the Champion sparking plugs is another fiddling job although it is easier than in early Silver Clouds. Plugs should be changed at 12,000-mile intervals.

I think I would give the job of changing fan belts — they actually drive the coolant pump and fan — to a Rolls-Royce service station. A five-groove pulley — six on US market cars — on the nose of the crankshaft drives the auxiliaries directly or indirectly. The handbook tells you how to adjust tension of the belts but wisely avoids telling you how to change them.

Electrics

The Silver Shadow has a complex electrical system and the fact that it is thoroughly executed is an essential rather than a refinement. Incidentally, it is rewarding to see what fine electrical equipment Lucas can turn out when they are not being beaten down by the buying department of a profit-hungry volume producer.

A belt-driven alternator charges a 71 ampere hour Dagenite battery, located in the boot, to provide electrical power. It is distributed by way of a fuse box containing no fewer than 20 fuses. The six main ones are bakelite mouldings with replaceable fuse wire. The others are Bulgin glass fuses and you are warned to use only this type. In typical motherly fashion, a bobbin of fuse wire is housed in a little hole in the side of the fuse box and there are four spare glass fuses. Circuit breakers guard the gearchange, door lock and window-lift circuits. The whole fuse box unit is beautifully made and carried on a plated chassis which hinges down from below the dash for ease of access. More fuses, for the hazard warning lights and the rear window heater, are positioned under the panel.

There are more than 30 clear-glass bulbs, apart from the headlamps, in and around the car. All of them can be replaced without fiddling, access to them is mostly by unscrewing the screws retaining the lens. Headlamp adjustment and removal is easy once the lamp bezels have been removed. All is revealed in the instruction book. A good safety feature is that should a headlamp circuit fail at least one lamp on each side is illuminated.

Special Lucas lightweight contact-breaker points are fitted. A spare set would be a useful addition to the tool kit.

Professional service

During the past few years Rolls-Royce have pruned their service network with the object of stimulating growth. As a result there are very efficient depots in all the main cities to which owners can take their cars with due confidence. Service charges are itemized in the table on p15. The regular services are not detailed for the reason that the price quoted is an all-in price for the service and includes any routine replacements and minor non-routine ones which are found necessary. These are what Rolls-Royce define as essential services. There are other special services, such as refilling the refrigeration system, which would be quoted for by the garage concerned. □

Travel Talk

Marseilles and back in a day

Proving a point in comfort (and a Corniche)

By Stanley Sedgwick

Ready to go with full petrol tank — so off to bed for an early start the next morning

Inspiration

ON returning home from the South of France in my (then) Lincoln Continental nearly two years ago having driven 864 miles in the day (*Autocar* 18 November 1971) my feeling of self-satisfaction was quickly deflated when I opened the current issue of *Autocar* and read the following statement in the Road Test of the Ferrari Daytona — "Across Europe it would undoubtedly prove the fastest touring car that traffic conditions would allow, easily coping with a trip from the Channel to the South of France and back in a day".

This struck me as an assertion worthy of demonstration and I resolved to see what could be done about it when the opportunity arose. Realization proved less easy than I had hoped, none of my Daytona-owning friends appearing to be in need of a co-driver to put hundreds of miles on the clock with the object of finishing up where they had started, so the idea went into the memory bank.

Some months later I drove my open 8-litre Bentley 1,000 miles in a day in England (*Autocar* 28 September 1972) in a total time of 17 hours 37 minutes. I then perceived that the same length of time at the wheel of a fast, modern, closed car — on roads free from speed limits — would encompass a considerably greater distance and it followed that I could drive all the way across France and back in a day myself given a suitable car. On further consideration I came to the conclusion that such a journey would be more meaningful if accomplished in a full-size saloon instead of an out-and-out sports-car. Thus it was that I asked a mildly surprised John Craig — Marketing Director of Rolls-Royce Motors — if I could borrow a Corniche for the trip. I was even more surprised when he said "Yes", and I proceeded to make plans.

Preparation

When? Who with? What route? The answer to the first question was easy. It made sense to undertake the drive when the daylight hours were longest, but before the holiday rush, and to choose a day in the middle of the week to avoid week-end congestion. I decided upon Tuesday, 22 May, and tackled the next question — who should I invite to accompany me? The occupant of the passenger seat had to be prepared to take three days' leave; to match his

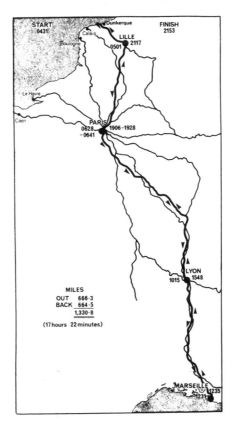

bodily needs to the mandatory stops for petrol; to keep a detailed record of time and distance throughout the trip and be prepared to be driven uncomplainingly at high speed in my sole company for 18 hours or so. I was fortunate in recruiting a Canadian friend, Hugh Young, to this (mobile) office which, in the event, he fulfilled untiringly and meticulously. What is more, he enjoyed every minute of it.

The choice of route to be followed gave me food for thought. Obviously the maximum use was to be made of *autoroutes* and this presupposed using A1 to Paris, A6 to Lyon and A7 beyond. Hitherto I had joined the A1 at Arras having driven from Calais, Boulogne or Le Touquet, but this is a tedious drive of about 70 miles passing through many villages and several towns which takes anything between 1¼ and 1¾ hours according to conditions. I was pleased, therefore, to discover* that a new stretch of *autoroute* had been completed from Dunkerque to Lille which entailed only a short distance through the outskirts of the latter city to reach the start of A1. I decided to use this new route feeling sure that the additional mileage would be more than justified from a time-saving point-of-view — and so it proved. I selected Marseilles as the turn-round point, for this met the fundamental aim of crossing France from Channel to Mediterranean, and set about building up a target timetable.

From a study of maps, A.A. and R.A.C. routes and my own records of previous journeys, I computed the mileage from Dunkerque to Marseilles to be 657.6. The target timetable, i.e. a yardstick of possible progress —

Footnote

* *Without (to our regret) having had occasion to emulate the rest of Mr Sedgwick's entertaining journey. Autocar endorses the usefulness of this route to A1, which is appreciable even in a weekend trip to Paris in a slower car (Autocar 8 February)* against which to assess the actual, which I devised was as follows:

	Miles	Hrs. Mins.
Urban areas — Lille	2.5	4
Paris (Boulevard Périphérique)	8.5	12
Lyon	3.0	5
Vienne	5.6	8
	19.6	— 29
Autoroutes — 638 miles at 90 mph	638.00	7 6
	657.6	7 35
Stops for petrol (and refreshment) 2 x 20 mins.		40
		8 15

Allowing for a 45-minute meal break in Marseilles, and the same time for the return journey, the total time (including stops totalling 2 hours 5 minutes) added up to 17¼ hours, thus a 4.30 a.m. start indicated a return to the starting-point by 9.45 p.m.—just about the duration of daylight at the time of year.

Realization

Another happy coincidence was the opening of "La Motellerie" at Armbouts-Cappel on the outskirts of Dunkerque just off the "Voie Express" (dual carriageway road) which becomes the *autoroute* A25 after eight miles. So we crossed from Folkestone to Calais on the 5 p.m. ferry *Hengist*, drove to Dunkerque and turned in early at "La Motellerie" having filled the petrol tank to the brim and requested a 4 a.m. wake-up call. (I had booked rooms for two nights, but we took the precaution of paying our bills for the first night to an astonished night-porter who couldn't believe his ears when I said that we were going to Marseilles for lunch.)

After self-made cups of coffee we left the motel at 4.20 a.m. and drove back into Dunkerque to the roundabout at Petite Synthe, three miles westward from the ferry terminal, where the "expressway" leaves the coastal road N40. I stopped there to take a photograph of the Corniche against the dawn sky before we set off at 4.31 a.m. The cloudless sky augured a fine day and the sun was soon shining brightly. As the engine warmed up I increased my cruising speed, mindful that the 45 miles to Lille should be covered in half-an-hour, and was surprised to find that 4,000 rpm corresponded with an indicated speed of 110 mph. Knowing that the true speed at 4,500 rpm was 118 mph I assumed that the rev.-counter was correct and that the speedometer exaggerated. I speeded up to find that 4,200 rpm equated to 120 mph on the speedometer and, at 4,500 rpm, (which marked the start of a yellow segment covering the next 500 rpm before changing to red at 5,000 rpm), the needle was going "off the clock" at 130 mph. Thus, it was evident that an indicated 130 mph corresponded to a true ground speed of 118 mph — an error of approximately + 10%. I knew, therefore, that so long as the instrument reading did not indicate more than 130 I would not be stressing the engine in any way. (Actually, my mandate from Rolls-Royce Motors had " . . . there should be no need to place any restrictions on the maximum speed for sustained cruising, other than those dictated by road and traffic conditions.")

Just before reaching Lille we overtook our first car — it was a light grey 3½-litre Bentley saloon pursuing its gentlemanly progress, presumably having crossed on the night ferry. (We didn't see another U.K.-registered car for another 400 miles and only five all the way across France.) The traffic was light and the trip recorder registered 139.7 miles at the end of 1½ hours. Odometer readings, only 0.2 per cent out, were taken as accurate. The distance to the toll-gate at Arras was 75 miles and we passed through 51 minutes after leaving Dunkerque — only two minutes later than planned — thus endorsing the choice of route.

We reached the end of the toll section at Survilliers (159.2 miles) at 6.13 a.m. having averaged 98 mph from Lille. The large number of cars approaching Paris at this early hour surprised us and traffic was running in all lanes on the *Boulevard Périphérique*. It was, however, maintaining a speed of 55 mph so that the drive round to the *Porte d'Italie* took only a minute longer than I had allowed. We sped off down A6 mentally calculating where we should replenish the tank and take our first sustenance of the day. Rolls-Royce Motors had indicated that I could expect to do not fewer than 11 mpg. This meant a range of about 250 miles — hence my plan to make five stops for petrol. We stopped at the Aire de Service Principale de Nemours at 7.14 a.m. and put in 94.6 litres (20.81 gallons) of *Super* to replace that used in covering the 233.5 miles to this point plus 5.7 miles the previous evening. This showed a consumption of 11.49 mpg at an average speed of 86 mph. We partook of a "continental" breakfast at a stand-up counter and the wheels were turning again after only 23 minutes. The next stage would take us to Lyon and we could not have had better conditions for the journey — dry road, sunshine, moderate traffic not too frequently obstructing the overtaking lane and only short stretches of single-working due to road works. I was cruising with the speedometer needle rarely dropping below 120 — a true speed of about 110 mph — and was conscious of the readiness of the car to exceed this gait without effort — but more of this later.

The first hour after breakfast saw another 96.4 miles behind us and then we were overtaken for the first time — by a 280CE Mercedes — as my speedometer showed 130, i.e. 118 m.p.h. We repassed it 10 minutes later at the same speed and did not see it again. At 9.31 a.m., five hours after leaving Dunkerque, I had driven 418.7 miles (= 83.74 m.p.h.) and 23 minutes later, i.e. after five hours driving time, 455.6 miles, which made the average driving speed 91.12 m.p.h. We reached the toll-gate at Villefranche at 09.52½ (dead accurate, this Young fellow) and five minutes later at Les Chères (458.5 miles) I put in nearly 22 gallons in exchange for 120 francs which, at 11.33 to the £, equalled £10.60, i.e. 48 pence per gallon. We decided not to eat — it was not yet 10 o'clock in the morning — and the stop took only seven minutes. As we pulled out on to the *autoroute* I pressed the wrong button to check the level of oil in the sump and for the second time opened the petrol-filler flap by mistake. Stopping to shut it, I remembered that this could not happen on the standard Silver Shadow, the button for testing being on the opposite side of the steering column to the petrol filler opener — a space taken up by the rev.-counter on the Corniche. We reached the entrance to the Tunnel de Fourvière at Lyon at 10.12 a.m. (469.3 miles). This tunnel has eliminated the slow, twisty, downhill run through the "agglomération" to the river bank and takes all the strife out of crossing Lyon. Five minutes after entering the tunnel we had resumed high-speed cruising on A7.

It always surprises me how rarely one is held up in Vienne where the *autoroute* runs between the town and the River Rhône and narrows to a single lane over a temporary flyover at the main intersection. We were on top of this flyover at 10.28 a.m. (489.3 miles) and passed through the *Péage* at Reventin four minutes later intending to run non-stop into Marseilles. At this point we were 18 minutes ahead of our target timetable, fuelling stops having occupied only 30 minutes against a budget of 40. The 137½ miles of *autoroute* to the *Péage* at Salon Nord, including 8.2 miles of single-lane working, took exactly 1½ hours (= 91.67 m.p.h.). After Orange we had been experiencing a cross-wind from a westerly direction strong enough to keep tell-tale windsocks in a horizontal attitude.

We arrived at the large roundabout encircling the triumphal arch which marks the end of A7 at Marseilles at 12.29 pm and pulled up at the *Cathédral* on the Quai de la Joliette at 12.31 p.m., exactly eight hours after leaving Dunkerque and a quarter of an hour ahead of our target time. The point-to-point average speed for the 666.3 miles had been 83.29 m.p.h. and, deducting the two stops aggregating 30 minutes, the driving time of 7½ hours (5 minutes less than planned) represented a driving average speed of 88.84 m.p.h.

Instead of stopping in Marseilles for lunch as programmed we decided to wait until we reached a "Jacques Borel" restaurant which we had noted 26 miles out on the *autoroute*. So, after a stay of four minutes in the South of France to photograph the Corniche against the background of *Cathédral* and docks, I turned the car and headed back to our starting-point. The *mistral* was blowing strongly, but the weather was still fine and sunny.

We stopped for lunch and petrol as planned. Two litres of oil went into the engine as a safety measure for none had been added for nearly 900 miles. The press-button oil-level indicator

Marseilles, and " . . . a stay of four minutes in the South of France to photograph the Corniche against the background of the Cathédral . . . "

was not designed to give a correct reading at the speeds I had been doing, when most of the oil is doing its job elsewhere than in the sump, and the engine hadn't had a chance to cool sufficiently to obtain a valid reading on the dipstick. In the restaurant we made for the snack bar, but the receptionist said it would be quicker in the *Gourmet Restaurant*. This seemed to me to be a contradiction in terms, but we had a simple *plat du jour* and, considering that the French don't know the meaning of the words *"tres pressé"* in the context of food, were lucky to be on the move again after 49 minutes. This meant that our stops now added up to within a minute of my original target timetable. As Hugh put it, we rejoined the *autoroute* as our ghost went past.

Just after 3 p.m. with over 800 miles behind me I sensed what I refer to as a lowering in my threshold of awareness. This is not really sleepiness, but I become conscious of blinking and of taking fractionally longer to refocus my vision when transferring the eyes

Marseilles and back in a day...

from road to instruments. After five minutes "zizz" in appropriately named *"Aire de Repos"* I was completely refreshed and resumed this most enjoyable drive. (Inured by this time to making calculations from recorded data my indefatigable Clerk of the Course wrote in the log: 40 winks in 5 minutes = 8 winks per minute!)

The traffic was lighter in this direction and the Corniche purred happily along in the manner of the fast, well-mannered touring car that it is, as is evidenced by the fact the distance of 196.4 miles between the toll-gate at Villefranche and the Courtenay/Sens Junction was covered at an average speed of 99 m.p.h. I see from the log that I averaged 100.17 m.p.h. over the last 117.7 miles of this stretch.

At 5.23 p.m. we had clocked up 1,000 miles which had entailed 11 hours 18 minutes driving at an average speed of 88.5 m.p.h. Just after 6 o'clock we were overtaken by our third car — a Citroen SM going like the clappers — and the weather deteriorated rapidly. Clouds gathered, lightning danced round the horizon and by the time we passed the exit for Courtenay and Sens it was raining very hard indeed. There was a brief respite at Fontainebleau and then a torrential downpour commenced which was to continue unabated for more than an hour and a half. The traffic was heavy approaching Paris and was reduced to a crawl and stationary jams by shunts and breakdowns. The *Périphérique* was congested and it took 23 minutes to drive from the A6 to the A1. At Le Bourget the thunderstorm continued, it was as dark as night though only 7.35 p.m., and the torrential rain was bouncing high off the road. We took advantage of a covered petrol station to refuel for the last time.

The weather cleared before Lille and we found the rather tricky route round the town using the partly-constructed *Boulevard Périphérique* whereon happened the only "untidy moment" of the whole trip when I found myself faced by an unlit barrier marking the end of the completed surface and an unsignposted deviation to the other carriage-way. I stopped without touching anything and the biggest surprise was the cloud of self-raised dust which enveloped the car after it had come to rest on the unmade surface! It is not difficult to imagine the thoughts which went through our minds at this happening so near to our goal. As we joined the A25 *autoroute* west of Lille for the final run-in Hugh told me that according to our target timetable we were due in Dunkerque, 40.4 miles away, in a quarter of an hour! The road was dry and traffic-free and we made it in 23 minutes, thus producing a sprint finish to our 1,330-mile run at 105.5 m.p.h. We stopped at the Petite Synthe roundabout at 9.53 p.m., eight minutes later than our target arrival, shook hands and drove back to the motel for supper and to honour our reservations.

"The traffic was heavy approaching Paris (on the return run) and was reduced to a crawl and stationary jams by shunts and breakdowns"

The return trip had taken longer than the outward run which had been made in as favourable conditions as one could ever hope to get.

The following is a summary of the journey:

Dunkerque-Marseille (666.3 miles)
Total Time : 8 hours = 83.29 m.p.h.
Driving Time : 7 hours 30 minutes = 88.84 m.p.h.

Marseille-Dunkerque (664.5 miles)
Total Time : 9 hours 18 minutes = 71.45 m.p.h.
Driving Time : 8 hours 14 minutes = 80.74 m.p.h.

Round Trip from Dunkerque to Marseille and back to Dunkerque (1,330.8 miles)
Total Time : 17 hours 22 mins. = 76.57 m.p.h.
Driving Time : 15 hours 44 mins. = 84.54 m.p.h.

I reckon that, on the basis of the above-reported progress, and remembering that 25½ miles were driven on single lanes where speeds were often restricted by heavy trucks to 40 m.p.h. or less, I must have driven more than 1,000 miles at speeds in excess of 100 m.p.h.

The following breakdown of the times taken to drive the various stages on the outward journey will serve as a basis for those interested enough to compare their own progress on the way to the Cote d'Azur without going to the unlikely destination of Marseille:—

Miles		Hrs.	Mins.
178.0	Dunkerque to Paris (Porte de la Chapelle)	1	57
9.3	On the *Boulevard Périphérique*		13
282.0	Paris (Porte d'Italie) to Lyon (Tunnel de Fourvière)	3	1
20.0	Lyon (Tunnel de Fourvière) to Vienne (top of flyover)		16
153.9	Vienne (top of flyover) to where the (new) autoroute to Aix-en-Provence and Nice forks off 8 miles south of Salon	1	42
643.2		7	9

Evaluation

That neither of us was really tired at the end of this long run says a lot for the Corniche. We wore the seat-belts all the time and, although we expected to be uncomfortable because of the relatively low top anchorage point (dictated by the construction of the two-door body) necessitating the strap coming up from behind the shoulder, we were not, in fact, inconvenienced. It is true to say, however, that the belts are more comfortable when not wearing a jacket.

Once placed in the right position the seats were so good as to call for no further adjustment and neither of us found it necessary to change our postures; nor were our joints stiff on getting out of the car. Two other factors played a tremendous part in making such a long journey unfatiguing — the air-conditioning and, because this enabled one to drive with all the windows closed, the insulation from outside noise. The controls for the air-conditioning are very complicated and the system is more difficult to "drive" than the car, but it is a good system and I hope it won't be long before it is automated and thermostatically controlled as has been the case in quality American cars these many years.

The Speed Control device is not designed to function at speeds over 90 m.p.h., so it was not much use on this drive. I did use it once, however, when we were confined to a single lane by road works, to check the speedometer reading at 100 k.p.h. against a timed kilometre — this distance took 38 seconds which is a speed of 94.74 k.p.h. indicating that the instrument was over-reading by about 5½ per cent at this speed. (It should be said that it is almost impossible to time accurately from a moving vehicle — especially at high speed — even using a split-second chronograph. At around 100 m.p.h. an error of only one-fifth of a second in reading the time between one kilometre marker and the next makes a difference of more than 1½ m.p.h.)

The Corniche performed impeccably. The fact that this particular example was built prior to the introduction of the major changes to the front suspension made little difference in this type of motoring. Only one fault manifested itself — in addition to the annoyingly optimistic speedometer — and that was the interior mirror being loose on its stalk. The right size Allen key would have fixed it in a moment, but it was aggravating having to reposition it at frequent intervals. When I handed the car back with 33,882 miles on the clock it was as smooth and quiet as when delivered to me.

Altogether I consider that the Corniche proved itself to be a very, very good fast touring car. Not quite fast enough, mind you, and I make this open plea to those at Crewe to offer a higher axle ratio as an option for those who habitually drive on the Continent. I feel that the car has reserves of power and that the resulting slight falling-off in acceleration, if any, would be a small price to pay for an available top speed nearer to 140 m.p.h. and a cruising capability in the region of 125/130. (If not an option on the Rolls-Royce version, I suggest that it should be standard equipment on the Bentley!) Such a modification would meet the oft-heard criticism that the top-end performance of the Corniche does not match up to the standards set by certain other manufacturers of very good, but less costly cars.

To those who cringe at the price of the Corniche — I am one — and who, having read this are inclined to say . . . "So it should", my reply would be "Well, it did, didn't it?" Of course, no-one in his right mind would wish to drive across France and back in a day, but it shows that a four-figure mileage on the Continent in daylight is well within the capability of this car. A few more miles of *autoroute*, and destinations way down in Italy (normally thought of as being at least two days away) could well be reached between an early breakfast and dinner — especially if two people share the driving.

Justification

I offer none. I undertook the drive solely for the pleasure I would derive from it. ("One man's meat . . . "). I was not disappointed. It cost a good deal of money and didn't last very long, but I can't think of anything which would have been more rewarding to me in terms of sheer enjoyment and I don't anticipate ever again using nearly 60 pounds' worth of petrol (119 gallons) and paying out over £12 in tolls in a day.

Thank you R.R. for a lovely day and thanks to H.J.T.Y. for bearing with me.

Rumination

It is interesting (to me, anyway) to note that the 1,330 miles took 15 minutes fewer than the 1,000 miles in the Vintage Bentley in England. Just to think of it — the average speed across France and back, including all stops, was nearly 10 per cent above the maximum speed allowed in this country?

My average speed for the round trip (76.57 m.p.h.) was slightly higher than that of the winning Speed Six Bentley of Woolf Barnato at Le Mans in 1930 (75.88 m.p.h.). I realise that the Bentley's average was maintained for 24 hours, but there were two drivers! There *has* been some progress in automobile and road engineering in the intervening years.

Was this the longest U-turn in history?

Speculation

I wonder how my new Double-Six Daimler would . . . . ? Better not!

Road test

by John Bolster

The depth and colour of the paint gives a splendid air of quality.

Rolls-Royce Silver Shadow: finer than ever

Who hath not owned, with rapture smitten frame. The power of grace, the magic of a name?

To start a road test report with poetry may be unconventional but I am dealing here with a very unusual car. The name of Rolls-Royce has denoted a car of incomparable excellence since Edwardian times, but let's skip a history which is already copiously documented and consider the latest production in terms of today and tomorrow.

In the course of my duties as a road test driver, I live with cars ranging from the rare and exotic to the small and popular. In a few short years, it has ceased to be a penance to drive cheap cars and indeed they are often superior in many important respects to the more costly machines. Some very expensive cars are remarkably bad and one might come to the conclusion that only the vast budgets of mass-production permit the full development of a really modern vehicle.

Yet, the Rolls-Royce is the exception that proves the rule. It is not the fastest or the biggest or anything like that, but it has such a well-balanced set of virtues that it makes everyday journeys memorable. I was lucky enough to be lent the latest Royce for an extra-long period and though the present dictates of economy denied me the long journeys and high speeds which I had promised myself, I liked it better every day. At the conclusion of the test, I asked myself a question. Would I, if I had that sort of money, buy a Shadow for my own use? The answer must be, "yes!"

When I collected the car from Dennis Miller-Williams at Conduit Street, I was amazed to see that it had a registration number which has figured on earlier test cars. Though I have previously tested a model with the enlarged 6.7-litre engine, this one had the important steering and suspension modifications which were introduced not long ago. Small changes are made as required to cars of this make and these had already included quicker steering and the deletion of self-levelling from the front suspension. However, a fairly radical redesign became possible when the latest generation of radial ply tyres came on the scene.

In brief, the changes concern more positive location of the front sub-frame, elimination of bump-steer effects under even the most extreme conditions, and particularly the transfer of road noises into the body, the reduction of which becomes a problem after mechanical and wind noises have been virtually eliminated.

The test car was a Silver Shadow saloon, which I prefer to the more expensive Corniche because it has four doors. Indeed, were I buying a Rolls-Royce I would order a long-chassis Shadow because, to me, this is essentially a car to be shared with friends. In a delightful shade of blue, the big machine looked smaller than it was, without any hint of vulgar opulence. That immortal radiator and bonnet too, look all the better in their later and lower shape.

When I entered the car, I set my seat both horizontally and vertically by touching a switch. Other switches arranged the door-locks and windows to my requirements, the electrically raised panes moving both quickly and silently, which is rare. Electric relays set the heating and air conditioning, too, with separate controls for the upper and lower parts of the interior. When I opened the door, a red light warned other traffic, and when I closed it the courtesy lights stayed lit for sufficiently long to let me insert the key and get the engine started. An interior control focused the outside mirror.

The transmission selector is really a switch, too, relays again doing the work. The engine has enormous torque and one can make a very impressive getaway in spite of an all-up weight of just over two tons, the independent rear suspension preventing wheelspin. The transmission is now so smooth that it is almost impossible to feel the changes or even to know which gear is in use.

The present very smooth transmission unexpectedly gives better acceleration figures than the earlier and less refined setup. Maximum performance was not the object of the present test and the performance figures in the data panel were taken by me in happier times, in a car with a similar specification though different steering and front suspension. Under prevailing conditions, I found that the Rolls-Royce speed control was a godsend, as it's difficult to keep a powerful car down to 50 mph without constantly looking at the speedometer. On a motorway, you simply switch it on, drive at the exact speed you want, and then press the "engage" button. The car will then cruise at that speed, irrespective of gradient. If it becomes necessary to accelerate or brake, the speed control at once goes out of action, until the driver presses the "resume" button, when it will speed up or slow down until it is cruising at the previously chosen speed. If a city has to be crossed, including stops at traffic lights, the device "remembers" and will resume the cruising speed when the open road is reached.

Only switching off the control, or the ignition, will erase the chosen speed, which can be selected again or replaced by any other. This device is a real petrol saver for it gives more correct carburetter control than the wobbly foot of the average driver. At 50 mph, the Rolls-Royce uses hardly any more fuel than a typical medium-sized saloon. It is above 100 mph that the big engine really becomes thirsty.

The new steering and suspension arrangements give more precise control (for example, the very light steering) yet gives sufficient "feel" for carefree driving on wet roads. The anti-drive suspension geometry avoids the occupants being thrown forward during hard braking and there is quicker recovery when S-bends are taken fast. The new front end does not cause any startling difference but it makes the car feel less ponderous to drive. Similarly, the revised insulation of road noise still admits some sound on certain non-skid surfaces but in general there has been an appreciable reduction of tyre noises.

The ride is absolutely outstanding, especially in the back seat, and the self-levelling system keeps the car level irrespective of passenger and luggage load. The Shadow moves with great ease over any surface and the absence of wind and engine noise is most noticeable in comparison with the majority of cars. Curiously enough, the vehicle seems less

Road test

silent to passers-by, the engine or fan noise, which is insulated from the interior, being audible through the radiator matrix. No doubt this tiny problem, too, will be solved, for nothing is more impressive than a car gliding away in total silence.

Obviously the size of a Rolls-Royce makes it less nippy than a smaller car in traffic or on winding roads, but the precision of the very light steering and the ease of judging the width do much to reduce this disadvantage. The driver's view is superb, with that most famous bonnet ahead, and the interior, with its leather, walnut, and lambswool carpets, it is in a class of its own. The depth and colour of the paint give a splendid air of quality and I washed the car regularly myself, for one does not risk the tender mercies of "Bert's Car Wash" with such a fine finish.

It's hard to put into words the pleasure and delight of living with a Rolls-Royce. There was a period of my life when I owned several in succession—all second-hand or even tenth-hand, I hasten to add—for I held the view that they were the only cars which I could afford to run. Now they are rather beyond my purse, apart from an Edwardian Silver Ghost which I have cherished for many years. However, those glistening Royces which carry the Royal Family on state occasions are all well over a dozen years old, I am told. These cars have always been a good investment and anybody who buys one nowadays can have a year of fabulous motoring and still sell it at a useful profit. Unfortunately, the waiting list for new ones is now so long that only young men need apply!

Car Tested: Rolls-Royce Silver Shadow four-door saloon, price £11,785.59 including tax and VAT.
Engine: Eight cylinders 104.1 mm × 99 mm (6745 cc). Compression ratio 9 to 1. Power output "sufficient." Pushrod-operated overhead valves. Twin SU carburetters.
Transmission: Torque converter driving 3-speed automatic gearbox, ratios 1.00, 1.48 and 2.48 to 1. Hypoid final drive, ratio 3.08 to 1.
Chassis: Combined steel body and chassis. Independent front suspension by wishbones, coil springs, and anti-roll tension bar. Recirculating ball power-assisted steering. Independent rear suspension by trailing arms, coil springs with automatic height control, and anti-roll bar. Telescopic dampers all round. Servo-assisted disc brakes on all four wheels with eight pads per disc and triple hydraulic circuits. Bolt-on disc wheels fitted Dunlop 205-15 radial ply tyres.
Equipment: 12-volt lighting and starting with alternator and ammeter; speedometer; fuel and oil level gauge; clock, heating, demisting, and ventilation system with refrigerated air conditioning and heated rear window; flashing direction indicators with hazard warning; two-speed and intermittent windscreen wipers and washers; unilateral parking lights; reversing lights; delayed-action courtesy lights; electric seat adjustment; speed control cruising system with engage and resume buttons; electric window lifts; electric door locks; radio with electrically-raised aerial and stereo-tape player.
Dimensions: Wheelbase, 9 ft 11½ in; track, 4 ft 9½ in; overall length, 16 ft 11½ in; width, 5 ft 11 in; weight, 2 tons 1½ cwt.
Performance: Maximum speed, 120 mph; standing quarter-mile, 17.0 s. Acceleration—0-30 mph, 3.2 s; 0-50 mph, 7.2 s; 0-60 mph, 9.8 s; 0-80 mph, 16.5 s; 0-100 mph, 28.4 s.
Fuel consumption · 12 to 18 mpg.

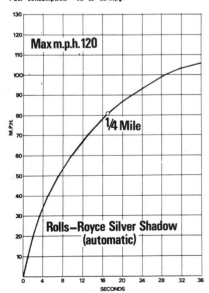

Speed control (left of wheel) helps improve the fuel consumption.

The standard wheelbase Shadow tested. Below: 6.7-litre V8 engine with SU carburetters.

Rolls-Royce Corniche

6,750 c.c.

You have to wait a long time for a new Corniche, but the latest models have many improvements and more standard equipment. The quiet, smooth engine, matched to an excellent automatic transmission makes the Corniche faster than the Shadow, with only marginally worse fuel consumption. The ride and handling have been improved, and the road noise levels reduced. It is expensive —but we think the high price is justified.

IN MOST people's eyes the Rolls-Royce Corniche is the absolute ultimate in cars. Not just a Rolls-Royce, it is a coachbuilt two-door coupé commanding a current list price of just over £15,000 and at present in such short supply that delivery is quoted at over three years. At pre-crisis auctions prices of as high as £28,000 have been reached for secondhand examples, although this is a long way from the real market value of around £17,000. Like a piece of select property or a rare antique, a Rolls-Royce is an appreciating investment and as such can be readily justified by company accountants.

When the Corniche was launched three years ago it followed the lines of the previous Shadow-based H. J. Mulliner, Park Ward two-door model, but was given more of a separate identity by using a special engine developing 10 per cent more power. The actual output is still undisclosed, but since the capacity was increased from 6·2 to 6·7 litres in 1970, the claim of "adequate" has been fully justified.

Somehow with a car like a Rolls-Royce, actual performance figures seem beneath its considerable dignity. Since the Corniche is a more sporting version and obviously aimed very much at the owner driver, we feel there is more justification for discussing its speed and acceleration in detail.

Compared with the Silver Shadow we tested in November 1972, the Corniche reached a mean maximum 3 mph higher at exactly 120 mph, with a best one-way run at 122 mph. At this latter speed the rev counter needle was just a whisker below the start of the red sector at 4,700 rpm, which is about as near as practice can get to theory whatever the car. On a slight down grade we had to lift off to prevent over-revving.

From a standing start the Corniche shot away with about a yard of wheelspin to record 0 to 50 mph acceleration time of 6·8sec (0·8sec quicker than that of the Shadow) and a 0 to 60mph time of only 9·6sec (10·2sec for

the Shadow). From rest to 100 mph took exactly 30sec (3·6 sec quicker than with the Shadow), which is no mean achievement for a saloon the size and weight of the Corniche.

Most of these gains have been brought about by improvements in the efficiency of the engine (re-timed camshaft, large-bore exhaust system, revised air cleaner) so fuel consumption suffers very little in consequence despite more eager feet on the pedals. Overall for close to 1,400 miles, which included a trip to northern France for maximum speed measurements and a lot of UK driving before the present fuel-saving campaign, we recorded 11·9 mpg. This is less than a five per cent drop on the 12·4 mpg we obtained with the Shadow in 1972. On a typical run away from London we returned over 13 mpg with ease, but during a spell of thick fog in heavy city traffic the consumption fell to less than 10 mpg, which is about the same as should be expected on a fast Continental trip with cruising speeds around 100 mph, speed limits permitting. The tank holds 24 gallons, so the useful range is poor, at just over 200 miles.

The engine starts first turn of the key but engagement is a noisy, out-of-character affair which surprises by-standers and sounds very loud inside a garage. When running, the engine is quiet and completely without temperament, but on the test car the hot idle was rather lumpy. An automatic choke takes care of all temperature variations.

Compared with the last Shadow we drove with the 6·7 litre engine, smoothness and refinement have been improved and there was none of the harshness we complained of in our last test. When the driver of the Corniche opens up to accelerate, there is a detectable and obviously intentional increase in engine noise which amounts to no more than a pleasant and very subdued power hum, like an electric motor speeding up.

A badge reading Corniche identifies the car at the rear, but although following the general lines of the Silver Shadow, the styling is quite distinctive

It is louder than on the regular Shadow, but in no way offensive. The rest of the time the near silence of the engine is a continual marvel and its sweetness throughout the range was delightful. These characteristics in a V8, with its secondary balance problems, are a remarkable achievement.

Rolls-Royce engineering policy is to make small but continuous improvements so that no version is immediately out of date but an owner changing his car after, say, three years will notice many worthwhile gains. We were impressed with how much progress has been made recently with the suspension, this nearly new 1974-specification demonstrator incorporating all the latest changes.

In the past we have felt that in the Rolls-Royce book ride took a decisive precedence over handling and we went to great lengths in our 1972 Shadow test to analyse the behaviour over all manner of surfaces and commented in detail on the limitations. In the final score we had to admit to being disappointed overall. The story now has changed considerably and the balance between ride and handling is a much better compromise. Steering response, for a start, is quicker and although understeer is still predominant, with power on the Corniche has much nearer neutral characteristics. In a brutal steering-pad test we got the car drifting sideways with front and rear tyres smoking equally, just to prove the point.

On the road there is still some vagueness in the steering about the straight-ahead position, accentuated to a large extent by the ultra-light servo assistance, but it is much easier than before to position the car accurately on sweeping bends or in narrow lanes. The small, 15in. dia wood-rimmed steering wheel which was fitted to the early Corniches has been replaced by the standard 16in. Shadow plastic wheel, which does not help, yet we soon settled in the big car and had enough confidence to flick it through the twists at remarkably high speeds.

Rolls-Royce Corniche

The biggest deterrent to this kind of driving is the body roll which throws occupants across the large slippery seats and tilts the horizon alarmingly. Actual cornering power is good, but for fast driving there are too many attitude changes, both laterally and in the fore and aft direction; although very effective anti-dive is built into the front suspension geometry, rear-end squat under hard acceleration is a real nuisance.

Improvements in ride are just as noticeable, road noise levels being much reduced (although whether this is due to engineering development or more a characteristic of the two-door bodywork, we cannot say for sure). On some coarsely-dressed surfaces, tyre roar begins to intrude, but on the kind of smooth tarmac that covers most main roads one can only marvel at the overall interior quietness. Abroad on French undulations we would have preferred stiffer damping, the rear suspension occasionally contacting the bump stops (despite no back seat passengers) and the general motion being rather like crossing a gentle swell at sea. On level main roads and the *autoroutes* the Corniche stormed along in perfect comfort with superb stability and no detectable wind noise.

For a few years now, Rolls have been using General Motors automatic transmission imported from the USA, the system employing a conventional three-speed epicyclic gearbox with torque convertor. This unit meets the exacting Rolls-Royce standards without modification and it behaved impeccably on the test car. Shift quality was an excellent match for the silky engine, and with sensitive part-throttle downshifts it gives the car a very lively response. At full throttle the gearbox changes up automatically at around 4,000 rpm, which must be close to the peak of the power curve because we could not improve on the acceleration times by holding to higher revs. At 4,000 rpm, maximum speeds are 42 and 73 mph in low and intermediate and the transmission will not kick down above 66 and 29 mph respectively. Low is prevented from engaging above 40 mph, to protect the engine. There is an electric selector on the steering column which is isolated in Park by the ignition switch, thus eliminating the need for a steering lock.

With ventilated disc brakes all round and a powerful servo system, pedal efforts are light and there was virtually no fade once the linings had reached a stabilized temperature. The handbrake is impressively powerful, recording 0.4g on its own from 30 mph and holding the heavy car securely on a 1-in-3 gradient. It is mounted in a slightly awkward position under the right of the facia and needed its full 9in. travel on the test car. Hydraulic circuits are duplicated for the servo system and backed up by a third non-servo installation, so any sort of brake failure is statistically out of the question.

Comfort and Equipment

For £15,000 the buyer has a right to the highest standards of luxury and in most respects he is unlikely to be disappointed with the Corniche. Full air conditioning is standard, best Connolly hide is used for the seats and trim, and wood veneer for the facia and door cappings is accurately matched with spare panels from the same part of the same tree held and indexed at the factory. Being painted and trimmed at the H. J. Mulliner, Park Ward works at Willesden, the Corniche received even more individual treatment than the Shadow, a justification in part for its premium price.

Front seats are large and very sumptuous affairs with folding armrests on the inside edges of the backrests and an adjustable armrest on each door. The inboard armrests quarrel with the seat belt buckles if lowered after fastening and we found it very difficult to get the buckle tongues past the door armrests. Seat belt arrangements are very poor in the Corniche and something the factory should attend to without delay.

Apart from being rather hard and slippery, as is so often the case with leather, the seats are comfortable and reasonably well shaped. You tend to sit on them, rather than in them, but support is quite well arranged. The multi-way joystick power adjustment levers are on the door trims with a much too coarse rake adjustment at the base of the backrest in line with the hinge. The steering column is fixed and several drivers found it too close to their chests once they had set the seat for comfortable pedal reach.

Getting into the back seat requires a stiff button to be pressed at the top of the backrest, to release it, and the rather heavy assembly to be swung forward. It is then a case of squeezing past the seat belt and exposed inertia reel before settling back in superb luxury and with a surprising amount of legroom. Assistance is needed to get out again as the front door must be opened first and its release is out of the reach of someone in the back. These criticisms aside, the Corniche must surely be the roomiest two-plus-two coupé on the market.

Facia layout is similar to that of the Shadow, except for the addition of a rev counter on the extreme right and a slightly different centre console arrangement. The glass of the rev counter picks up reflections from the side window and is often impossible to read. In a car like this with such an excellent automatic transmission it serves little purpose anyway. At night we found the instrument lighting patchy.

Heating and Air Conditioning

On all Bentley and Rolls-Royce models full air conditioning is standard and once the rather complex controls have been mastered, it works very effectively. Temperature is selected by a progressive rotary knob which is pulled out through four positions to regulate the volume of heated air, a multi-speed booster fan working independently. With practice the system can be set and controlled easily, but we have to question whether something simpler on the lines of that used by Cadillac and now Jaguar would not be more in keeping with this sort of ultimate car. The positions of the various outlets and the balance of flow rates has been well thought out. Far from being something for the occasional heat-wave, the air conditioning was useful on nearly every trip even in winter.

A considerable number of items which normally cost extra are included in the standard price, and the quality has not been skimped to permit this. There is still a wide choice available to the customer and some of these alternatives were fitted to the test car. One of the most satisfying was a Bosch Frankfurt FM stereo radio instead of the standard pushbutton AM unit, with four speakers. It adds £77.46 to the price. Apart from poor installation of the rear speakers which was not noticed in the front, the quality was most impressive, especially when stereo was being broadcast by the commercial radio stations. In contrast the eight-track Radiomobile cartridge player was disappointing and we would have preferred the alternative Bosch four-track cassette machine which costs £25 extra.

Also standard is an automatic cruise control which maintained the speed to within 1 mph of the setting chosen. It is engaged by pressing a button when the required speed is reached, switching it out or touching the brake cancelling this command immediately. It can be overridden in use by depressing the accelerator against extra pressure.

Living with the Corniche

When the total invoice of the car delivered is close to £15,500 and the wait can amount to around four years, one expects a lot from the Corniche. We found the test car stood up to the closest scrutiny of every detail and felt that a buyer would be well pleased with his possession.

After the first 3,000-mile service, the Corniche needs attention every 6,000 miles, when the cost alternates between £40 and £50, assuming a typical labour charge of £3.50 per hour. The warranty period lasts for an impressive three years and after that replacement of components like brake linings, dampers, starter or alternator is no more expensive than on much cheaper cars. The stainless-steel exhaust system, which should last several years without trouble, is expensive at just under £260, but that is a fair price for long life and quality materials.

Owner checks are made extra simple by the traditional sump level gauge that works through the fuel gauge and in many ways the Rolls can be run without attention, like a washing machine or television. Circuits for the several warning lamps and seat belt buzzer can be tested by pushing a button on the facia.

Another of the standard luxury items is centralized locking

The Corniche is available only with two-door bodywork – saloon (as tested) or convertible. The huge doors open wide for easy access, and the luxurious thick squabs of the front seats tip forward for entry to the rear compartment, after the chrome button at the top of the backrest has been pressed. The engine compartment is a magnificent, if daunting, sight. Below, the world famous mascot is spring-loaded for pedestrian safety. A further refinement is a drop-down panel beneath the facia carrying the fuses and relays

Rolls-Royce Corniche

operated by a two-way button on the trim panel of the driver's door. It locks both doors and the boot, the latter being released only with the key or by pushing a button inside the glove locker which can itself be locked with the boot lid key. The luggage can thus be left secure while the car is driven by an attendant. After closing both doors the interior lamp stays on with a time switch for about 10sec.

Conclusions

In the light of advances made recently by Mercedes, we would like to see a few more engineering features on this excellent British product. High on our list would be anti-squat suspension to prevent the tail dipping under acceleration, power hydraulics for the brakes and aerodynamic guttering to prevent the

Comparisons

MAXIMUM SPEED MPH
Jaguar XJ12C ....(£5,243) 146
Citroen SM ....(£6,369) 139
Mercedes-Benz 450SEL.. (£9,582) 134
Rolls-Royce Corniche (£15,104) 120
Fiat 130 Coupé ....(£5,970) 116

0–60 MPH, SEC
Jaguar XJ12C 7.4
Mercedes-Benz 450SEL 9.1
Citroen SM 9.3
Rolls-Royce Corniche 9.6
Fiat 130 Coupé 10.6

STANDING ¼-MILE, SEC
Jaguar XJ12C 15.7
Mercedes-Benz 450SEL 16.7
Rolls-Royce Corniche 17.1
Citroen SM 17.1
Fiat 130 Coupé 17.7

OVERALL MPG
Fiat 130 Coupé 20.6
Citroen SM 17.9
Mercedes-Benz 450SEL 14.7
Rolls-Royce Corniche 11.9
Jaguar XJ12C 11.4

Performance

ACCELERATION SECONDS

True speed mph	Time in secs	Car speedo mph
30	3.2	30
40	4.9	39
50	6.8	49
60	9.6	58
70	12.7	68
80	16.9	79
90	23.2	92
100	30.0	103
110	—	113

Standing ¼-mile
17.1sec 81 mph

Standing kilometre
31.8sec 101 mph

Mileage recorder: accurate.

GEAR RATIOS AND TIME IN SEC

mph	Top (3.08–6.28)	Inter (4.56–9.30)	Low (7.64–15.59)
10–30	—	—	2.5
20–40	—	3.7	3.0
30–50	—	4.2	3.8
40–60	6.6	5.0	—
50–70	7.6	6.1	—
60–80	9.0	7.8	—
70–90	11.2	—	—
80–100	13.1	—	—

GEARING
(with 205VR–15in. tyres)
Top ....26.2 mph per 1,000 rpm
Inter ....17.7 mph per 1,000 rpm
Low............10.6 mph per 1,000 rpm

MAXIMUM SPEEDS

Gear	mph	kph	rpm
Top (mean)	120	193	4,600
(best)	122	196	4,660
Inter	83	134	4,700
Low	50	80	4,700

BRAKES
FADE (from 70 mph in neutral)
Pedal load for 0.5g stops in lb

1	20	6	30–40
2	25	7	30–40
3	25–30	8	35
4	28–32	9	35
5	30–35	10	35

RESPONSE (from 30 mph in neutral)

Load	g	Distance
20lb	0.45	67ft
40lb	1.00	30.1ft
50lb	1.05	28.7ft
Handbrake	0.40	75ft
Max Gradient	1 in 3	

Consumption

FUEL
(At constant speed—mpg)
30 mph 19.1
40 mph 18.7
50 mph 18.1
60 mph 17.2
70 mph 15.8
80 mph 14.1
90 mph 12.3

Typical mpg..12.5 (22.6 litres/100km)
Calculated (DIN) mpg 14.4
(19.6 litres/100km)
Overall mpg 11.9 (23.7 litres/100km)
Grade of fuel Super, 5-star (min. 100RM)

OIL
Consumption (SAE20W50) 500mpp

TEST CONDITIONS
Weather: Fine Wind: 0 mph
Temperature: 10 deg C. (50 deg F).
Barometer: 30.1in. hg. Humidity: 93 per cent.
Surface: Dry. concrete and asphalt
Test distance: 1,366 miles.

Figures taken by our own staff at the Motor Industry Research Association proving ground at Nuneaton and on the Continent.

Dimensions

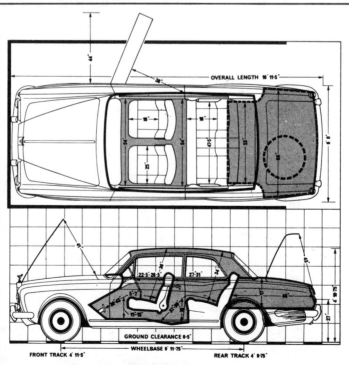

STANDARD GARAGE
16ft × 8ft 6in.

TURNING CIRCLES
Between kerbs
L, 38ft 9in.; R, 38ft 0in.
Between walls
L, 40ft 11in.; R, 40ft 2in.
Steering wheel turns, lock to lock: 3.4.

WEIGHT
Kerb weight 430cwt
(4,816lb–2,190kg)
(with oil, water and half-full tank).
Distribution, per cent
F, 53.2; R, 46.8.
Laden as tested: 46.0cwt
(3,156lb–2,340kg).

82

side windows from becoming obscured in wet weather. Two additional details would be a push-button trip reset, like that used on various cheap Fords, and a switch in the door jamb to energize the electric windows with the ignition off and the door open. At the moment they are totally isolated by the ignition key.

These few points apart, we feel the latest Corniche sets very high standards at the top end of the luxury car market and is well deserving of the considerable esteem in which it is held. Whilst its price is extremely high, it is easy to see where the money has been spent and as a non-depreciative asset it is unique in the modern car field. □

MANUFACTURER:
Rolls-Royce Motors Ltd., Pyms Lane, Crewe, Cheshire.

PRICES		EXTRAS (inc. VAT)	
Basic	£12,675.00	*Everflex roof covering	£184.71
Special Car Tax	£1,056.25	*Bosch Frankfurt AM/FM stereo radio in lieu of standard	£77.46
VAT	£1,373.13	*Bosch cassette player in lieu of 8-track cartridge unit	£24.64
Total (in GB)	**£15,104.38**	*Fire extinguisher	£11.81
Seat Belts	Standard	*Fitted to test car	
Licence	£25.00		
Delivery Charge (London)	Varies		
Number plates	£5.65		
Total on the Road (exc. insurance)	**£15,135.03**	**TOTAL AS TESTED ON THE ROAD**	**£15,409.01**
Insurance	Group 7		

Specification

Rolls-Royce Corniche

FRONT ENGINE REAR WHEEL DRIVE

ENGINE
- Cylinders: 8, in 90-deg. vee
- Main bearings: 5
- Cooling system: Water; pump, thermostat and viscous-coupled fan
- Bore: 104.1mm (4.1in.)
- Stroke: 99.1mm (3.9in.)
- Displacement: 6,750 c.c. (412 cu. in.)
- Valve gear: Centre camshaft, hydraulic tappets
- Compression ratio: 9.0-to-1. Min. octane rating: 100RM
- Carburettors: 2 × SU HD8
- Fuel pump: 2 × SU electric
- Oil filter: Purolater full-flow, renewable element
- Max. power: Not disclosed
- Max. torque: Not disclosed

TRANSMISSION
- Gearbox: GM automatic, 3-speed with torque converter
- Gear ratios: Top (auto) 1.0–2.1 / Inter 1.50–3.15 / Low 2.50–5.25 / Reverse 2.00–4.20
- Final drive
- Mph at 1,000 rpm in top gear

CHASSIS AND BODY
- Construction: Integral steel body and chassis, aluminium alloy doors, boot lid and bonnet

SUSPENSION
- Front: Independent; coil springs, double wishbones, anti-roll bar, telescopic dampers
- Rear: Independent; coil springs, semi trailing arms, anti-roll bar, telescopic dampers, automatic ride height control

STEERING
- Type: Power-assisted recirculating ball
- Wheel dia.: 16.0in.

BRAKES
- Make and type: Rolls-Royce/Girling ventilated disc front, plain disc rear. Three independent hydraulic circuits, operating twin front calipers and dual-piston rear.
- Servo: Two independent engine-driven pumps
- Dimensions: F 11.0in. dia. / R 11.0in. dia.
- Swept area: F 227 sq. in., R 286 sq. in. Total 513 sq. in. (223 sq. in./ton laden)

WHEELS
- Type: Pressed steel disc, 5-stud fixing 6.0in. wide rim.
- Tyres—make: Avon, Dunlop or Firestone
- —type: Radial ply tubeless
- —size: 205-15in.

EQUIPMENT
- Battery: 12 volt 71 Ah
- Alternator: Lucas 55 amp
- Headlamps: Lucas 200/75 watt (total)
- Reversing lamp: Standard
- Electric fuses: 20, plus thermal cutouts for headlamps, gear selector and door locks.

- Screen wipers: 2-speed, plus intermittent delay
- Screen washer: Standard electric
- Interior heater: Standard, with air-conditioning
- Heated backlight: Standard
- Safety belts: Standard, inertia reel
- Interior trim: Connolly hide seats, cloth headlining
- Floor covering: Wilton carpet, with nylon rugs
- Jack: Screw pillar
- Jacking points: 2 each side, under sills
- Windscreen: Laminated
- Underbody protection: Zinc-coated steel, phosphate dip primer, bitumastic compound.

MAINTENANCE
- Fuel tank: 23.5 Imp. gallons (107 litres)
- Cooling system: 28.5 pints inc. (heater)
- Engine sump: 14.5 pints (8 litres) SAE 20W50. Change oil every 6,000 miles. Change filter every 6,000 miles.
- Gearbox: 18.6 pints. SAE ATF. Change every 12,000 miles.
- Final drive: 4.5 pints. SAE 90EP. Change every 24,000 miles.
- Grease: 6 points every 12,000 miles.
- Valve clearance: Self adjusting
- Contact breaker: 0.014in.–0.016in. gap; 26–28 deg. dwell.
- Ignition timing: 5 deg. BTDC (stroboscopic at 800 rpm)
- Spark plug: Type: Champion N14Y. Gap 0.023–0.028in.
- Compression pressure: N.A.
- Tyre pressures: F 28; R 28 psi (normal driving) / F 28; R 32 psi (full load).
- Max. payload: 1,000lb (454kg)

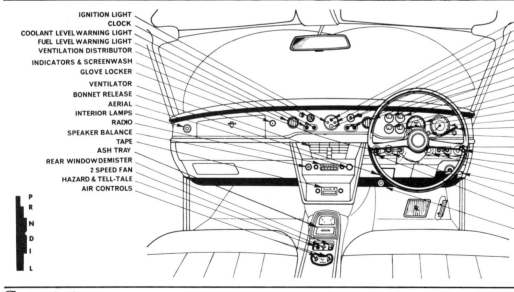

Servicing

	3,000 miles	6,000 miles	12,000 miles	Routine Replacements:	Time hours	Labour	Spares	TOTAL
Time Allowed (hours)	2.75	9.00	9.50	Brake Pads – Front (set)	1.00	£3.50	£20.19	£23.69
Cost at £3.50 per hour	£9.62	£31.50	£33.25	Brake Pads – Rear (set)	0.60	£2.10	£10.09	£12.19
Oil Change	£2.84	£3.25	£9.06	Exhaust System	6.25	£21.88	£258.50	£280.38
Oil Filter	—	£3.30	£3.30	Dampers – Front (pair)	5.00	£17.50	£24.64	£42.14
Breather Filter	—	—	—	Dampers – Rear (pair)	7.75	£27.13	£24.64	£51.77
Air Filter	—	—	—	Replace Half Shaft (exchange)	1.00	£3.50	£41.80	£45.30
Contact Breaker Points	—	£1.57	£1.57	Replace Generator (exchange)	0.80	£2.80	£25.41	£28.21
Sparking Plugs	—	—	£2.94	Replace Starter (exchange)	2.00	£7.00	£29.70	£36.70
Total Cost:	**£12.46**	**£39.62**	**£50.12**					

83

ROAD IMPRESSIONS

R-R has gone through some rough patches — but the new models are much more driveable than before . . . .

From GORDON WILKINS
(in Britain)

With the hood down, the Corniche shows the handbuilt craftsmanship of the London coachbuilders, H.J. Mulliner-Park Ward.

Rolls-Royce Shadow and Corniche

For some time past the Rolls-Royce philosophy on handling and road-holding seemed to be diverging from the mainstream of European thought. The Silver Cloud — with massive understeer scrubbing the treads off the front tyres at quite moderate cornering speeds — did not encourage fast driving on winding roads, and lack of road feel in its power steering made it difficult to drive confidently on snow or ice.

Nor (with power doing the work) was it easy to see why the steering needed 4¼ turns from lock to lock.

Company thinking as expressed by then Chief Engineer, Harry Grylls, was influenced by the "sneeze factor". American buyers, it seemed, could not be trusted with quick, direct steering lest a sneeze caused them to steer themselves into the ditch. Such things have happened, but if the steering were not so extremely light the possibility of such a ludicrous disaster might be reduced.

So for the benefit of Americans who had hay fever and couldn't drive, all had to suffer impaired control. It certainly helped to foster the image of the Rolls-Royce as an old man's car, and it was not surprising when market research showed that the average age of buyers was over 50.

In 1965 the Silver Shadow came, bringing self levelling hydro-pneumatic suspension for a super-smooth pitch-free ride, and a new triple braking system with discs (though it shares one weak link with the previous model — if some clod fails to secure the clevis pin at the base of the pedal properly, contact with all braking systems except the handbrake is lost). The steering still lacked feel, and on corners the body roll was spectacular.

FIXED STEERING WHEEL

We never did agree about feel in the steering. Grylls maintained that you don't drive by feel, but by the position of the wheel. He even had a car built with a fixed steering wheel which turned the front wheels according to the pressure applied, and no one could drive it.

But after you have spun the wheel back and forth a few times to correct a

Interior of the Corniche: A/T selector is on the right of the column, and a rev-counter (right) is standard.

series of slides, the position of the wheel tells you nothing — except that you are in trouble.

Being one of the most complex cars ever built, the Shadow suffered a series of youthful ailments which took some time to sort out. It was perhaps fortunate for Rolls-Royce that the Mercedes Benz 600 — an equally complicated but better-handling car — also suffered dire mechanical and hydraulic afflictions which so tarnished its image, that it is now only made to order — and then at a heavy loss on every car produced.

Rolls-Royce were helped through their time of travail by the extraordinary loyalty of the customers. I had one traumatic drive from Rome to Milan with the driver's door window stuck open because the electric motor had failed, seized-up suspension-levelling jack, a main oil pump out of action,

Sculptured lines of the Corniche convertible. Engine is an 8-cylinder of 6 230 cm3.

and a hydraulic system that had thrown away most of its oil.

Later cars were given emergency manual operation for the windows, new bore treatment kept the hydraulic pumps working, and the leaks were cured. The ride remained superb, and though an engineer with a sensitive ear might claim to detect traces of crank rumble at 100 mph and over, the Shadow remained probably the quietest car on the road, in mechanical and wind noise, and in road noise.

RADIALS AS AN OPTION

Radial-ply tyres later brought an improvement in steering and wet weather grip, plus nearly twice the tyre life, but at a cost of a slightly harder low speed ride and a noticeable thump at road joints or cats' eye road markers so they remained an option, not standard equipment.

In response to criticisms, the steering ratio was eventually changed to give lock-to-lock in 3¾ turns instead of 4¼ and something was done about the body roll. This was not easy, because suspension geometry, particularly at the rear, gave a roll centre that was down near the deck. However, an anti-roll bar was added at the rear and the front one was made thicker, which was about all that could be done without introducing a lot of lateral roll shake, which is tiring to the neck muscles.

Thus equipped, the two-door Corniche was launched. The engine had already been increased from 6,2 to 6,7-litres and for Corniche it was modified to give a further 10 per cent in power. But the steering and the roll still caused criticism from fast drivers, and the extra power began to show up the limitations of the brakes.

The consequences were spelled out by a Shadow owner, Dr Alex Moulton, inventor of the British Leyland Hydrolastic suspension and the Moulton bicycle, in a paper he read at a conference organised by Britain's Transport and Road Research Laboratory in January 1972. Carefully-instrumented chicane tests, simulating accident avoidance and recovery, showed that the Shadow made slowest time of ten cars tested, and had the biggest roll angle for a given lateral acceleration.

In a Clotoide test simulating one of those treacherous corners which get tighter as you go round, the Shadow came tenth out of 12 cars tested, the only worse performers being an Anglo-American 1,7-litre and an American sedan.

This sort of thing, added to the financial crisis brought about by the RB 211 aero engine made the future look bleak, but the Car Division was quickly turned into a separate entity with new management. Young David Plastow became managing director, and John Hollings took over as chief engineer when Harry Grylls retired. Soon they had boosted production, and made some radical improvements to the product, which give better handling and steering response with less roll, while more "compliance" in the front suspension quells the harshness and thump from radial-ply tyres.

"STEERING THE SUB-FRAME"

Sub-frames carrying front and rear suspension were attached to the body shell by coil springs stabilised by masses of metal wool, which soon became known as the "pan scrubbers". They damped out road noise very effectively, but permitted a slight amount of lateral movement which had exaggerated effects.

As one engineer remarked: "You were steering the sub-frame, not the car." So the pan scrubbers found their way to the dustbin and were replaced by rubber mountings, made more resistant in the lateral plane that affects the steering.

Radial-ply tyres are now standard, and the job of damping out the radial thump is now done by more compliant front suspension in which the top wishbone is replaced by a cranked rubber-mounted arm linked to a short strut.

There are other benefits too. The front roll centre is raised by 5 cm, front track is increased slightly, and at this point a little has a big effect.

Steering now needs only 3¼ turns lock to lock.

Steering response is much better, and though the reduction in roll angle is small, the car feels more stable, probably because the roll centre is now more accurately located instead of swimming about with the sub-frame.

Low-speed ride is back to the old standard, even with radials, and the thump has been tamed. Even so, minor departures from true roundness, which can occur in manufacture of big radial tyres, can create minor vibrations, so Rolls-Royce get the tyre manufacturers to grind the tyres truly circular after they have been fitted to the wheels — just one more example of the relentless attention to detail which gained them their reputation.

SET BRAKES ON FIRE!

If you let the steering wheel spin back through your fingers after a sharp corner, the earlier Shadows (like the Clouds) would waddle out in a series of roll oscillations amazing to behold, and drivers learned to feed the wheel back from hand to hand, but the latest Shadow and Corniche come out smoothly with no more than a slight shake as they hit the straight.

For the first time in years I found myself enjoying the silence and comfort and the impeccable finish of every detail, and enjoying the driving at the same time.

Soon I was throwing a Shadow down an Alp at speeds I had never attempted before, and eventually I set the brakes on fire. They cope with this on the Corniche by fitting ventilated discs at the front.

They have already got the average age of Rolls-Royce buyers down to 45, but if they get it any lower, they're going to need ventilated discs on the Shadow, too! ●

The Cubic-Dollar Comparison:
Cadillac Seville Vs. Rolls-Royce Silver Shadow

BY DON SHERMAN

If you mercilessly strip away all the illusions, there can only be one standard of excellence.

• Once your course is set and the automatic controls are programmed, you can forget about driving. Just keep the lane markers flowing through the silver wreath on the hood with one finger's worth of pressure on the wheel and Cadillac takes care of the rest. Your speed is frozen. Your headlights flash on as the sun sets and politely dim at the sight of oncoming traffic. The climate in your sealed capsule is constant. Peace prevails, interrupted only by the soft murmurings of the stereo FM electronically locked to its signal. It's your leather-cushioned den gliding down the freeway—technology's finest expression of the automatic motor car.

Society has schooled us to expect no less than effortless comfort of a Cadillac. After all, it has long been the biggest, most expensive and therefore best American car, the gleaming flagship of a nation dedicated to indulgence.

Yet the tremors of change have cracked the very bedrock of the country's tenets, leaving Detroit teetering on the edge of one particularly deep fault. Motor City's sustenance, the big car, is being buried under an avalanche of crowded roads and

windling resources. So the baton has been passed. Cadillac's Seville will now bear the crest of Detroit's first family, as well as its reputation as the "standard of the world."

But is it really? To find out, we turned for a look at the world's most expensive production sedan, a timeless example of cost-no-object machinery: the Rolls-Royce Silver Shadow. At 10 times the price of today's basic transportation, the Silver Shadow is the sole survivor of an era when cars were crafted by men rather than machines. Any similarity it has to an assembly-line automobile is a coincidence of function—because both the Silver Shadow and the Cadillac Seville are designed to carry their passengers in the highest attainable level of comfort and luxury.

What we uncovered in this comparison is how truly impressive a piece of equipment the Seville is. The Rolls is a "better" car—more capable in the hands of a demanding driver—and outright luxury penetrates its soul far deeper than it does the Cadillac's. But the Seville has the advantage of newer technology, which makes it both more comfortable and at the same time a surprisingly responsible member of a modern transportation system.

The Rolls-Royce is a purely functional design. The mechanicals, passengers and luggage are accommodated in a fitting manner, sealed away from the rude world about them, protected by five-mph bumpers . . . and that's it. The Seville, however, came to life only after calculated scrutiny of the competition by Cadillac planners. Their target was not so much Rolls-Royce as Mercedes-Benz, Jaguar and BMW—manufacturers experienced in building affordable luxury without bulk. The Seville came cut six inches too long to fit into the "precision-size" mold, but it is the smallest Cadillac in 36 years—more than two feet shorter and eight inches narrower than the current Sedan de Ville. And definitely prime evidence that the harsh light of reality is filtering through cracks in the bigger-is-better philosophy. You'll never, however, see the Seville described as small, mini or baby in advertising, because words like that connote a scaled-down price, which, at $12,470 for starters, this new Cadillac doesn't have.

PHOTOGRAPHY: GENE BUTERA

From the front seat of a conventional Cadillac, the pavement lies somewhere beyond an expanse of sheetmetal so vast it makes you feel you're conning an aircraft carrier. But in the Seville, Cadillac buyers can ride in luxury without an escort fleet to scout a navigable route for the biggest vessel on the road. The idea was to pack every attribute for which a Cadillac is known—space, comfort, long lists of standard power equipment and a pervading air of luxury—into a smaller package. The Seville's hood is manageable in size and low in profile, and the tall greenhouse positions you upright for clear outward vision. Wide rear roof-pillars and tapered rear fenders make it a little difficult to pinpoint the back corners, but any driver rated for big-car operation will stroke around in the Seville with a foot of extra clearance in every direction. In sum, what Cadillac has here is the right-size plug to stem the flow of defectors toward the imports.

The Silver Shadow is virtually the same length and width, but it emits an illusory feeling of bulkiness on the highway,

The revolutionary thing about the Seville is that it is a Cadillac you can enjoy driving without feeling naked because you've given your battalion of outriders the weekend off.

quite possibly a subconscious effect of the prestige radiated in every direction. You look down on normal traffic from a leather seat you have to step up to. The *Spirit of Ecstacy* guides your progress from her perch on the monolithic radiator shell as she has on every Rolls-Royce since 1911. A legend trails respectfully behind.

The tale of a Rolls-Royce Silver Shadow is automotive lore in its richest form. The body represents three months of handwork to manufacture the aluminum and steel panels before they are ready for 20 coats of paint. The interior is lined with materials that thwart any attempt at mass production. Italian walnut sweeps across the instrument panel and window sills in such a delicately interlocking veneer that it represents two weeks of one very skilled man's labor. The leather from eight hides cloaks the door panels and seats that tower like weighty furniture. The chrome is as deep as buffed silver. Nylon carpeting covers the floor and over that lies a plush blanket of lamb's skin.

During each step of the six-month-long manufacturing process, the Silver Shadow is coddled by an unhurried work force. On file are the results of metallurgical inspection of a specimen from every crankshaft. Vaults protect intentionally reserved walnut veneer so that any trim damaged in service can be matched perfectly. Every window frame is an artistic hand fabrication with the finish of fine pewter. The radiator shell is a stainless-steel sculpture shaped to suit the eye of a craftsman rather than the rigid contours of a stamping dye. The bright moldings that surround the windshield and back window are each one continuous band with no laps or joints to interrupt their sweep.

The Seville makes no pretense at matching Rolls-Royce standards. In Detroit, the assembly line is the only workable way to build a car, and the same construction techniques are used for the Seville as for the Pinto, although the Cadillac line's speed has been slowed to a mere 14.5 cars per hour instead of the normal 50 to 60. Nor are the materials used in the Seville particularly unusual. It missed out on the crash-diet attitude currently sweeping Detroit design circles, so no attempt to replace steel and iron with aluminum and plastic derivatives has been made in the Seville, which weighs a hefty 4400 pounds. The woodgrain is the chemical variety, and leather is reserved as an option for the seats only. Smooth vinyl covers the instrument panel and armrests, with hard plastic held to a bare minimum masking roof pillars. It amounts to a trim level as good as anything offered recently by Detroit—but no better.

The advantage the Seville has over every car of its size in the world lies exactly where Cadillac engineers felt it would be most appreciated: ride and quietness. This is what American cars are all about, and Cadillac has consistently set the pace with its cruisers. Scaling down the size intensifies this challenge because vast bulk and a block-long wheelbase aren't available to soak up road noise and dampen ride motions.

The Seville is patterned after GM's K-body (Nova) line of cars, though only one sheetmetal panel—the trunk floor—is shared. The basic construction is a two-piece unitized body. From the firewall back, the car is a welded-together structure, and to this bolts a very stiff front sheetmetal assembly (fenders, radiator yoke, hood, grille). A half-length frame ties the two structures together from below and also supports the engine, transmission and steering/suspension components. A live rear axle is mounted directly to the body with semi-elliptic leaf springs, even though the general assumption that anything less than a full frame and all-coil suspension demands penalties in ride and noise has held sway in Detroit for dec-

ROLLS-ROYCE SILVER SHADOW

Importer: Rolls-Royce Inc.
P.O. Box 189
Paramus, New Jersey 07652

Vehicle type: front-engine, rear-wheel drive, 5-passenger 4-door sedan

Price as tested: $34,355.00
(Manufacturer's suggested retail price, including all options listed below, dealer preparation and delivery charges, does not include state and local taxes, license or freight charges)

Options on test car: base Rolls-Royce Silver Shadow, $33,500.00; vinyl roof, $840.00; federal excise tax, $15.00

ENGINE
Type: V-8, water-cooled, cast aluminum block and heads, 7 main bearings
Bore x stroke .... 4.10 x 3.90 in, 104.1 x 99.1 mm
Displacement .... 412 cu in, 6740 cc
Compression ratio .... 9.1 to one
Carburetion .... 2 x 1-bbl SU
Valve gear ... pushrod-operated overhead valves, hydraulic lifters
Power (SAE net) .... N.A.
Torque (SAE net) .... N.A.

DRIVE TRAIN
Transmission .... 3-speed, automatic
Final drive ratio .... 3.08 to one

Gear	Ratio	Mph/1000 rpm	Max. test speed
I	2.48	10.4	45 mph (4350 rpm)
II	1.48	17.4	76 mph (4350 rpm)
III	1.00	25.8	112 mph (4350 rpm)

DIMENSIONS AND CAPACITIES
Wheelbase .... 119.8 in
Track, F/R .... 59.5/57.8 in
Length .... 203.5 in
Width .... 71.0 in
Height .... 59.8 in
Ground clearance .... 6.5 in
Curb weight .... 4950 lbs
Weight distribution, F/R .... 52.7/47.3 %
Battery capacity .... 12 volts, 68 amp-hr
Alternator capacity .... 475 watts
Fuel capacity .... 28 gal
Oil capacity .... 10 qts
Water capacity .... 17 qts

SUSPENSION
F: ...ind, unequal-length control arms, coil springs, anti-sway bar
R: .... ind, semi-trailing arm, coil springs, auto. level control

STEERING
Type .... recirculating ball, power assisted, idler arm damper
Turns lock-to-lock 3.2
Turning circle curb-to-curb .... 38.5 ft

BRAKES
F: .... 11.0-in. dia. vented disc, power assisted
R: .... 11.0-in. dia. solid disc, power assisted

WHEELS AND TIRES
Wheel size .... 6.0 x 14-in
Wheel type .... stamped steel, 5-bolt
Tire make and size .... Dunlop Formula 70, HR70-15
Tire type .... fabric cord, radial ply, tubeless
Test inflation pressures, F/R .... 28/28 psi
Tire load rating .... 1890 lbs per tire @ 36 psi

PERFORMANCE
Zero to	Seconds
30 mph	3.2
40 mph	4.9
50 mph	7.6
60 mph	10.8
70 mph	15.3
80 mph	21.4
90 mph	29.7
100 mph	42.2

Standing ¼ mile .... 17.9 sec @ 75.1 mph
Top speed (observed) .... 112 mph
70-0 mph .... 203 ft (0.80 G)
Fuel economy, C/D mileage cycle .... 11.5 mpg, city driving
                                          12.5 mpg, highway driving

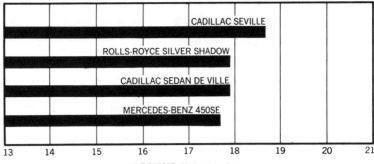

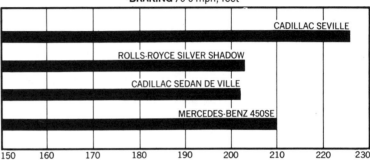

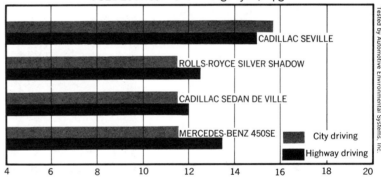

Tested by Automotive Environmental Systems, Inc.

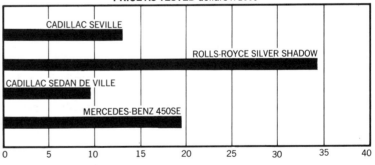

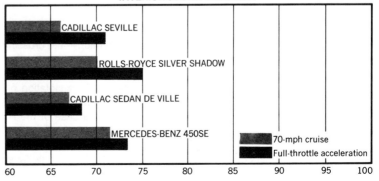

CADILLAC SEVILLE

Manufacturer: Cadillac Motor Car Division
General Motors Corporation
Detroit, Michigan 48233

Vehicle type: front-engine, rear-wheel-drive, 5-passenger 4-door sedan

Price as tested: $13,178.00
(Manufacturer's suggested retail price, including all options listed below, dealer preparation and delivery charges, does not include state and local taxes, license or freight charges)

Options on test car: base Cadillac Seville, $12,479.00; AM/FM stereo tape player, $229.00; Guidematic headlamp control, $52.00; Twilight Sentinel, $45.00; rear window defogger, $73.00; cruise control, $100.00; vanity mirror, $43.00; illuminated entry system, $50.00; door guards, $11.00; carpeted floor mats, $36.00; special wheel discs, $43.00; trunk mat, $10.00; license frame, $7.00

ENGINE
Type: V-8, water-cooled, cast iron block/heads, 5 main bearings
Bore x stroke . . . . 4.06 x 3.28 in, 103.0 x 85.9 mm
Displacement . . . . 350 cu in, 5730 cc
Compression ratio . . . . 8.0 to one
Carburetion . . . . Bendix electronic fuel injection
Valve gear . . . pushrod-operated overhead valves, hydraulic lifters
Power (SAE net) . . . . 180 bhp @ 4400 rpm
Torque (SAE net) . . . . 275 lbs-ft @ 2000 rpm

DRIVE TRAIN
Transmission . . . . 3-speed, automatic
Final drive ratio . . . . 2.56 to one

Gear	Ratio	Mph/1000 rpm	Max. test speed
I	2.48	12.4	60 mph (4850 rpm)
II	1.48	20.7	100 mph (4850 rpm)
III	1.0	30.7	110 mph (3600 rpm)

DIMENSIONS AND CAPACITIES
Wheelbase . . . . 114.3 in
Track, F/R . . . . 61.3/59.0 in
Length . . . . 204.0 in
Width . . . . 71.8 in
Height . . . . 54.7 in
Curb weight . . . . 4406 lbs
Weight distribution, F/R . . . . 56.9/43.1 %
Fuel capacity . . . . 21 gal
Oil capacity . . . . 5 qts
Water capacity . . . . 18.9 qts

SUSPENSION
F: . . . ind, unequal-length control arms, coil springs, anti-sway bar
R: . . . . rigid axle, semi-elliptic leaf springs, automatic level control, anti-sway bar

STEERING
Type . . . . recirculating ball, centerlink damper
Turns lock-to-lock 3.1
Turning circle curb-to-curb . . . . 40.0 ft

BRAKES
F: . . . . 11.0-in dia vented disc, power assisted
R: . . . . 11.0 x 2.0-in finned cast iron drum, power assisted

WHEELS AND TIRES
Wheel size . . . . 6.0 x 15-in
Tire make and size . . . . . . Firestone Steel Belted Radial, GR78-15
Tire type . . . . steel-belted radial ply, tubeless
Test inflation pressures, F/R . . . . 24/24 psi
Tire load rating . . . . 1620 lbs per tire @ 32 psi

PERFORMANCE
Zero to   Seconds
30 mph 3.9
40 mph 5.7
50 mph 8.2
60 mph 11.5
70 mph 16.4
80 mph 22.7
90 mph 30.8
100 mph 42.7
Standing ¼-mile . . . . 18.7 sec @ 73.8 mph
Top speed (observed) . . . . 110 mph
70-0 mph . . . . 226 ft (0.72 G)
Fuel economy, C/D mileage cycle 15.5 mpg, city driving, 15.0 mpg, highway driving

---

ades. According to Robert Templin, Cadillac's chief engineer, that's a theory of the past. He claims to have been able to utilize computer analysis to optimize structural and rubber bushing designs to refine a unit body to full-frame ride levels and save weight in the process. The only limitation—for Cadillac, at least—is assembly-plant space. The assembly line turn-arounds are too tight for a body even as short as the Seville's, so Cadillacs will continue to have a detachable front sheetmetal section and therefore a subframe.

The computer pointed out structural weaknesses of Seville designs that were under consideration and determined where modifications could contribute most effectively to creating a rigid base for the rest of the car—for example, the addition of a simple tubular strut across the open bottom of the Seville's transmission tunnel greatly reduced floorpan drumming effects. Rubber isolators for suspension, engine and subframe mounting were also computer-selected, and each of the six front subframe mounting cushions has a different stiffness in all three directions to minimize shake and maximize isolation. The harsh rap you usually get over expansion joints is softened by a very compliant horizontal rate. Since this could

allow a continuing body-shake motion in the horizontal plane long after the disturbance is crossed, a pair of conventional shock absorbers (placed nearly horizontal with their forward end attached to the subframe and the trailing end tied to front sheetmetal) are used to damp out this oscillation.

All of this results in a ride any Cadillac owner will be at home with. The big cars in the family stable are probably slightly better, but the increment of improvement in no way justifies the excessive sheetmetal baggage you have to tote around. On smooth roads, driving in a Seville is like riding an air cushion—and crossing rough pavement is little worse.

The Silver Shadow is clearly outranked when it comes to ride. It has firm, entirely enjoyable poise over the road but not the supreme isolation of the Cadillac. The basic design approach of a unit body and partial frame is common to both, but in the case of the Rolls, a subframe is used both in the front *and* the rear. And instead of rubber isolators, there are spring-loaded steel-mesh mounts (unusual in the automotive industry but frequently specified for aerospace applications) that look a lot like cylindrical Brillo pads. Road isolation suffers somewhat, but the metal mounts don't deteriorate with

> Because Cadillac knows its customers, the Seville offers infinite detail rather than sweeping imagination.

time—a major problem with Detroit's soft-rubber approach.

The Cadillac demonstrates its comfort advantage quite clearly in the one parameter we can quantify: interior noise level. With 66 dBA at 70 mph, the Seville is the quietest car ever to murmur into our sound-level meter at speed. The Rolls trails by four dBA, a substantial amount, and that relationship also holds during acceleration, where induction noise stirs the Cadillac's quiet with 71 dBA. The Silver Shadow is best at idle, when its big aluminum-block V-8 ticks over with a beat that makes the clock on the instrument panel sound like a jack hammer by comparison. After all, this is the car that sparked the classic smoothness test: balancing a coin on edge on the radiator with the engine running. It was a claim to fame of the 1907 Silver Ghost—and the Silver Shadow does equally well today. At a stoplight, the Rolls takes the record for the quietest car we have tested; at only 45.5 dBA, it is substantially quieter than the average room in your home and

(Continued on page 92)

PHOTOS: MARC MADOW

The legend of how Rolls-Royces are built is automotive lore in its richest form—but even more dazzling is the hands-on proof that the myths are based on solid fact.

In Search of Seville Country

· MONTEREY, California—A landmark day in a legendary town. It is the evening of April 30, the day before the public introduction of the car which will change automotive history: Cadillac's "international-size" Seville, the American Answer.

The local dealership, Butts Cadillac, seems to have been created in the General Motors styling studios just for this preview. It lies in the far corner of a 60-acre automotive Disneyland called Heitzinger Plaza. The whole walled area has been designed for the sole purpose of showing and selling cars: a GM Futurama realized, a selling city in a totally self-contained park-like enclave.

Tonight Jake Butts, with Cadillac Motor Car Division looking carefully over his shoulder, has invited almost all the big-buck retired executives who live in the area. Perhaps more importantly, he has also gone through Polk's car registry and sent an invitation to every Jaguar, Continental, BMW, Mercedes and Rolls owner in town—and there are a lot of them.

Monterey, it seems, is a statistical anomaly, just the kind of domestic-versus-import puzzle that General Motors hopes to solve with the Seville. Almost a quarter of the high-priced foreign cars sold in the U.S. are sold in California, and there are more of those cars per capita in Northern than in Southern California. For the last 20 years, imports have registered at least 40 percent of all cars sold on the Monterey Peninsula—roughly twice the current national average. While Cadillac outsells Mercedes-Benz by about six to one in the U.S. (220,000 to 37,000 last year), they have been put to shame by Mercedes in Monterey for the last five years.

"Cadillac wants conquest registrations with the Seville," says Jake Butts. And it is clear that the car and the dealer he wants to conquer is Mercedes. So the reaction to these two Sevilles on the floor of Butts Cadillac at this preview is going to mean as much to Cadillac as all its motivational studies, market tests and gut feelings combined.

Jake, the Sevilles, the engraved invitations and the high-roller canapés pulled about 250 people and sold three cars. Two were bought by a lumpy man with a Brooklyn accent, a partner in a restaurant called the Sardine Factory, whose company already owned eight Cadillacs and who had been thinking of replacing a couple with a pair of Eldos. Another was bought by a lady with a 350SL.

The following morning, Butts Cadillac sold a fourth, taking as trade-in a Mercedes 280SL with 22,000 miles on it (at book value: $5200) after its elderly female owner drove the Seville exactly one mile. Her husband stood by beaming, allowing as how he had owned 20 Cadillacs in his time.

Jake Butts is optimistic about the Seville in Monterey. His body count after 24 hours is four warm ones: Half of them female (always a good indication), half of them former believers in German Engineering Infallibility. Monterey, the nut that Cadillac Division committed itself to crack, is showing definite signs of becoming Seville Country. *—Leon Mandel*

SEVILLE VS. SILVER SHADOW

3.5 dBA quieter than the Seville. For all practical purposes, however—which necessarily include moving down the highway—the Seville is the quietest car in the world and one of the best riding; the Rolls trails slightly behind. Big-budget Cadillac technology was aimed full force at this goal—and reached it.

The Rolls, however, is clearly roomier. Its stately six-foot roofline is not only a prestige factor but allows bolt-upright seating with plenty of headroom. Position of the massive front bucket seats is controlled by a jewelry-like toggle pivoting in a ball-and-socket joint, but there is no adjustment for the steering column.

In the Cadillac, only the driver's seat has a six-way power adjuster as standard equipment. And there is no way to alter the cushion-to-backrest angle on the driver's side, even though there is a power option to do this for the front passenger—at Cadillac, reclining seats mean sleeping, not an adjustable driving position. Compared to the throne-room feeling you get in the Rolls, the Seville interior seems slightly claustrophobic, but by American standards it's quite roomy. This is one car where stylistic manipulation of the roofline is not at odds with comfort. The top-hat formality of the rear window keeps the ceiling up and out of the rear passenger's hair, so the Seville is one of the few American-made compacts where the back seat isn't a second-class accommodation. Four fit inside with due consideration for their extremities; though Cadillac rates the car for five, that calls for some rear-seat sacrifices you don't have to make in the Silver Shadow, Mercedes-Benz 450SE or Cadillac Sedan de Ville, all wider in back by four to six inches.

Styling concerns did, however, outrank trunk space and undoubtedly will be a disappointment to any veteran Cadillac owner used to a trunk the size of a U-Haul trailer. Since the rear deck in the Seville tapers and is of modest proportions compared to the blocked-off roof, there just isn't much volume under it, and a collapsible spare takes up a large share of the floor area. This is a side effect of the demise of GM's rotary engine. The Seville was originally designed as a three-rotor, front-wheel-drive car. When the Wankel engine plans died, the conventional rear axle ended up cutting heavily into trunk room. But even though the trunk is small, it is lined like an oversized jewelry box with fuzzy carpet (even on the lid) and plastic covers over the hinges to guard against abrasion. The Rolls' luggage compartment is just the opposite: not so plush but voluminous. The average American family wouldn't have enough luggage to fill it.

The two different design philosophies represented by Cadillac's nominee for the world's standard of excellence and by the traditional favorite, Rolls-Royce, come into sharp focus in the driving. Cadillac treats the act as an unpleasant task to be foisted off on Jeeves if possible, performed by power assists and automatic controls if necessary. There is a power boost to do everything from releasing the parking brake to opening the decklid. It's as close as you can get to the ultimate push-button car.

Every measure of performance is clearly secondary to comfort in the Seville. The tall 2.56-to-one axle cuts engine noise during cruising and delivers a respectable 15 mpg. The force of flat-

The baton has been formally passed. The traditional large car is gasping its last; the Seville will now carry on.

out acceleration is—probably by design—barely detectable. The brakes are numb to your touch and uncooperative when you try to stop fast. Rear wheels lock much too early, with a light load stretching stopping distances to a lengthy 226 feet from 70 mph. (Cadillac engineers claim the strong rear brake bias is necessary to pass 1976 federal brake-performance standards. They have no explanation for the fade we found during the *first* stop from 70 with a three-passenger load.) And the steering wheel is strictly a one-way control: You can accurately position your path with it, but there's no feedback from the road to help you during quick transitions. The instruments are reduced to a bare minimum: speedometer, fuel gauge and an opaque panel hiding the warning lights until they have something to signal.

The Silver Shadow does a much more proficient job of satisfying a discerning driver, but you still get the distinct impression from the truck-like black plastic steering wheel that the task is meant to be performed by an hourly man—it's as if a luxury piece was deemed unnecessary for this one link back to the working class. There's not much more feel through the steering wheel than in the Seville, but the Silver Shadow's finely damped ride motions encourage more aggressive driving in spite of a much higher center of gravity. Its cornering limit just matches the Seville's at 0.67 G, but the Rolls assumes a much more neutral attitude with its tail slightly wide. Rear tires do their share of the cornering, so this car doesn't suffer from the Cadillac's grinding understeer. Poise is also maintained during hard braking: The Shadow merely settles down to the pavement while the Seville tries to stand on its nose. It's done with an uncommonly large amount of anti-dive geometry in the unequal-length control arm front suspension, using braking forces to cancel 68 percent of the forward load transfer. A similar anti-lift design in the rear keeps the tail from rising, so a practically level attitude is maintained during even the hardest stops.

Braking from 70 mph took 203 feet in the Rolls, due in part to premature lock-up of the rears. On paper, you'd expect a good deal more from this car's brake system. There is a large disc for every wheel with vented rotors and double calipers in front, a hydraulic boost system and plumbing to divide the brakes into three independent systems. All that's missing is an anti-skid device to avoid sliding one or more wheels in sharp-braking situations.

The sophistication of the Rolls's brake system is not matched in its powerplant. The large-displacement (412 cubic inches) V-8 is quite run-of-the-mill by Detroit standards. Except for twin SU side-draft carburetors and an aluminum block and heads, it's just like any pushrod American V-8. There are, in fact, those who insist it's a close copy of an early postwar Cadillac design.

Cadillac engineers had to come up with a new engine for the Seville because their 500-cu. in. monster-motors are hardly in tune with the car's conservative new image. Oldsmobile actually supplies the 350-cu. in. block for this application, and to that Cadillac adds their own cylinder-head design as well as the new Bendix electronic fuel injection. This makes a powerplant just as prestigious

SEVILLE VS. SILVER SHADOW

Continued

as the Rolls engine without the fussy attention to the manufacturing tolerances, tests and retests for which the British engine is famous. Fuel injection also eliminates drivability problems that accompany elaborate emissions controls, so Seville owners won't have to tiptoe through warmup on a cold morning. Just twist the key and it's ready, with an electronic brain to take care of the engine's every want and need. The American-made injection system is much like the equipment used for years on VW and Mercedes cars (built under a Bendix license by Bosch) with some subtle upgrades. The electronic brain uses late-technology integrated circuits, while the Bosch design consists of less reliable discrete components. Slowly but surely the world is turning to electronic injection to allow otherwise-conventional engines to meet the needs of low emissions and reasonable fuel economy. Fuel injection will delight the typical American luxury-car buyer, even if he doesn't have a clue how it works.

This set of consumers also insists on long lists of special features and elaborate power and automatic equipment to assure themselves that theirs isn't just another car. Mercedes can always boast about its overhead cams, fuel injection, all-independent suspensions and four-wheel disc brakes, but even it has had to add automatic speed control—a strange piece of equipment for the ultimate *driving* sedan—to satisfy the every-possible-option collectors. The Seville buyer won't feel neglected: Automatic controls operate the headlamps, parking-brake release, radio antenna, interior climate and rear ride height.

As for more esoteric chassis items, the Seville is a little light. Its rear suspension is a relatively unsophisticated live axle with leaf springs, and the drum brakes back there are not the substance of over-$13,000 cars, especially when even the Granada offers an optional all-disc brake system with anti-skid. So the Seville buyer is going to have to rely on ride and quietness to impress his friends, because the Cadillac philosophy ranks comfort all important.

Since Rolls-Royce is a small company and doesn't produce enough cars per year to do rapid development work, they have no choice but to purchase many off-the-shelf items. Dunlop is the exclusive supplier of tires—a rather conventional fabric-cord radial design in a currently popular low aspect ratio. The quadrasonic tape player is a Japanese Pioneer, while the stereo radio is a German Blaupunkt. All locks are Yale's finest. General Motors has the contract for transmissions (the same Turbo-Hydramatic used in the Seville); the only Rolls-Royce modification is the addition of an electric shifting mechanism to replace the conventional manual linkage. The air-conditioning compressor is also by GM (Frigidaire), widely acknowledged as the best the world has to offer. It's a key component of the Silver Shadow's air-conditioning system, which has two huge squirrel-cage blowers to force-ventilate the interior. Unlike the Seville's system, the flow is impressive whether or not the compressor is engaged. But the Rolls also has a climate-control flaw in comparison to the Cadillac: It lacks the Seville's ability to thermostatically maintain a preset temperature.

That's the nature of these two cars.

The Seville makes no pretense at matching Rolls-Royce standards. In Detroit, the assembly line is the only workable way to build cars.

The Cadillac is a totally up-to-date car, with every convenience the product planners can think of to isolate you from the hardships of driving. Yet it would make an ideal Cannonball Baker racer, for it has the comfort, speed and room to move a relay team of drivers Coast-to-Coast without wrinkling their driving suits or straining their smallest muscles. It's better than a Rolls in many important respects: ride, quietness and fuel economy. And the thought of making it a better driver's car is at least in Cadillac engineers' minds because they have hinted at the possibility of a "European suspension" option for the future.

The Silver Shadow is, of course, a precious yardstick by which to measure any car. And it pinpoints the Seville as a mass-produced automobile cultivated to an unsurpassed level of comfort. In a car of manageable proportions, this is one American accomplishment worthy of international esteem. •

Rolls-Royce Silver Shadow

Cheapest Rolls-Royce offering refinement and finish equalled by few manufacturers.
Quiet V8 engine giving improved economy with adequate performance.
Smooth automatic transmission and good brakes.
Much improved steering response and roadholding but softer ride.

As dignified as the House of Commons, the latest Silver Shadow has changed little outwardly though there have been a number of specification changes under the bonnet

IT IS NOW just over 10 years since Rolls-Royce launched the Silver Shadow, a departure from tradition in so many ways. Now, for some people, the Shadow and its relatives the Corniche and Camargue (which share the same floorpan) *are* the tradition. Part of that tradition is the Rolls-Royce policy of continuous development, so that the Silver Shadow of today is a very different animal from the earliest cars. Autocar last published a Silver Shadow test in November 1972, by which time the engine had already grown from 6,230 to 6,750 c.c., and the three-speed General Motors transmission had become standard. Since then the engine has been changed again, not in size but in compression ratio which has come down from 9-to-1 to 8-to-1, permitting the use of four-star petrol with a lower lead content. Lucas Opus electronic ignition is now used and helps to maintain a consistently low level of exhaust emissions, helped also by the thermostatically-controlled air intake. Carburation is still by two SU instruments, with the fuel pumped by twin SU electric pumps.

On the chassis side, several changes have been made to the suspension to obtain better compliance (for lower road noise) without compromising geometry and therefore handling. At the same time the tyres have become lower and wider, the 205-15 covers of the 1972 car being replaced by 235/70-15 size. The former triple braking system has been replaced by a simpler — still fully-duplicated — double system.

Rolls-Royce are no more insulated from the efforts of inflation than anyone else, and it is sad to note that the price of a Silver Shadow has risen from the £10,550 of our 1972 test car, to £17,813 today. This does, however, underline what many owners have found, that a Silver Shadow can turn out to be an investment.

Performance and economy

As always, Rolls-Royce quote no power or torque output for the V8 engine. By comparing the performance with that of other cars, however, one can make an intelligent guess. The peak power must be a good deal less than the 225 bhp of the much quicker Mercedes 450SEL for instance, and indeed it is doubtful if the Rolls-Royce engine produces an honest 200 bhp (DIN). What is more difficult to determine is the engine speed at peak power, and hence the intelligence of the overall gearing.

Such a low power output from an engine of this size, corresponding to less than 30 bhp per litre, argues that the designers have sought other qualities than sheer performance, in particular quietness and reliability. Rolls-Royce have always termed their power output "sufficient" and one can hardly argue with a maximum speed of 116 mph and a 0-60 mph time of 10.6sec. A higher maximum speed would be increasingly useless, and greater acceleration is difficult to use without disturbing passengers.

What seems certain is that little power has been lost with the lowering of the compression ratio. Compared with the 1972 test, the present car is just 1 mph slower at its maximum and has lost only 0.3sec in the standing start acceleration to 60 mph, despite being 80 lb heavier at the kerb and being tested in worse conditions. It all but held its own in acceleration right through to 100 mph, but the 110 mph figure was well down, indicating that some of the top-end eagerness has been lost.

The overall gearing gives a relatively modest 25.8 mph per 1,000 rpm, so that the 116 mph mean maximum is reached at 4,500 rpm which must be fairly close to the power peak. Certainly that peak, when reached, is followed by a rapid strangling of the engine. It is worth noting that while the Intermediate ratio will run to 94 mph, the 70-90 mph time is slower in Intermediate than in top. When running the standing-start accelerations, we found that any attempt to hold the transmission beyond the natural full-throttle change-up points (44 and 74 mph) resulted in a slower time, so there is absolutely nothing to be gained by fiddling with the transmission selector. In the wet conditions of the test, it proved possible to spin the wheels briefly when leaving the line, if the torque converter was brought up to the stall point against the brakes.

Rolls-Royce claim the engine changes have made the Silver Shadow appreciably more economical, and this was borne out by almost all our figures, whether at steady speeds or in normal driving. The steady-speed figures show the "hole" at 30 mph typical of almost all today's emission-controlled cars, so that it is actually more economical to cruise at a steady 40 mph (which is the only speed at which the big car bettered 20 mph). After that there is a steady decline

Above: The latest engine has an 8.0 to 1 compression ratio and Lucas Opus electronic ignition. The underbonnet area looks less accessible than most routine items are. The refrigeration compressor is mounted on top of the engine at front

all the way to 11 mpg at a steady 100 mph. The only speed at which the latest car proved marginally inferior to the 1972 model was at 80 mph (14.7 against 15 mph).

In overall terms, the latest car achieved 13.6 mpg against the 12.4 mpg of the previous test. This may not sound much, and it certainly doesn't lift the Shadow much higher in the economy league table, but it *is* a 10 per cent improvement and as such, well worth having.

Naturally, it is a figure that reflects our usual harder than average driving, and a more usual result would be about 15 mpg which incidentally coincides with the predicted DIN touring comsumption. Given gentle driving, up to 18 mpg should be possible but clearly, nobody is ever going to see the right side of 20 mph except in freak circumstances.

Handling and brakes

The first thing that must be said under this heading is that Rolls-Royce know their market. It is a market that does not demand Ferrari-style handling and steering, and it doesn't get it. One is actually driven to question the fairness of one's own criticisms, since the *Autocar* test staff are — perhaps regrettably from their point of view — a long way from being typical Silver Shadow owners. Even so, there are some points where the car might be improved even in the most general view.

The steering is reasonably geared with 3.6 turns of the wheel between extremes of a lock which, for this size of car, is by no means bad: well under 40ft. On the other hand the steering is still extremely light, the assistance very powerful, and it takes a gentle hand at the wheel to detect even a suggestion of feel. It might be argued that this would be a good thing for a chauffeur, driven back to relying on the same senses as those of his passengers. At the same time, one wonders if it is really a good thing to be so divorced from any idea of what the front wheels are trying to do, and how close to the limit of adhesion they may be. If it is permitted to take the Shadow's two obvious competitors as yardsticks, the steering is much more Cadillac than Mercedes — not surprisingly perhaps, since the steering gear like the transmission comes straight from General Motors.

If the lack of feel does not matter as much as it once did, that is because the latest Shadow is massively stable. We commented in 1972 that the car was much improved in stability compared with the earliest examples (which were decidedly deficent in this respect, at least by today's standards). The further changes to the suspension, and the adoption of low-profile tyres, have made things better still so that the Silver Shadow of today sits on the motorway, running arrow-straight with no need for the driver to correct its course more than occasionally. Even strong, blustering sidewinds do little to push the car off line, and this is all to the good.

Cornering behaviour depends very much on the driver's approach. If the Shadow is driven smoothly, as we trust all chauffeurs will, there is a steady build-up

Above: The wide low profile 235/70HR-15 radials are the only outward change on the latest car

of understeer, smooth and predictable but quite quick, scrubbing off speed unless a lot more power is applied. Rougher driving — which requires some confidence, given the inertia involved — can result in the tail swinging mildly out of line, in which case speed is scrubbed off even more quickly. Increasing the tyre pressures by 6psi all round, to those recommended for driving at over 110 mph, has little effect on the stability but improves the cornering behaviour quite noticeably. At high speed, however, a slight tendency to oversteer can then be detected (and became moderately disturbing when the car was entering the steep MIRA banking at 120 mph). The roadholding is good at all times, and excellent in the wet, thanks to the Avon tyres fitted to the test car.

AUTOTEST
Rolls-Royce Silver Shadow

Above: A 120 in wheelbase makes for a roomy passenger compartment. The side marker lamps are illuminated with the sidelamps. Right: Hefty overriders and a substantial bumper protect the bodywork from minor scrapes. The rear fog warning lamps are standard

The brakes are rather lighter than before, and those familiar with older Shadows may notice the lower brake pedal – now nicely levelled with the accelerator – and reduced pedal travel. Given the lower effort, it would have been easy to over-servo the brakes, but this has not happened. Instead the action is so beautifully progressive that the car is one of the easiest to stop smoothly with no suggestion of a final jerk – doubtless something that was accorded high priority during development. We were expecting a disappointing ultimate performance on the wet track but the wide Avons hung on superbly to record 0.96g before all four wheels slid with considerable reluctance. Stability and control under heavy braking are far better than average.

Our brake fade test showed no problem whatever. To begin with the brakes proved a little speed-sensitive, but once they were fully warmed their performance was absolutely consistent with no increase whatever in pedal travel or effort. Clearly the four big discs are well up to their exacting task of stopping well over two tons of car. The handbrake also did well, recording 0.3g when used alone before locking the back wheels on the wet surface; it held securely on the 1-in-3 MIRA test hill, on which the Silver Shadow restarted – as one would expect – with contemptuous ease.

Comfort and convenience

This is the area in which the Silver Shadow might be expected to excel, and generally it does so, though it is not without its weaker points. The ride is helped by the self-levelling arrangement which ensures that the rear springs are always working around their mid-point whatever the load condition. The self-levelling can sometimes be felt working when the car has been brought to a quick halt, but most of the time it goes quietly about its task and some owners might never be aware of its existence.

In other ways, the suspension is fairly conventional with a front linkage arrangement which corresponds to double wishbone geometry, and semi-trailing arms at the rear. Suspension travel is long and the springing correspondingly soft; the damper settings are also on the soft side, though not so much as to induce the upsetting vertical heave so often encountered in the full-sized American car. Instead the softness is felt more when the car reaches the bottom of a ramp, or when it traverses a single hump, when it lifts or dives more than one might expect. On rough surfaces the damping works well, while the big wheels stand the car in good stead. Roll angles are not high, but are probably more than many designers would think desirable today. It should not be inferred from this that the Silver Shadow's ride is poor in any way; it is extremely good, though biased towards those who prefer soft rather than sporting feel. If it no longer seems exceptional, that is because so many manufacturers have managed to improve the ride of their cars in the past few years.

The front seats are power-adjusted by minature "joysticks" in the centre console, and their range of adjustment is sufficient to cater for a very wide range of drivers. Although the steering column is not adjustable, most drivers seem able to arrive at a comfortable and efficient sitting position, though really short people may have more trouble. The seats themselves are less satisfactory. Both cushion and squab are very thick slabs of beautifully leather-upholstered springing. They lack sufficient shaping to give proper sideways support – even our heaviest driver found he didn't sink very far into them – and several testers thought the squab was badly shaped to support the small of the back. It helps that one can adjust the tilt of the seat, but even so there is a tendency on long trips to slide

Specification

ENGINE
	Front; Rear drive
Cylinders	8, in 90 deg vee
Main bearings	5
Cooling	Water
Fan	Viscous
Bore, mm (in.)	104.1 (4.10)
Stroke, mm (in.)	99.0 (3.90)
Capacity, cc (in.)	6,750 (411.9)
Valve gear	ohv
Camshaft drive	Chain
Compression ratio	8.0-to-1
Octane rating	97RM
Carburettor	2 SU HD8

TRANSMISSION
Type	General Motors three-speed automatic with torque converter

Gear	Ratio	mph/1000rpm
Top	1.0	25.80
Inter	1.48	17.43
Low	2.48	10.40
Final drive gear	Hypoid bevel	
Ratio	3.08-to-1	

SUSPENSION
Front—location	Double wishbones
springs	Coil
dampers	Telescopic
anti-roll bar	Yes
Rear—location	Semi-trailing arms
springs	Coil
dampers	Telescopic
anti-roll bar	No

STEERING
Type	Recirculating-ball
Power assistance	Yes
Wheel diameter	15.7 in.

BRAKES
Front	11.0 in. dia. disc
Rear	11.0 in. dia. disc
Servo	Hydraulic, engine-driven

WHEELS
Type	Pressed steel disc, 5-stud fixing
Rim width	6.0 in.
Tyres—make	Avon (on test car)
—type	Radial-ply tubed
—size	235/70-15in.

EQUIPMENT
Battery	12 volt 71 Ah
Alternator	55 amp
Headlamps	Four-lamp system, 270/75 watt (tota
Reversing lamp	Standard
Hazard warning	Standard
Electric fuses	21
Screen wipers	2-speed plus intermittent
Screen washer	Electric
Interior heater	Air blending
Interior trim	Leather seats, Ambla headlining
Floor covering	Carpet
Jack	Screw pillar type
Jacking points	1 each side under sill
Windscreen	Laminated
Underbody protection	Zinc plating plus bitumastic overall

MAINTENANCE
Fuel tank	23.5 Imp galls (107 litres)
Cooling system	28.5 pints (inc. heater)
Engine sump	14.75 pints SAE 20W/50
Gearbox	5 pints Dexron
Final drive	4.5 pints SAE 90 EP
Grease	6 points
Contact breaker	Lucas Opus ignition
Ignition timing	15 deg BTDC (stroboscopic at 1,200 rpm)
Spark plug—type	Champion N14Y
—gap	0.030 in.
Tyre pressures	F24; R28 psi (normal driving)
Max payload	1,000 lb (454 kg)

Maximum Speeds

Gear	mph	kph	rpm
Top (mean)	116	187	4,500
(best)	120	193	4,650
Inter	94	151	5,400
Low	56	90	5,400

Acceleration

True mph	Time secs	Speedo mph
30	3.8	30
40	5.6	40
50	7.6	50
60	10.6	60
70	14.1	70
80	19.0	80
90	25.8	91
100	36.5	103
110	60.7	115

Standing ¼-mile:
18.1 sec, 78 mph
kilometre:
33.3 sec, 97 mph

mph	Top	Inter	Low
10-30	—	—	2.3
20-40	—	—	3.4
30-50	—	4.9	4.2
40-60	—	5.6	—
50-70	—	6.7	—
60-80	10.0	8.8	—
70-90	12.5	14.9	—
80-100	21.1	—	—
90-110	32.8	—	—

Consumption

Fuel
Overall mpg: 13.6
(20.8 litres/100km)
Calculated (DIN) mpg: 15.0
(18.8 litres/100km)

Constant speed:

mph	mpg
30	19.3
40	20.1
50	19.6
60	18.3
70	16.5
80	14.7
90	12.8
100	11.0

Autocar formula
Hard driving, difficult conditions
12.2 mpg
Average driving, average conditions
15.0 mpg
Gentle driving, easy conditions
17.7 mpg
Grade of fuel: Premium, 4-star (97RM)
Mileage recorder: 0.8 per cent over reading

Oil
Consumption (SAE 20W/50)
2,000 miles/pint

Brakes

Fade (from 70 mph in neutral)
Pedal load for 0.5g stops in lb

start/end		start/end	
1	27/24	6	30/30
2	29/27	7	30/30
3	30/30	8	30/30
4	30/30	9	30/30
5	30/30	10	30/30

Response (from 30 mph in neutral)

Load	g	Distance
20lb	0.45	67ft
30lb	0.60	50ft
40lb	0.75	40ft
50lb	0.90	33ft
60lb	0.96	31ft
Handbrake	0.30	100ft
Max. gradient 1 in 3		

Test Conditions

Wind: 10-17 mph
Temperature: 8 deg C (46 deg F)
Barometer: 29.5 in Hg
Humidity: 80 per cent
Surface: wet asphalt and concrete
Test distance 2,120 miles

Figures taken at 17,000 miles by our own staff at the Motor Industry Research Association proving ground at Nuneaton.

All Autocar test results are subject to world copyright and may not be reproduced in whole or part without the Editor's written permission

Regular Service

	Interval		
	3,000	6,000	12,000
Engine oil	Yes	Yes	Yes
Oil filter	No	Yes	Yes
Gearbox oil	No	No	Yes
Spark plugs	No	No	Yes
Air cleaner	No	No	Yes
C/breaker	No	No	Yes

Total cost £15.48 £42.37 £74.69
(Assuming labour at £4.30/hour)

Parts Cost

(including VAT)

Brake pads (2 wheels)—front	£19.63
Brake pads (2 wheels)—rear	£17.92
Silencers (stainless)	£331.35
Tyre—each (typical advertised)	£38.15
Windscreen	£60.35
Headlamp unit	£3.37
Front wing	£83.24
Rear bumper	£86.40

Warranty Period
36 months/50,000 miles (mechanical)
12 months (bodywork)

Weight

Kerb, 42.4 cwt/4,752 lb/2,156 kg
(Distribution F/R, 53.9/46.1)
As tested, 45.7 cwt/5,117 lb/2,332 kg

Boot capacity: 22.5 cu. ft.

Turning circles:
Between kerbs L, 39ft 8in; R, 39ft 2in
Between walls L, 41ft 3in; R, 40ft 9in
Turns, lock to lock 3.6

Test Scorecard

(Average of scoring by Autocar Road Test team)

Ratings: 6 Excellent
5 Good
4 Better than average
3 Worse than average
2 Poor
1 Bad

PERFORMANCE	5.00
STEERING AND HANDLING	4.67
BRAKES	4.60
COMFORT IN FRONT	4.58
COMFORT IN BACK	5.14
DRIVER AIDS	4.25
(instruments, lights, wipers, visibility etc)	
CONTROLS	4.00
NOISE	5.17
STOWAGE	5.67
ROUTINE SERVICE	3.10
(under bonnet access: dipstick etc.)	
EASE OF DRIVING	5.00
OVERALL RATING	**4.49**

OVERALL LENGTH 16' 11·5"
OVERALL WIDTH 5' 11"
OVERALL HEIGHT 4' 11·75"
GROUND CLEARANCE 6"
FRONT TRACK 4' 9·5"
WHEELBASE 10' 0"
REAR TRACK 4' 9·5"

Comparisons

	Price £	max mph	0-60 sec	overall mpg	capacity c.c.	power bhp	wheelbase in.	length in.	width in.	weight lb	fuel gall	tyre size
Rolls-Royce Silver Shadow	17,813	116	10.6	13.6	6,750	—	120	203½	71	4,752	23.5	235/70-15
Fiat 130	5,721	113	11.4	15.7	3,235	165	107	187	71	3,560	17.5	205-14
Jaguar XJ12	7,496	146	7.4	11.4	5,343	285	113	195	70	4,152	20	205/70-15
Mercedes 450SEL	11,312	136	9.0	14.1	4,520	160	105	181½	68	2,690	15.2	195-14
Opel Commodore GS/E	5,068	115	10.7	18.1	2,784	225	116½	199	73½	3,870	21	205/70-14
Volvo 264	5,896	104	12.7	18.6	2,664	140	104	193	67	3,195	13.2	185/70-14

Rolls-Royce Silver Shadow

forward into a slouched position which eventually gives backache. In fact, while the seats look tremendously impressive, they are far from being the best seats available from an ergonomic point of view. The leather upholstery is standard, but Rolls-Royce will of course supply almost any alternative (at a price).

By contrast with the front seats, the back seat is a model of good shaping and comfort. There is plenty of head and knee room for very large passengers, as indeed there should be with a 120in. wheelbase. Entry and exit are easy and if anything, the ride feels better than it does in the front. One is forced to the overall conclusion that it is nicer to ride in the back of the car than in the front: a relic of a former order of priorities?

The major controls are an odd mixture of good and bad. Many people may be surprised to find that the standard steering wheel is a thin-rimmed plastic affair, which seems small in relation with the bulk of the car. The two pedals are well offset to the right, leaving plenty of space to rest the left foot (which has the responsibility of operating the dipswitch). The transmission selector is mounted in the right of the steering column and is very easy to operate, the only danger for a newcomer being a tendency to confuse it with the indicator stalk mounted opposite. At least one staff member frightened himself by selecting neutral when intending to signal a left turn. The transmission itself is almost, but not totally, unobtrusive. We noticed an occasional slight thump when engaging Drive at rest, and the kick-down from Intermediate to Low can certainly be felt. For the rest, it is possible to

Above: Instrumentation appears more comprehensive than it actually is, though there are numerous warning lamps. Push-pull knobs control the central fresh air outlets. Surrounding the clock are (l to r): Panel light switch, radio balance control, electric aerial switch, wiper switch. The main lighting switch and the ignition/starter switch are to the right of the steering column. Right: A Pioneer quadraphonic cartridge player (top) and a Bosch Frankfurt AM/FM stereo radio are both standard. Ahead of the ashtray are the electric seat adjustment "joysticks." Controls for the upper and lower heating systems and refrigeration are central on the console. Below: The boot release switch is positioned in the glove locker so that this can be locked leaving luggage in the boot safe when the car has to be handed to others for parking at hotels

detect ratio changes if one is concentrating and listening for them; but passengers with their minds on higher things will inevitably be unaware of the transmission. If the driver hangs on to Low for an unreasonable time, a safety change up takes place at 56 mph, which corresponds to 5,400 rpm. The handbrake, while effective, is poorly placed, tucked away under the facia on the right. We noticed this especially when having to lean right forward to reach it when restarting on the 1-in-3 hill.

The minor controls are "traditional" and poor by modern standards. The lights switch on the right is a rotary knob selecting side, head and fog lamps in turn. It is matched by a smaller switch on the left which selects the three wiper speeds (two steady, one intermittent). The wiper switch especially calls for the driver to lean well forward to reach it. Screen washing is controlled by a much handier switch on the end of the indicator stalk, which also brings in the wipers for several strokes.

The heater controls, mounted in the centre console, are by contrast easy to reach and operate. The system is not of course a simple heater, but also offers a recirculating circuit with refrigeration (as distinct from the complete fresh-air conditioning system seen on the Camargue, and now also the Corniche). The heater controls select upper and lower temperature independently, flow being controlled by pushing the knobs in (less flow) or pulling them out. Both hot and cold air are available in copious quantity, and the separation of upper and lower flows makes it easy to control distribution. Fan speed is separately controlled. The only possible criticism here is that the coarsely-spaced detents of the temperature selectors can make it tricky to achieve exactly the right degree of warmth.

For the driver, one pleasant and mildly surprising aspect of the Silver Shadow is the excellent visibility. The front corners of the car can be clearly seen, and the tail is visible for reversing without any undue craning of the neck. A weaker spot is the wipers, which clear small arcs and on the test car

Above: Four switches on the driver's door control the electrically-operated windows and there is remote adjustment of the door mirror. Above the adjustable armrest is the switch for the central locking system

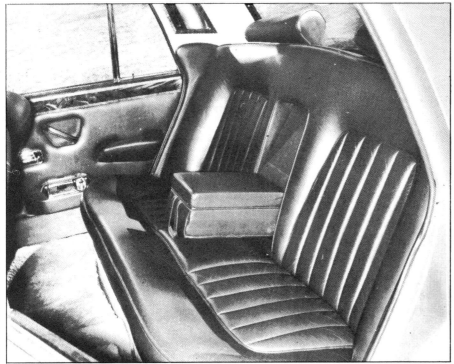

Back seat legroom is generous and there are moveable footrests. Headrests, a reading lamp and mirror in each rear quarter, and an ashtray and cigar lighter in each rear door are provided

at least, did not adhere all that well to the screen at high speed. The headlamps are extremely good on main beam, and well-controlled on dip, though owners may specify halogen headlamps if they need even more light.

Where silence is concerned, the Silver Shadow remains mildly disappointing. Rolls-Royce are not alone in discovering that very wide radial-ply tyres create and transmit a lot of road noise, not only through the suspension mountings but also by air-transmission to the wheel arch interiors. In its latest form the Shadow is certainly better than it was but it is still far from silent. It is quiet enough to fool many passengers they are riding in silence until they notice a change in "background quality" when the car crosses from one surface to another. On smooth asphalt, the Shadow is indeed *very* quiet; on ribbed concrete, it is noticeably less so. This is all a matter of degree, but thrown into more prominence by the Shadow's excellence in other respects. The engine *can* be made to emit a characteristic V8 throb when accelerating flat-out at low speed, but for the most part one is not aware of it. The transmission is quiet at all times. At high speeds in the test car there was a whisper of wind noise, untypical in our experience, from the top of the driver's door – a less than perfect seal, perhaps. In these circumstances it is inevitable that the road rumble will be noticed, and there is certainly quite enough of it to drown the ticking of a clock. There is also, in these latest tyres, an element of bump-thump over transverse road joints, while a large pothole or raised drain cover may give rise to a single loud crash as the car passes across.

Living with the Silver Shadow

One does not, of course, "live with" the Silver Shadow as one might with a lesser car. Few owners will be concerned with washing it – though except for its sheer size, the job is easy – or with doing their own servicing (which is assuredly much more tricky). When it *has* to be lived with, as on a long Continental journey, it has all the equipment needed to ease the strain of life. The long list of standard equipment includes a Bosch Frankfurt AM/FM radio and electrically-retractable aerial, and a Pioneer quadrophonic cartridge player (the much more sensible cassette system may be specified at a small extra cost). Other, lesser items such as the heated rear window and the fog lights one takes for granted.

There is unlikely to be any problem with stowing luggage. The boot is very large, and loading is helped by a feature missing in so many cheaper cars – the absence of a sill over which cases must be humped. Interior stowage for odds and ends is also much better than average, a greatly-appreciated point. The fuel tank, with its remotely-operated filler cover, holds 23½ gallons. This is sufficient for a range of well over 300 miles, though it is disconcerting to discover that the fuel gauge calibration is far from linear, and that the needle wavers when the car is driven on twisting roads.

Speaking of the fuel gauge highlights one odd point of the Silver Shadow: its lack of instruments. It could be argued that this is no bad thing, but there must be keen owners who long for more information than is provided by a speedometer, ammeter, fuel contents gauge and clock. Under the bonnet, things are not as bad as they might at first appear. No big V8 engine is the easiest unit to maintain, but the Rolls-Royce unit is not as bad as it looks. Many of the apparent complications are due to items not immediately associated with the operation of the car, such as the refrigeration compressor. Others, like the Lucas Opus distributor, are mounted high and clear of encumbrances. In other words, although the initial picture is daunting, the Silver Shadow by no means conspires to defeat the keen owner-driver.

In conclusion

It is not easy to sum up the Silver Shadow. A hard-bitten road tester may assess it in terms of performance or ride or handling; but how is he to take into account the feeling which overcomes him when he steps into the back of the car, in full public view, after it has called for him in the morning? Of such feelings are reputations made. In cold-blooded terms, one is entitled to ask if any car is worth nearly £20,000. Included in our comparison tables are three cars costing less than a third as much, though with comparable performance and in each case, at least something of their own sophisticated aura. Yet the Silver Shadow sells. It not only sells, it maintains its value in a unique way. Personal feelings apart – and there are many whose personal feelings will transcend what *Autocar* has baldly to say about specific output or handling or road noise or minor control layout – we should all be grateful that one manufacturer continues to work to a different, and higher, set of standards. And we British should be grateful above all that that manufacturer continues to be one of our most successful exporters. Is it wrong that such a car, so well developed and carefully assembled, should continue in the face of all econoic adversity to be a symbol of success in Britain too? □

The boot is large and luggage loading is helped by the lack of a sill. Small tools and spare bulbs are housed in a fitted tray on top of the battery. The spare wheel winds down from under the boot floor

---

MANUFACTURER:
*Rolls-Royce Ltd.,
Pyms Lane,
Crewe, Cheshire*

PRICES
Basic	£15,225.00
Special Car Tax	£1,268.75
VAT	£1,319.50
Total (in GB)	**£17,813.25**
Seat Belts	Standard
Licence	£40.00
Delivery charge (London)	£38.60
Number plates	£6.48
Total on the Road	
(exc. insurance)	**£17,898.33**
Insurance	Group 7

EXTRAS (inc. VAT)
Halogen headlamps	£22.81
Cassette player	£29.83
Non-standard carpet	£22.81
Non-standard paint	£119.92
Everflex roof covering	£310.05
Woolcloth upholstery	£64.93

**TOTAL AS TESTED
ON THE ROAD** £17,898.33

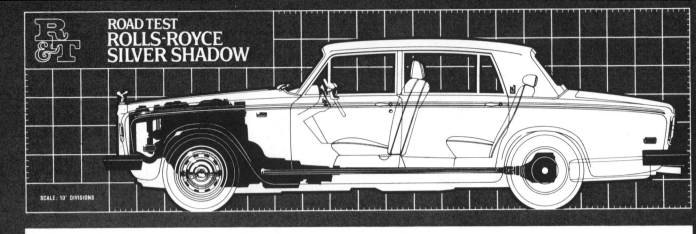

ROAD TEST ROLLS-ROYCE SILVER SHADOW

PRICE
List price, all POE $44,000
Price as tested $44,225
Price as tested includes standard equipment (air cond, AM/FM stereo, tape deck, leather upholstery, self-leveling suspension), matching leather on dash & console ($225)

IMPORTER
Rolls-Royce Motors, Inc
Century Rd, PO Box 189
Paramus, N.J. 07652

GENERAL
Curb weight, lb 5005
Test weight 5425
Weight distribution (with driver), front/rear, % 53/47
Wheelbase, in 123.5
Track, front/rear 57.5/57.5
Length 211.5
Width 71.8
Height 59.8
Ground clearance 6.5
Overhang, front/rear...........33.2/54.8
Usable trunk space, cu ft 19.4
Fuel capacity, U.S. gal 26.0

ENGINE
Type ....ohv V-8
Bore x stroke, mm ....104.0 x 99.1
Equivalent in. ....4.10 x 3.90
Displacement, cc/cu in. 6750/412
Compression ratio 7.3:1
Bhp @ rpm, net ......est 190 @ 4000
Equivalent mph........................105
Torque @ rpm, lb-ft ....est 290 @ 2500
Equivalent mph..........................64
Carburetion.........................two SU (1V)
Fuel requirement......unleaded, 91-oct
Exhaust-emission control equipment: catalytic converter, exhaust-gas recirculation, air injection

DRIVETRAIN
Transmission ....automatic; torque converter with 3-sp planetary gearbox
Gear ratios: 3rd (1.00) 3.08:1
2nd (1.50) 4.62:1
1st (2.50) 7.70:1
1st (2.50 x 2.0) 15.40:1
Final drive ratio 3.08:1

CHASSIS & BODY
Layout ....front engine/rear drive
Body/frame........................unit steel
Brake system ....11.0-in. vented discs front, 11.0-in. discs rear; vacuum assisted
Swept area, sq in. 471
Wheels ....steel disc, 15 x 6JK
Tires ....Dunlop Formula 70, HR/70HR-15
Steering type ....recirc ball, power assisted
Overall ratio..........................19.3:1
Turns, lock-to-lock 3.2
Turning circle, ft......................39.6
Front suspension: unequal-length A-arms, coil springs, tube shocks, anti-roll bar
Rear suspension: semi-trailing arms, torque arm, coil springs, tube shocks, anti-roll bar; self leveling

ACCOMMODATION
Seating capacity, persons.................5
Seat width, f/r ....2 x 23.0/49.5
Head room, f/r....................36.0/35.0
Seat back adjustment, deg.............30

INSTRUMENTATION
Instruments: 130-mph speedo, 99,999 odo, 999.9 trip odo, ammeter, combined fuel & oil levels, clock
Warning lights: oil press.,brake fluid level, brake press. 1, brake press. 2, parking brake, alternator, coolant temp & level, low fuel, low washer fluid, stop lamp failure, rear fog lights, seatbelts, hazard, high beam, directionals

MAINTENANCE
Service intervals, mi:
Oil change.............................3000
Filter change.........................3000
Chassis lube.......................12,000
Tuneup................................12,000
Warranty, mo/mi................36/50,000

CALCULATED DATA
Lb/bhp (test weight)............est 28.6
Mph/1000 rpm (3rd gear) 25.5
Engine revs/mi (60 mph) 2350
Piston travel, ft/mi......................1530
R&T steering index.....................1.27
Brake swept area, sq in./ton........174

ROAD TEST RESULTS

ACCELERATION
Time to distance, sec:
0-100 ft 4.0
0-500 ft 10.3
0-1320 ft (¼ mi) 18.7
Speed at end of ¼ mi, mph 73.5
Time to speed, sec:
0-30 mph................................3.9
0-50 mph................................8.4
0-60 mph..............................11.8
0-70 mph..............................16.6
0-80 mph..............................23.3
0-100 mph............................33.6

SPEEDS IN GEARS
3rd gear (4050 rpm).....................106
2nd (3200) 65
1st (4000) 42

FUEL ECONOMY
Normal driving, mpg...................10.5
Cruising range, mi (1-gal. res) 263

HANDLING
Speed on 100-ft radius, mph.........na
Lateral acceleration, g..................na
Speed thru 700-ft slalom, mph......na

BRAKES
Minimum stopping distances, ft:
From 60 mph............................159
From 80 mph............................313
Control in panic stop ....excellent
Pedal effort for 0.5g stop, lb..........22
Fade: percent increase in pedal effort to maintain 0.5g deceleration in 6 stops from 60 mph .....nil
Parking: hold 30% grade? ....yes
Overall brake rating............very good

INTERIOR NOISE
All noise readings in dBA:
Idle in neutral 49
Maximum, 1st gear.........................74
Constant 30 mph............................61
50 mph 66
70 mph 71
90 mph 73

SPEEDOMETER ERROR
30 mph indicated is actually 29.0
50 mph................................50.0
60 mph................................60.0
70 mph................................70.0
80 mph................................78.0
Odometer, 10.0 mi................10.0

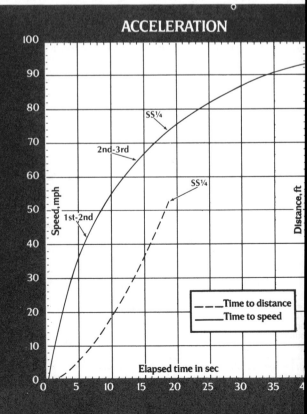

ROLLS-ROYCE SILVER SHADOW LONG WHEELBASE SALOON

"The quality will remain long after the price is forgotten." –
Sir Henry Royce

BY JOHN DINKEL
Engineering Editor

PHOTOS BY JOE RUSZ

 Is a Rolls-Royce the best car in the world? That depends upon your definition of "best" but if it includes such things as using the finest materials money can buy, the ultimate in hand craftsmanship, attention to the most minute detail, and time-and-money-no-object workmanship, then you'd have to place the Rolls at the top of the list.

In a world gone plastic and throwaway, a Rolls is one of the few remaining links with the quality of the past. Name another car that takes from 12 weeks to six months to go through the whole production process with checks, double checks and triple checks that most manufacturers don't make even once. For example, every crankshaft is checked for balance and is made overlength so a piece can be snipped off for metallurgical analysis. Engine valves are a combination of three separate materials because the stresses imposed on heads, stems and wearing surfaces are different in character. And each valve undergoes at least two fluoroscopic examinations during manufacture to detect minute cracks. All connecting rods are also crack-tested and their bolts await final assembly, packed separately in protective plastic jackets. Every single component that must withstand pressure is pressure tested.

To justify the claim that the Rolls-Royce is the quietest car in the world, Rolls-Royce engineers go to painstaking lengths. Hydraulic lifters are "assembled in a bath of kerosene lest a speck of dust should cause wear or noise." To minimize fan and alternator noise the blades are unevenly spaced. A spring steel band is set into the periphery of each rear disc brake to prevent squeal. To reduce noise from electric window winding mechanisms, even such seemingly mundane components as nylon sprockets are made by Rolls-Royce rather than farmed out to an outside supplier, and the teeth on the sprockets are precision cut, not molded. To further minimize engine noise the passenger compartment is shielded from the engine compartment by a double bulkhead.

Rolls-Royce really means it when they say nothing but the finest materials attainable are used throughout. Can you think of another automaker that annually sends a 2-man expedition to Italy to choose the tree of the year? Not just any tree will do for the instrument panel and window moldings of a Rolls-Royce. Only Circassian walnut from trees at least 100 years old is deemed worthy because the grain of older woods has much more depth and character. Following the precepts of the finest furniture makers, the dash and door trims are veneered in two halves, each a mirror image of the other. The woodwork is lacquered and buffed by hand to a finish that looks like glass and is almost as hard. And Rolls is proud to state without being the least bit facetious that the wood is

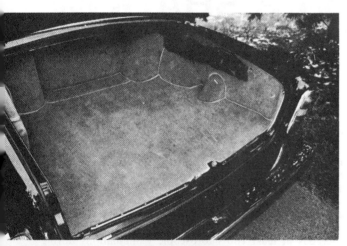

Sumptuous rear seat accommodations include foot rests and lighted vanity mirrors set in walnut veneer in rear pillars. Trunk is better appointed than the interiors of most cars. Tool kit is located on left side of trunk.

also termite proof!

Only the finest English leather from Connolly Brothers is used for the upholstery and trim in a Rolls-Royce. Each car requires eight hides and upward of 500 are inspected for every one chosen. "Rejects," Rolls says, "are likely to end up as expensive handbags. Those that win through, invariably come from clean-living animals grazed in fields protected by electric fences. Lacerations produced by barbed wire and lesions caused by parasites are equally unacceptable." Incredible.

Only 10 men in the world can make the distinctive stainless steel radiator for a Rolls-Royce, that along with the Flying Lady that tops the radiator ranks as one of the most recognizable symbols in all automobiledom. The slender panels are only about 2.5 mm thick and the columns look straight only because they are slightly curved. The Greeks called this effect *entasis* and built it into the Parthenon. At Rolls-Royce, craftsmen build it into the radiator the same way the Greeks did, by hand and eye alone.

There's more. Every engine is run-in for the equivalent of at least 150 miles on natural gas on a test bed and every car gets up to 150 miles of on-the-road testing. The radial tires were especially designed for the car and arrive on their wheels as individually balanced units and are secured by brass nuts to eliminate rust or jamming. Where chrome plating is used, it is 10 times thicker than normal. And more than 400 parts are stove enameled because in some areas even the best paint isn't good enough. These are some of the reasons why when many auto companies have been mired in the throes of one of the worst sales slumps to ever hit the world automakers, Rolls-Royce could announce a new model, the Camargue, and have the first three years production sold to customers who had signed an order without ever having seen the car or knowing its ultimate price: $90,000.

What this is all leading up to is that in conjunction with the U.S. introduction of the Rolls-Royce Camargue, I was invited to England to spend a few days in London and at Rolls-Royce Motors headquarters in Crewe, a small picturesque town located about 200 miles northwest of London in the county of Cheshire.

Famous figure, known officially as the Spirit of Ecstasy, is cast of stainless steel by the ancient lost wax process. Driver's door panel contains switches for electric window lifts, a master switch that allows only the driver to operate the windows plus a switch for the central door locking system. Armrest adjusts up and down and fore and aft.

The trip started with a whirlwind tour of the Mulliner Park Ward coachworks where the bodies for the Corniche and Carmargue are hand built by skilled craftsmen. Bodies for the "bottom of the line" Silver Shadow are built by British Leyland. If you're a bit surprised and shocked by such an admission, don't be. Just remember that grape juice and the finest Dom Pérignon champagne start out the same way—with the squeezing of the grapes! Following the factory visit, I was taken for a brisk (and I do mean brisk) sightseeing ride around London in a Carmargue capably driven by Public Relations Officer Evan Morgans (who happens to be related to Morgan the pirate but that's another story) and if I'm a bit hazy on the details on what I actually saw (Parliament and Hyde Park were two of the places I think) it's because it takes my brain three or four days to convince the rest of my body that a sleepless all-night plane ride and the effects of jet lag are just what the doctor ordered.

Following lunch (I think it was lunch but my stomach wanted either breakfast or dinner and I was too tired to figure out which) Morgans and I caught a train for the 2-hour ride to Crewe. I took this opportunity to grab some much needed sleep so the trip went by quickly with just brief recollections of mile upon mile of yellow mustard flowers that I later found out are used for feed and a few drying streams and lakes, the result of an extended drought.

I was driven to my "motel," a typically English Inn called The Wild Boar Motor Lodge Inn in the picturesque farming town of Beeston, not far from Crewe. That night I supped in the Inn's delightful l'Aperitif Restaurant Beagle Cocktail Bar and Penthouse (if you don't believe me just check the photo) where I dined on some of England's world famous Dover sole (it's even better than the N.Y. fluke and flounder I had been brought up on) and topped off the meal with some delicious peach mousse.

The next morning Morgans picked me up in a Silver Shadow and we headed for Crewe. There I made the rounds of the various assembly areas. As you might expect there is nothing hurried about the construction of a Rolls-Royce. Each worker can virtually take as much time as needed to get it right and the pride each worker takes in building the "best car in the world" is evident wherever I looked. (Managing Director David Plastow said it best when the question, "How fast does the Rolls-Royce assembly line move?" was put to him by an

American news journalist used to the frantic pace of Detroit. "I think it moved last Thursday," was Plastow's droll reply.)

Morgans says that strikes, the bane of every other British automaker, are virtually nonexistent at Rolls-Royce and that despite wages lower than in the more industrialized areas of England, the labor turnover is almost nil. Building a Rolls is a true labor of love.

The Rolls-Royce philosophy of introducing a change only when it is an improvement and never simply for the sake of change was quite evident in the talks I had with several of the top engineers at Crewe. That's why you won't find opera windows in a Rolls or hidden windshield wipers but why Rolls is currently studying electrically operated outside mirrors like those used by Porsche (Rolls engineers are impressed with the precision adjustments such a mirror allows) and why you won't see fuel injection for at least three to four years (it hasn't proved itself from cost, noise and throttle response considerations to Rolls-Royce satisfaction). An ultra-conservative company, Rolls-Royce Motors doesn't see itself in the forefront of any coming revolution or evolution in automobile design but prefers instead to let other companies do the research and then wait until such advances are fully developed before considering them for inclusion in Rolls-Royce cars.

Although I spent some time driving a Silver Shadow along the motorways and narrow, twisty 2-lane roads that snake through the area around Crewe, it's difficult getting a valid impression of a car when you are not only intimidated by driving on the wrong (left) side of the road and wrong (right) side of the car for the first time in your life but also by having an important personage from Rolls-Royce sitting next to you watching for your every mistake in his $38,750 motorcar. There's nothing like driving and testing a car on your home turf so upon my return to the states R&T arranged through the west coast distributor to borrow a Rolls-Royce from our local dealer, Roy Carver in Newport Beach. Following a lengthy briefing on the various controls, Sales Manager Joe Davison handed over the keys to a Silver Shadow Long Wheelbase Saloon.

We can't think of anyone who has ever looked at a Rolls-Royce who hasn't come away thoroughly impressed with its incredible quality and detailing and our Oxford blue with beige roof Silver Shadow was no exception, standing up to the closest scrutiny by the R&T staff and scoring a solid 99 percent (no car is perfect, not even a Rolls).

Introduced in December 1965 (the 4-in. longer wheelbase version came out in 1969) the Silver Shadow has stood up to the test of time very well indeed. There were cries of anguish from traditionalists that it didn't "look like a Rolls" but surely there were similar cries when its predecessor, the Silver Cloud, was introduced in 1955. People are saying the same thing about the Camargue but Pininfarina who designed the Camargue refutes those who take umbrage at the styling by saying, "If you took to it straightaway it would already be dated."

The Silver Shadow, like any Rolls, is an engineering *tour de force*. The long-wheelbase version is built on a 123.5-in. wheelbase and is 211.5 in. long. It occupies about the same road space as a Cadillac Seville, but the Silver Shadow is considerably taller and much roomier inside. Its unit steel body, which has aluminum doors, hood and rear deck, carries large subframes isolated from the main structure by unusual steel gauze doughnuts. There's independent suspension front and rear with coil springs and anti-roll bars at both ends of the car and self-leveling at the rear.

There's little chance of running completely out of brakes because of the Rolls' unique 3-part brake system. Disc brakes are fitted to all wheels with two twin cylinder calipers at each front wheel and one 4-cylinder caliper attached to each rear wheel. In addition to these two separate and independent hydraulic circuits, a mechanical foot-actuated parking brake system operating on a set of lever-actuated calipers on the rear discs is fitted.

The V-8 engine displaces 6750 cc and is an all-aluminum design with steel liners and twin SU carburetors. Otherwise the pushrod overhead-valve unit is of thoroughly conventional design. Rolls-Royce doesn't publish power or torque figures but we estimate the output as 190 bhp at 4000 rpm and 290 lb-ft of torque at 2500 rpm. The transmission is a 3-speed Turbo Hydra-Matic unit Rolls purchases from General Motors. The only Rolls-Royce modification (obviously GM matches the shift points to the output of the Rolls engine) is the addition of an electric shifting mechanism in place of the conventional linkage arrangement.

Those are the basic facts and specifications. But to dwell upon numbers and comparisons with other cars is to miss the whole point of a Rolls-Royce. When you drive a Rolls you aren't driving a car but a tradition, a philosophy and a way of life that in no way can be related to the mass production techniques used to build other cars. Instead you talk about the incredible attention to detail inside! Things like fold-down armrests for driver and front-seat passenger, door armrests that adjust up and down and side to side, the button on the dash that allows you to check the oil level without ever opening the hood or touching a dipstick, the cigarette lighters for each of the three ashtrays, carpets that are edged in leather from those same eight hides used to cover the seats.

It's true that a Rolls-Royce can't be thrown around curves with the abandon of a 450SE but the overall ride and handling are really impressive. It's as comfortable as the softest riding American luxury car without the rocking and rolling motions that find favor only with the seafaring crowd. It stops and it goes and the steering, though light, has reasonable road feel and is quite responsive. The Blaupunkt AM/FM stereo radio and the Pioneer quadraphonic 8-track, 4-channel tape deck provide exquisite listening pleasure. And it really is a quiet car. Our previous noise champ, the Cadillac Seville, registered 52 decibels when idling in neutral; the Silver Shadow was 3 dBA quieter. The Rolls was one dBA noisier at 70 mph but one dBA quieter at 90 mph. And what was most impressive was that when I drove a similar car at 120 mph on an English motorway there was hardly a discernible increase in road, engine or wind noise from the level noted at 60 mph.

How do you justify paying $38,750 to $90,000 for a car? The same way you justify spending $200,000 for a yacht or $1,000,000 for a thoroughbred race horse. Some people will never settle for second best and are willing to spend as much as required to acquire what they consider the ultimate. There's legend to a Rolls of course: Lawrence of Arabia with an armored Rolls-Royce in the desert, so reliable he said it was more valuable than rubies; Rolls-Royce as the car of Hollywood stars, and diplomats, aristocrats and royalty around the world; Rolls-Royce, a car as close to perfection as man can make it; Rolls-Royce, the epitome of the American dream both as a symbol of success and what success can bring.

Don't you think you deserve a Rolls?

PHOTO BY JOHN DINKEL

new cars

As "The Best Car in the World" sits in majesty note the new inconspicuous air dam which gives improved stability.

The Silver lining

Known for 70 years as "The Best Car in the World," the Rolls-Royce is more popular than it has ever been, in spite of unavoidable price increases. Especially on export markets, the sales are steadily rising, and many millions of pounds are being spent on plant and buildings in order to satisfy the demand. It is in this extremely healthy atmosphere that the new Silver Shadow II is announced.

The latest car is a logical development of the existing Shadow. Of recent years, the Rolls-Royce has become much more of a driver's car and it is in this respect that further advances have been made. The front suspension has been redesigned, the geometry following racing practice in giving more swing-axle effect. This tends to keep the wheels more upright in relation to the road, reducing noise and tyre wear and increasing roll resistance. The rear anti-roll bar has been reduced in diameter to retain the correct degree of understeer for stability.

For the first time on a Rolls-Royce car, rack and pinion steering has been adopted, with power assistance, and the size of the steering wheel has been reduced. An inconspicuous air dam is built into the front of the body to give improved stability.

The big light-alloy V8 engine has been the subject of much research, especially in the fields of part-throttle fuel economy and exhaust emission control. Twin SU HIF7 carburettors have now been adopted, which have linear ball-race dash pots and bi-metal control of the main jet position, to compensate for fuel temperature and viscosity. A new dual exhaust system with six stainless steel silencers relieves back pressure and gives more vivid acceleration in the upper speed ranges. A quieter and more efficient plastic engine fan is driven by a viscous coupling, that is torque-limited to a pre-determined maximum revolution speed and an electric booster fan cuts in automatically if the temperature ever reaches 105 deg C.

An electronic speedometer has been adopted, the deletion of the usual cable drive eliminating a possible source of noise transmission. The odometer has an extra digit, reading up to 999,999 miles, as befits a car that has an immense life expectation, and this type of speed indicator is fundamentally more reliable than the mechanical pattern. The new instrument panel, incorporating ergonomic research, resembles that of the Camargue.

Also developed from the Camargue, the new automatic air conditioning system is a highly sophisticated installation that cannot be described in detail owing to limitation of space. It differs from the previous design principally in its elaborate sensing and control system, which renders the operation fully automatic, calling for no skill on the part of the user or re-adjustment under changing conditions. Air is introduced at two levels and entirely different temperatures can be maintained in the upper and lower parts of the saloon.

The air conditioning system uses 0 deg C as its datum, and all incoming air is first cooled to this temperature before being either heated or cooled further. This de-humidifies the air and prevents the windows misting up. The heating of the rear window is turned on automatically when required and has no manual switch. After a stop, when air in the ducts may warm up and become humid, there is a 13 second delay on the fan motors while the air is frozen, to avoid the windows misting up. Another arrangement prevents the lower ducts from directing cold air into the occupants' feet before the engine has warmed-up.

The Silver Shadow II can be identified by a name plate on the rear panel, a slightly deeper radiator and the two chromium-plated exhaust pipes.

With a life expectancy stretching far into the future, the body shell undergoes anti-corrosion measures of a thoroughness that would be pointless on an ordinary car of planned obsolescence. The door, bonnet and boot-lid are of light-alloy for ease of handling, the finishing processes being long and elaborate to avoid deterioration of the paintwork.

In the past one of my few criticisms of the Rolls-Royce car has concerned the fan noise, for whereas the engine was silent the fan was not. Though its mild commotion could not be heard by those inside the car, it was audible to people on the pavement, slightly spoiling the effect of a magic carpet gliding away. When Rolls-Royce Motors invited me to drive the new car in the sunshine of Spain, this was the first point which I checked. I am delighted to say that exterior fan noise has been eliminated and the Silver Shadow II moves off in a religious hush. The exhaust silencing is also more effective and such sound as there may be is of a very deep timbre.

I chose a winding route through Spanish mountains as I was more interested in the handling of the car than in mere maximum speed. The new steering feels totally different, being altogether quicker and more responsive. One can now fling the big machine around in a manner befitting a small sports car. The steering is, as always, very light, but it now transmits messages from the road to the driver's hands.

I have driven all the variations on the Silver Shadow theme since I first met the car in 1965. It is impossible to generalise, because there have been so many stages of development but, whereas the earlier series tended to roll and make a lot of tyre noise when handled with some brutality, the Silver Shadow II remains remarkably upright and the scrubbing of tortured rubber never degenerates into a scream. Although such driving was necessary to determine the car's limits, I hasten to add that I know my manners better than to behave like that in a Rolls-Royce on normal journeys!

Though there is long travel on the suspension, it is well damped and there is no suspicion of pitching or wallowing, while the automatic self-levelling system avoids a tail-down attitude when passengers and luggage are added. Road noise used to be a problem with the early Silver Shadows, but this has steadily improved as the suspension and tyre technicians have worked together. In the Silver Shadow II this progress has gone a stage further and wind noise is totally absent, thanks to the perfect fit of doors and windows.

The 6750cc V8 engine can hardly be heard and such sound as there is denotes that the crankshaft is turning unusually slowly, even when the car is accelerating briskly. Though the brakes are very light in operation, the pedal action is progressive and one soon feels at home with them. All the controls have a sensitive touch, which gives an almost sensual pleasure to the experienced driver. The car is extremely easy to drive, but it responds best of all to the pilot with a little finesse at his finger tips.

The Silver Shadow has been the most successful Rolls-Royce model ever and the results on the export markets have been outstanding. The Silver Shadow II, or Bentley T2 if you prefer the other radiator shell, is even more enjoyable to drive, has many small improvements and uses appreciably less fuel. Even at its price of £22,809, it is certain to break all the sales records of its predecessor.

If you look at the dashboard below you may see the odometer which will read up to 999,999 miles.

105

NEWCOMERS:
THE SHADOW'S SECOND COMING

LJK Setright on some welcome changes to Rolls-Royce.... but are they enough?

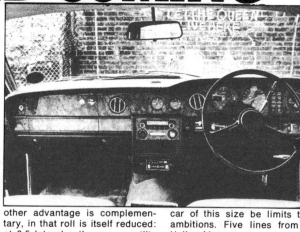

WHEN ROLLS-ROYCE INTRODUCED the Camargue in 1975, Mr Engineering Director Hollings told me that he wished he had been the man who thought of subternasal air dams, since they could not fail to improve the behaviour of virtually any car. It is therefore no surprise to find such a dam beneath the nose of the latest Silver Shadow which, after a long series of alterations since the model first appeared in 1965, is now graced by a plate on the boot declaring it to be the Mark II. The dam is easily the most noticeable feature of the exterior, and although it has no effect on the car's drag coefficient it does have a great stabilising influence, especially in cross-winds, when the car is travelling quickly. The maximum speed remains at about 115mph, but at speeds well below this the earlier Shadows demanded quite a lot of steering correction.

It is in the quality of the steering that the Mark II Shadow is most sensibly improved. Gone is the mushy erratic over-powered system that used to make the Shadow feel like a stepped-input jelly. In its place is a precise rack-and-pinion apparatus with beautifully consistent power assistance and much more alacrity of response than one might expect in a car of such heft. Burman make the rack, but of course it is a very special one, with the hydraulic cylinders at its extremities and the output to the track-rod halves at the middle. The need for this layout was imposed by the desirability of setting the inboard steering joints on the same line as the inner pivots of the lower suspension wishbones, so that suspension movement might introduce no spurious steering movements; and this in turn is rendered more important by suspension revisions that have markedly improved the cornering power of the front tyres.

The revisions to suspension geometry are of a simple nature, but they have two very important and welcome results. One is that the swing axle effect is increased, so as to reduce the detrimental change of camber towards positive when the car rolls in cornering: by keeping the outer wheel more nearly perpendicular to the road in these conditions, the cornering power of the tyre is maximised and tyre wear (Shadows used to be great chewers of tread shoulders) is minimised. The other advantage is complementary, in that roll is itself reduced: at 0.5 lateral g the car now tilts through 4¼ degrees, half a degree less than before, because the geometric revisions have raised the instantaneous roll centre from 3 to 3.8ins above the ground.

The result is a relevation. The Mark II Shadow has far better steering than any other Rolls-Royce ever had. This new-found precision does not make the Shadow a sporting car, however: the greater cornering power of the front tyres reduced the car's tremendous understeer so much that there was some danger of converting it to oversteer, which the average Rolls-Royce customer would not like, so the rear anti-roll bar has been weakened in compensation. The effect is that, although the new Shadow is much more agile than any of its predecessors, it still discourages hard cornering. No longer does the car feel betrayed by its front wheels, but rather let down by those at the back. Force the car hard into a corner and you get an immediate sensation of roll oversteer as the rear wheels assume the odd contortions that all trailing-arm independent suspension systems seem to encourage. This is but a transient stage, after which the car seems to settle down into a stable condition of roll and understeer that (given reasonable judgement of the entry speed) can be maintained through the rest of the corner. It is probably not as hazardous as it at first feels, but it is somewhat discomfiting: the Rolls-Royce is not yet a car in which one can hurry decorously. This is surely a pity: a Rolls-Royce should be supreme in doing everything decorously, but presumably there must in a car of this size be limits t[o] ambitions. Five lines from Hollow Men seem to sum it up[:]
 Between the idea
 And the reality
 Between the motion
 And the act
 Falls the Shadow.

Good old T S Eliot. Good Rolls-Royce, for that matter: are getting nearer to producin[g the] best car in the world, and i[f an]body beats them to it they [won't] yet succeed.

The quality traditionally as[soci]ated with Rolls-Royce is as a[bun]dant as ever. The quality o[f life] within the car has been impr[oved] by the adoption of a fully a[uto]matic air-conditioning system [of]fering the same completely [in]dependent temperature contr[ol at] upper and lower levels as [was] previously unique to the C[am]argue, but it has now been re[vised] to make its control simpler. S[ome] of the temperature sensing [and] automatic control is electr[onic] and so is the new speedom[eter,] more reliable and less poten[tially] noisy than the old cable-dr[iven] type, and allegedly tamper-[proof] for the first million miles or k[ilom]etres recorded on its odomete[r.]

The instrument is part of a [new] display incorporating a recta[ngu]lar bank of warning lights loo[king] strangely at odds with the cir[cular] dials and ventilation nozzles [but] the veneers and trimmings o[f the] fascia are as handsome as e[ver] and rather more practical [than] before. Above the front pas[sen]ger's knees is a compartment [hou]sing the fuse board, and g[iving] immediate access not only t[o the] fuses but also to the reset bu[ttons] for the thermal interr[upter] switches that safeguard som[e of] the car's complex circuitry. H[igh] quality electrics were alway[s]

106

Royce speciality, and in the Mark II Shadow they are more extensive than ever: the speed-hold device is now electronic, with a memory and a reset/accelerate facility, previously unique to Bristol and Porsche and better than before.

As in the Bristol, if not quite so effectively (in the Chrysler-engined car, the speed-hold accelerator commands only the primary chokes of the four-barrel carburettor) the speed-hold can be used to ensure economical driving. The Shadow has a pair of SU carburettors, those elegantly engineered HIF7 jobs with linear ball-bearings in the dash-pots to ensure really accurate motion of the piston and thus make possible close control of mixture for emission requirements as well as fuel economy. There is even a bi-metal thermostat to control the main jet position according to fuel temperature (and viscosity), giving a much weaker carbon monoxide concentration when the engine is idling. All these emission controls, together with a somewhat reduced choke size for the carburettors, combine to reduce very slightly the power of the 6.75litre engine, but this deficit is restored by a new twin-exhaust system (with six stainless steel silencers) which imposes less back pressure. The engine-driven radiator fan is smaller and more efficient, and thus wastes less fuel, being supplemented by a heat-switched electric fan.

None of this husbanding of power makes the new Shadow a fast car. Its performance is virtually the same as its predecessors, a casual 0 to 60mph dash taking 10.7secs on a Spanish road that was not quite level. There is no improvement to be found by delaying upward changes through the GM automatic transmission: the gear-hold lever, beautifully gated to define its movements in a scheme that should be an example to all except Daimler-Benz, is used only to heighten one's joy in controlling the car. That joy is no less real for being based on intangible things such as serenity, quality, reputation and all the other non-material things that contribute as much as the car's 80,000 component parts to the enviable state of being conveyed by Rolls-Royce. May the Ghost of James Shirley forgive me, but 'The glories of our blood and state are Shadows not-substantial things ' ●

Star Road Test

ROLLS-ROYCE SILVER SHADOW II

The latest version of Rolls-Royce's best seller embodies a number of significant improvements. It's not the fastest or best handling of luxury saloons, or even the quietest. But we've yet to try anything better

THE SILVER SHADOW is by far the most successful car Rolls-Royce have ever made. Over 20,000 of these rather conservative looking but magnificently constructed cars have now been built at Crewe since 1968, though the latest versions are very much better than the originals. The 2000-odd modifications made to the car in its nine-year production run (likely to continue for several years) have kept the stately carriage abreast of the times and, if anything, enhanced the marque's unique standing and reputation.

With the Shadow II designation, justified by fundamental design modifications, came some obvious visual changes. For instance, there are now rubber-tipped bumpers front and rear without over-riders, a deeper radiator shell and different door handles with more deeply recessed buttons. Still more evident is the bib spoiler under the front bumper, claimed to have considerably improved high-speed stability, especially in a cross wind. There is also a new-style door-mirror of spring-back design which is adjustable from inside the car.

However, most of the latest improvements are considerably more significant you can't see are the switch from recirculating ball to rack-and-pinion power steering with far better "feel" than before, and revised front-end suspension geometry to improve the handling and adhesion, and help reduce tyre wear. Fuel consumption is said to have been improved by fitting different carburettors, less power-sapping cooling fans and by reducing back-pressure with a new dual-pipe, six-silencer exhaust system.

Inside, there is a revised facia with better instrumentation, switch-gear that is more ergonomically arranged, a smaller-diameter steering wheel, and the incredible fully automatic, split-level air conditioning system first seen in the Camargue.

Riding in a Rolls-Royce has always been an experience to savour, especially by those unaccustomed to such opulence. Driving one, however, has in the past betrayed a number of shortcomings — particularly in the handling department — that have not always endeared the car to everyone here. The latest changes have dramatically improved the Shadow, though even now not to the point where on-the-road results yet match the unrivalled standards set by Rolls-Royce for the car's construction and execution. It is the way the Shadow is made, finished and equipped rather than what it does which still sets it apart from any other car.

By normal standards, as you would expect for an outlay of £23,000, it is extremely comfortable, quiet and refined — but the Jaguar XJ12, costing very much less, is superior on all counts and the big 6.9 Mercedes faster and better handling. Even so, despite Rolls-Royce's failure to make the Shadow a match for *any* rival in *every* department, the majority view here is that no other car we know cossets you quite so well or provides quite so much in the way of sophisticated creature comforts in such luxurious and rich surroundings — even if not all you see and hear (like the magnificent Japanese quadrophonic tape player) is as traditionally British as the purists would like.

From the hypercritical among us, there is still a note of dissent for that sort of money, they argue, the Shadow should be even better than it is. From the rest there was at long last recognition that, taking all things into account, perhaps Rolls-Royce really do now make the Best Car in the World, even if it's still far from perfect.

PERFORMANCE

★★★ Though V8 engines are largely the prerogative of the American motor industry (even the ubiquitous Rover unit was spawned Stateside) Britain has two such engines of which to be proud — that which powers the Aston Martin and Lagonda cars and the all-alloy 6.75-litre unit which purrs behind the mascots of all current Rolls-Royces bar the Phantom VI which uses a short-stroke version of the same unit.

A mass of plated pipery and polished enamel, the underbonnet ironmongery of a Shadow II is an impressive marriage of craftsmanship and advanced technology. The engine's output, though, remains undisclosed. You could make an educated guess at the power required to propel over two tons of car and to drive pumps for the brake and power steering systems and the air conditioning, not to mention a heavy duty alternator which feeds a plethora of electrically operated components. Yet for a car of the Shadow's calibre, the *way* in which it performs is of greater importance than *what* it can achieve. Suffice to say, as R-R always do, that the power is enough.

For the record, the mean maximum speed of our test car, in far from ideal conditions, was 113 mph. On the fastest leg it did 114 mph, figures comparable with those achieved by the Shadow I we tested in 1972.

Momentarily spinning its huge tyres on the damp surface of MIRA's mile straight, the Rolls-Royce accelerated peacefully and briskly to 60 mph in under 11 seconds and on to 100 mph in 37 secs. Though nothing special by absolute standards, and comfortably bettered by, say, the V12 version of Jaguar's sprightly XJ range, such performance makes the stately Shadow quite a quick car. Its overtaking ability is shown by the kickdown times, 30 to 50 mph requiring 4·1 secs, 50 to 70 mph 6·8 secs.

A flick of the ignition key is usually sufficient to fire the muted V8, the shimmy of the auxiliary gauges being the only clue that the engine is running unless you're sitting in silence. The automatic chokes of the emission SUs work perfectly and the Shadow will pull without fuss when stone cold. In fact the only proof of the choke's operation is the accompanying rise in idling speed which in turn momentarily dips the car's tail as "drive" is engaged.

While Jaguar surpass even Rolls-Royce when it comes to suppressing engine noise, the note of the Crewe V8 never rises above a distant thrum. A faint increase in induction noise is evident when opening up the throttle but the murmur is never obtrusive — with the tape-player or radio on, it is inaudible. An occasional tremor is the nearest the carefully balanced and heavily insulated engine ever gets to the booms and engine-excited vibrations excited by more mundane machinery.

The controls for the excellent automatic speed hold are now incorporated in the gear selector lever. An on/off and resume switch is mounted on the side of the stalk, the speed-set button forms its tip. Once a speed has been selected, the car will go up hill, down dale at exactly that speed without surging, taking much of the strain out of long distance driving. On a motorway, we found this "feet-off" driving very relaxing and of real practical value if you don't want to exceed the speed limit. A touch of the brake pedal cancels the control and normal progress is resumed.

ECONOMY

★ A Rolls-Royce cossets its occupants to a standard unattainable with lesser machinery, to a degree where vehicle weight and aerodynamics are of secondary importance to accommoda-

108

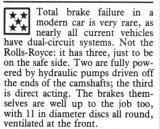

Above: wide doors open the way to the sumptuous interior. There is ample space on the hide-covered rear seat for three adults to lounge in comfort. Below: courtesy light, spot light, mirror; passengers are truly pampered

tion and creature comforts. So economy also takes a back seat. While Rolls-Royce are constantly at pains to improve the efficiency and therefore reduce the consumption of their cars, it is no surprise that the Silver Shadow II remains an extremely thirsty vehicle.

Changes affecting economy include a more efficient viscous-coupled fan (assisted by another, electrically operated, in conditions of high ambient temperature); more suitable SU carburetters; and a dual exhaust system of reduced back pressure. Even so, our test car was no more thrifty than its predecessors, returning a touring consumption of 12.5 mpg and 11.1 mpg overall. With such a thirst even the Shadow's 23.5 gallon fuel tank allows no more than 300 miles between fill-ups — too little for a car that's well suited to trans-continental trips. With its 8.0:1 compression ratio the Rolls-Royce performs happily on four star fuel.

TRANSMISSION

★★ Early in the life of the Silver Shadow the old four-speed Hydramatic transmission was dropped in favour of the three-speed GM400 unit, since when we have been running out of superlatives to describe its uncannily smooth operation. Rolls-Royce themselves must be reasonably satisfied as no changes have been made to the gearbox of the new car.

Those used to stirring a floor-mounted gear selector will find the Shadow's column-mounted electric selector switch a revelation. Under most circumstances it can be flicked in to "drive" and forgotten. Part-throttle upward changes are virtually imperceptible other than for a slurring change in engine note — if you can hear it. A full throttle change from first to second is accompanied by a slight and very satisfying surge as the second ratio comes in. The second to top change is as smooth as silk.

So sensitive is the kickdown, so ideally matched are the characteristics of gearbox and engine, that manual selection of the lower ratios is superfluous other than when dawdling in traffic. Only when manual downwards changes are executed does the transmission jerk.

At full throttle, upward changes occur at 40 and 74 mph respectively while less enthusiastic motoring will see second engaged by 10 mph and top before you have reached 20 mph. There is no advantage in hanging on to the gears manually; indeed, the handbook discourages it as there is no fail-safe device to prevent over-revving. However, there is an electronic over-ride that locks the transmission whenever the ignition key is removed — irrespective of where the lever is positioned.

HANDLING

★★★ Silver Shadows have never been endowed with the taut, sporting handling of, say, a Mercedes 450 SEL 6.9. Though safe and predictable, early Shadows discouraged fast driving by squealing their tyres, rolling a lot and understeering excessively. Continual development brought improvements but never more so than now with the changes made to the MkII.

Originally the car was equipped with power-assisted recirculating ball steering. An early change in ratio improved response but the system was always too light for our taste. The MkII, however, comes with a new rack and pinion set-up which, with a smaller diameter steering wheel, is better weighted and more precise. Apparently it has been well received by customers at home and abroad.

There have also been changes in front-end geometry, intended to keep the outside wheel more upright when cornering and to lower the front roll centre. To assist in front-end adhesion a slimmer rear anti-roll bar is fitted.

All this adds up to a car that rolls less and handles better which, with the improved steering, encourages more enthusiastic motoring. Even so, the Shadow is still no driving machine in the Mercedes mould: soft suspension and the sheer mass of motor car somehow temper the mood of the most press-on driver.

What is impressive, though, is the adhesion of the enormous Avon radials. In the dry it takes brutal provocation to push them off line, and even in slippery conditions the engine's considerable torque can be transmitted to the tarmac with little or no wavering of the weighty tail. Any incipient slide can be quickly and easily checked thanks to the better feel of the new steering.

BRAKES

★★★★ Total brake failure in a modern car is very rare, as nearly all current vehicles have dual-circuit systems. Not the Rolls-Royce: it has three, just to be on the safe side. Two are fully powered by hydraulic pumps driven off the ends of the camshafts; the third is direct acting. The brakes themselves are well up to the job too, with 11 in diameter discs all round, ventilated at the front.

Our standard 20-stop fade test brought a slight initial rise in pedal pressure that was barely detectable by the driver. The water splash had no effect whatsoever.

On the road the brakes always inspired confidence. At town speeds braking is light and progressive. Braking from high speed needs more pedal movement but the power and sensitivity are as good.

Though awkwardly placed, the umbrella handbrake is extremely efficient. It made light of holding the heavy car on a 1 in 3 incline and gave a remarkable 0.4g stop from 30 mph, locking the rear wheels with ease. Only a 0.25 deceleration is required by law.

ACCOMMODATION

★★★★ At a shade over 17ft (the rubber-tipped bumpers have added half an inch) the Silver Shadow is a large car that provides ample accommodation for five adults and their luggage. Head and knee room in the back are sufficient for three six-footers to lounge in extreme comfort. For two-up travel the large central armrest may be unfolded to provide welcome lateral support which the smooth surface of Connolly-trimmed seats don't provide. Comfortable head-rests are standard in the back.

Silver-plated contacts ensure hypersensitive control from the front-seat "joy-sticks". With them, a practised hand can position the armchair seats in exactly the right position: legroom, height and cushion tilt can be altered electrically. Even tall, fat people will find ample head, leg and shoulder space in the front of the Shadow, but we're sure that many, like us, would prefer more lateral and lumbar support

109

Above left: the burr walnut may pick up stray reflections, but its quality is unmatched. Top right: the new instruments are an improvement, though their glasses do reflect. Left: the hypersensitive seat "joysticks" and the self-emptying ashtray are excellent features, as is the speed control, below left. Below right: the switchgear is beautifully precise. Right: there are warning lights for almost everything.

from the seats. And why no headrests?

Stowage space within the car is adequate rather than generous. Shallow pockets in the front doors will hold Ordnance Survey maps or paperbacks while a leather trimmed, walnut-faced locker hides small valuables from prying eyes. The deep pockets in the seat backs are ideal for storing magazines or map books. The rear shelf, however, is too small for anything but bric-a-brac.

Either the master key or button located in the glovebox will unlock the bootlid which lifts high to reveal a beautifully trimmed hold of enormous dimensions. We persuaded no less than 14.5 cu ft of our Revelation cases into this easily loaded area.

RIDE

★★ ★★ Owners of ordinary family saloons would rightly be over-awed by their first taste of Rolls-Royce motoring, not least by its ride. At town speeds on smooth roads the Shadow's compliant suspension imparts an almost magic carpet effect; irregularities are smothered by the self-levelling system and only the continual albeit muffled bump-thump of the fat Avons betrays the hard-working suspension.

By absolute standards, however, the Shadow's ride is not exceptional for it is bettered by even the cheapest of Jaguar's XJ saloons. At speed on undulating surfaces the heavy car "floats" a bit, nodding its majestic prow. Furthermore, sharp irregularities can catch the soft suspension out altogether, causing the shell to shudder. Even on smoother surfaces the big tyres tend to relay the presence of irregularities with muffled thuds or damped vibration.

In short, though the ride is very comfortable, there are occasions when it fails to match that of some much cheaper cars.

AT THE WHEEL

★★ ★★ Juggling with the seat adjuster should result in a near perfect driving position for even the most finicky. Short people can sit high, tall people well back. Those who like a reclining position can use the manual backrest adjuster for fine tuning. However, you are aware of sitting on rather than *in* the seats, especially when cornering hard.

The large brake pedal is well placed for operation with either foot but the two-feet brigade will hit trouble at night as there is also a foot-operated dip switch to keep the left leg busy.

Though tidier than before the minor controls are still scattered. For instance, the electric washers are on the indicator stalk yet the two-speed-and-pause wipe control is mounted on the facia. Above it, and introducing yet another type of switch, is the master-light control. The remaining knobs and buttons are neatly installed on the lower edge of the facia and all the controls have a very satisfying action.

The gear selector is a hand-span from the rim of the new steering wheel and is beautifully light and simple to move. The umbrella-style handbrake, however, can catch your knee or trouser leg when pulled on.

Unlike Mercedes, Rolls-Royce do not shroud their seat belts; consequently they not only look untidy (the webbing is not even colour-matched like those of the new Audi 100) but recoil in such a way that the steel tongues frequently clout and mark the superb walnut door trims. The cheap-looking plastic buckles also look out of place.

VISIBILITY

★★ ★★ As the seat is adjustable for height, nobody should find themselves straining to see past the Silver Lady's flowing drapes. Though long and wide the regal snout of the Silver Shadow is easily aimed. At a stretch even the extremities of the boot are visible from the driver's seat so reversing is no problem. Between them the dipping interior and new-style remote-control door mirror provide a good view of following traffic and a virtually invisible heating element makes light of demisting the rear screen.

However, the wipers leave several unswept areas and the blades start to lift at anything over 100 mph. Moreover, the first sweep takes the blades only half way across the screen. The four-shot headlight system is fine on dip but nothing special on main beam.

INSTRUMENTS

★★ ★★ With the MkII comes a welcome revision of instrumentation, a reversion to the layout used on earlier Rolls-Royces. A large circular speedometer with trip and total mileage recorders is flanked by a compound dial incorporating gauges for oil pressure and water temperature as well as an ammeter and fuel gauge. The latter also doubles as an oil gauge, pressing a button on the facia causing it to register the level of oil remaining in the sump.

The other two dials in the centre of the facia are for a quartz clock and an ambient temperature gauge. This has its sensor mounted just below the Silver Lady mascot and is affected by bonnet heat when the car has been standing. Once on the move it works well. The new (silent) electronic speedometer was 2 per cent fast at 100 mph but thereafter became wildly optimistic. The distance recorder was 1 per cent fast.

Though far from reflection-free the traditionally styled dials are boldly calibrated and easily read through the upper segment of the new, small-diameter steering wheel.

To the right of the speedometer is a window of no fewer than 10 warning lights that register everything from brake failure to a lack of washer fluid. There is even a light to warn of freezing conditions outside.

HEATING

★★ ★★ Air conditioning has been standard equipment on all Rolls-Royces since 1969. The system previously fitted to the Shadow, though highly effective, was quite complex and needed occasional adjustment as ambient temperature varied. Its superb replacement is fully automatic and has just three controls.

Firstly, there is the "function" knob which has five positions: off, low, auto, high and defrost. All self-explanatory. Then there are two temperature controls, one for the upper, the other for the lower level. They allow any temperature from 63 to 91 deg F to be selected. If the control switch is then set on either "low" "auto" or "high" the chosen blend will be maintained irrespective of conditions outside. Make no mistake about the power of the system either; the heater's output is 9 kw, sufficient for a three-bedroomed house, while the cooling capacity of the air conditioning unit, according to Rolls-Royce, is equivalent to 30 domestic fridges.

Countless little refinements are built into the system. For instance, the rear window demister operates automatically whenever heated air is being directed to the screen outlets.

MOTOR ROAD TEST NO 28/77 ● ROLLS-ROYCE

The system has 0°C as its datum temperature and any incoming air is cooled to that temperature before being re-heated, thus ensuring almost total dehumidification and the elimination of interior misting. The fans (there is one at each end of the scuttle) will not function until the temperature is 44 deg C or above. Finally, to make absolutely sure no misting occurs, the fans are subjected to a 13-sec delay whenever the system is re-started, just enough time for the air to re-freeze in the ducts and thus dry out. They have thought of everything for this piece of world-beating equipment.

In practice the system is as efficient as it sounds, and is the only automatic one we know which really can be left entirely to its own devices. Our only complaint: there is no manual override for the fans, which would be useful at times.

VENTILATION

★★★
★★

In the centre of the facia there are two eye-ball vents, the flow through which is governed by the automatic air conditioning system. Below them is a pair of chip-cutter grilles the air for which is controlled by a manually operated flap. Cold air is fed to all three vents whenever the upper temperature control is set to cool the interior. Like the heating, the ventilation is extremely efficient and we could not fault it.

NOISE

★★★
★★

There are a few (very few) quieter cars than the Silver Shadow but its cosseted passengers will never fail to be impressed by their insulation from the outside world. Wind noise is imperceptible below 50 mph and even at 100 mph it is more of a distant whoosh than a roar. The uncannily smooth engine is barely audible at tickover and its distant hum is well suppressed on part throttle, whatever the speed. Only under hard acceleration or at high revs does it take on a more purposeful note — and it is then that the Jaguar V12 is notably superior.

Tyres of 235 section make a big footprint but also impart muffled thumps as they pass over cats'-eyes and man-hole covers; occasionally they also cause a trace of vibration. Still quieter sounds come from the air-conditioning as its boost is automatically brought into play or one of the electric motors adjusts an internal flap, from the body when it utters a faint creak over heavily undulating surfaces, and finally from the clock. Yes, the Kienzle clock as it intermittently draws current to keep itself fully wound.

FINISH

★★★
★★

Before judging the quality of a Rolls-Royce it is necessary to adjust your standard accordingly. There's no point in searching for ill-fitting panels, shoddy stitching or poor fitting carpets, for such flaws would never pass the Crewe inspectorate. At first, even the most critical eye will be side-tracked by the sheen on

PERFORMANCE

CONDITIONS
Weather — Wet and overcast; wind 0-22 mph
Temperature — 47°-50°F
Barometer — 29.3 in Hg
Surface — Damp tarmacadam

MAXIMUM SPEEDS
	mph	kph
Banked Circuit	112.9	181.6
Best ¼ mile	113.9	183.2
Terminal Speeds:		
at ¼ mile	76	122
at kilometre	96	154
Speed in gears		
1st	40	64
2nd	74	119

ACCELERATION FROM REST
mph	sec	kph	sec
0-30	3.5	0-40	2.8
0-40	5.3	0-60	4.7
0-50	7.6	0-80	7.5
0-60	10.7	0-100	11.2
0-70	14.4	0-120	17.0
0-80	19.7	0-140	24.0
0-90	26.3	0-160	36.4
0-100	37.1		
Stand'g ¼	17.5	Stand'g km	32.3

ACCELERATION IN KICKDOWN
mph	sec	kph	sec
20-40	3.4	40-60	1.9
30-50	4.1	60-80	2.8
40-60	5.4	80-100	3.7
50-70	6.8	100-120	5.8
60-80	9.0	120-140	7.0
70-90	11.9	140-160	12.4
80-100	17.4		

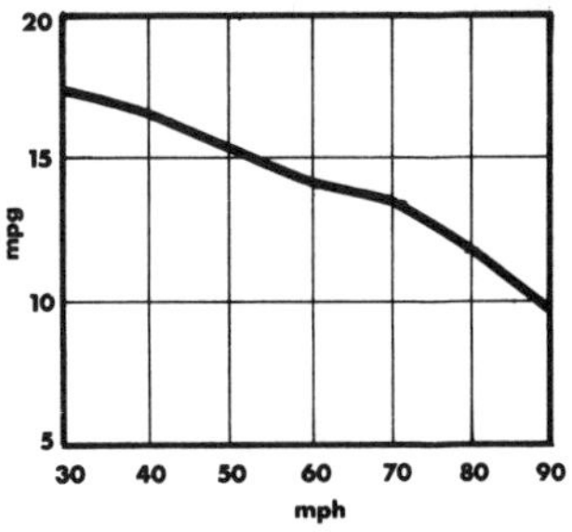

FUEL CONSUMPTION
Touring* — 12.5 mpg / 22.6 litres/100 km
Overall — 11.1 mpg / 27.4 litres/100 km
Fuel grade — 98 octane / 4 star rating
Tank capacity — 23.5 galls / 106.8 litres
Max range — 293.8 miles / 472.7 km
Test distance — 1250 miles / 2011 km
*Consumption midway between 30 mph and maximum less 5 per cent for acceleration

BRAKES
Pedal pressure deceleration and stopping distance from 30 mph (48 kph)
lb	kg	g	ft	m
25	11	0.56	54	16
50	23	1.00+	30	9
Handbrake		0.40	75	23

FADE
20 ½g stops at 1 min intervals from speed midway between 40 mph (64 kph) and maximum (76 mph, 122 kph)
	lb	kg
Pedal force at start	25	11
Pedal force at 10th stop	30	13
Pedal force at 20th stop	30	13

STEERING
Turning circle between kerbs
	ft	m
left	34.1	10.4
right	36.0	10.9
Lock to lock	3.5 turns	
50ft diam. circle	1.0 turns	

SPEEDOMETER (mph)
Speedo	30	40	50	60	70	80	90
True mph	28	38	48	58	68	78	88

Distance recorder: 1.0 per cent fast

WEIGHT
	cwt	kg
Unladen weight*	42.6	2164.2
Weight as tested	46.3	2352.1

*with fuel for approx 50 miles

Performance tests carried out by Motor's staff at the Motor Industry Research Association proving ground, Lindley.

Test Data: World Copyright reserved; no unauthorised reproduction in whole or part.

GENERAL SPECIFICATION

ENGINE
Cylinders — V 8; 90°
Capacity — 6750 cc (412 cu in)
Bore/stroke — 104·1/99·1 mm (4.1/3.9 in)
Cooling — Water
Block — Aluminium alloy
Head — Aluminium alloy
Valves — Ohv, push rods, hydraulic tappets
Valve timing
  inlet opens — 11° btdc
  inlet closes — 76° abdc
  ex opens — 53° bbdc
  ex closes — 34° atdc
Compression — 8.0:1
Carburettor — Twin SU H1F7
Bearings — 5 main
Fuel pump — Twin SU electric
Max power — Not disclosed
Max torque — Not disclosed

TRANSMISSION
Type — Three-speed automatic, GM type 400 with torque convertor
Internal ratios and mph/1000 rpm
  Top — 1.0:1/26.2
  2nd — 1.5:1/17.5
  1st — 2.5:1/10.5
  Rev — 2.0:1
Final drive — 3.08:1; Hypoid spiral

BODY/CHASSIS
Construction — Unitary steel monocoque with alloy panels
Protection — Zinc-coated steel in vulnerable areas and bitumastic under sealing

SUSPENSION
Front — Ind. by wishbones, coils, anti-roll bar
Rear — Ind. by semi-trailing arms, coils, anti-roll bar with automatic height control

STEERING
Type — Rack and pinion
Assistance — Yes
Toe in — 12' ± 5'
Camber — 0·5° ±15'
Castor — 2°30' – 3°30'
King pin — 11°30'

BRAKES
Type — Discs all round, ventilated at front
Servo — Two high pressure pumps supplying hydraulic reservoirs
Circuit — Triple: two high pressure, one direct acting
Rear valve — Yes
Adjustment — Automatic

WHEELS
Type — Pressed steel, 6JK × 15in
Tyres — Avon radial 235/70 HR 15
Pressures — 24 psi front; 28 psi rear

ELECTRICAL
Battery — 12V, 71Ah
Polarity — Negative earth
Generator — Alternator
Fuses — 20
Headlights — 4 × sealed beam, 50W (inner) 37½/60W (outer)

IN SERVICE

GUARANTEE
Duration...3 years or 50,000 miles for mechanical parts, 1 year for coachwork

MAINTENANCE
Schedule ....Every 6000 miles
First service ....at 3000 miles
DO-IT-YOURSELF
Sump ....14.5 pints, SAE 10W/50
Gearbox ....18.6 pints, Dexron
Rear axle........4.5 pints, SAE 90 EP
Steering gear .... 1·3 pints, Dekron
Coolant ....28.5 pints
Chassis lubrication ....Four points every 12,000 miles
Spark plug type ...... Champion RN 14Y
Spark plug gap ....0.030 in
Tappets ....Hydraulic, self-adjusting

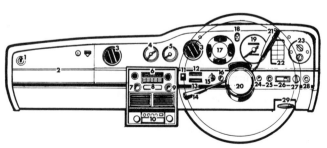

1 map light
2 glove locker
3 air vents
4 clock
5 ambient temperature gauge
6 radio
7 speaker balance
8 seat belt warning light
9 cigar lighter
10 quadrophonic tape-player
11 aerial lifter
12 upper-level temperature control
13 lower-lever temperature control
14 indicator/wash-wipe/ headlamp flash
15 air-conditioning control
16 hazard flashers
17 ammeter/temp, fuel and oil pressure gauges
18 main beam tell-tale
19 speedometer/trip and total mileage recorders
20 horn
21 gear selector/ speed-hold control
22 warning lights
23 light switch-ignition switch
24 fuel-filler release
25 panel light rheostat
26 rear fog light switch
27 wiper control
28 oil-level indicator button
29 handbrake

Make: Rolls-Royce
Model: Silver Shadow II
Maker: Rolls-Royce Motors Ltd, Crewe, Cheshire, CW1 3PL
Price: £19,495 plus £1624.58 car tax plus £1689.57 VAT giving total as tested of £22,809.15

Star rating guide

★★★
★★ excellent

★★
★★ good

★★
★ average

★
 poor

 bad

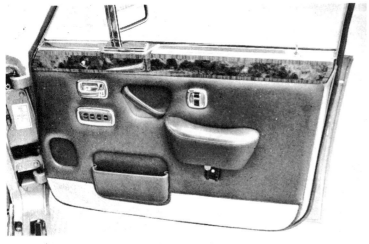

The driver has complete control, with electric window lifters, electric door lock and mirror "joy-stick" at his finger tips. Even his armrest is adjustable

the Italian burr-walnut facia. Even the most blasé will appreciate the expanse and bouquet of unmarked hide trim, the lamb's wool rugs and for that matter the beautiful short-pile carpet they protect. Even the positive, rewarding way in which the controls operate has to be experienced to be appreciated. Indeed the quality is more than skin deep, and detail inspection of the fully trimmed boot or highly finished engine compartments, where plating and enamel abound, will reveal the same degree of attention to detail and eye-pleasing finish.

But there were imperfections on our test car. Glue was oozing from the sun visors, and from the driving seat ripples were clearly visible on the upper surfaces of the long steel wings. There was also a blemish on the foremost fold of the classic radiator and cracks in the rubber surround of the rear window. Nevertheless, the standard of finish and craftsmanship is unsurpassed by anything else.

EQUIPMENT

★★★★★ To say the Silver Shadow is well equipped is a gross understatement. For your £22,809 you might well expect automatic transmission, air conditioning (in this case the unique Rolls-Royce split-level variety), electrically operated tinted windows, a heated rear pane, a clock, reversing lights as well as courtesy lights for bonnet and boot and remote control door mirror. But such items represent a fraction of the useful, and frequently pleasure-giving, devices included as standard on Shadow II. Listing electrical gadgets alone we must mention the front seat joy sticks, the speed-hold, the retractable radio aerial, buttons for remote release of petrol filler flap and boot catch, and the master lock which instantly secures all four doors and the boot lid.

Each door has an arm rest, those on the front being adjustable for height. In addition, rear passengers have grab handles and coat hooks within grasp. All three passenger seats are serviced by individual spot lights while a seven-second delay is built into the automatic courtesy ones.

Between them the Blaupunkt FM/AM radio and the fabulous quadraphonic cartridge player should keep even the most musical of ears satisfied, while smokers will be in their element with a cigar-sized ashtray on the console and one each for the driver and rear seat passengers. There are also three lighters too — one on the facia and two in the back. Vanity mirrors are incorporated in the rear pillars and on the reverse of the passenger sun-visor. Back seat passengers are provided with headrests and footrests.

IN SERVICE

Rolls-Royce expect to see your car after 3000 miles or three months, whichever comes first. From there on oil changes dictate the service intervals of 6000 miles.

Though few owners will attempt their own maintenance it is encouraging to know that most under-bonnet components are reasonably accessible and that the layout is not as complex as it seems. Unlike most modern cars, however, the Silver Shadow has four points on the chassis that also demand periodic lubrication.

The care with which these cars are assembled and the plethora of warning systems that keep the driver informed of what's happening should make involuntary stops so unlikely that they can be disregarded. However, in such a case the neat tray of tools and spares could save the day. There are two feeler gauges, an adjustable spanner, an open-ender, a 2BA Allen key for removal of the mascot, a sump plug spanner, a screwdriver with interchangeable ends, a pair of pliers and a tyre pressure gauge. The spares include two flasher bulbs, two stop/tail bulbs, two side-lamp bulbs and one side marker bulb.

This kit is housed in a plastic tray and is mounted above the battery in the nearside of the boot. The battery itself is away from all heat and only requires attention twice a year. The jack, tommy bar and box spanner for wheel changes are found in the offside of the boot. The spare can be inflated without removing it by way of the rubber plug in the boot floor.

The Rolls-Royce guarantee all the Shadow's mechanical parts for three years or 50,000 miles and is testimony to their faith in the product. So, perhaps, is the inclusion of a tamper-proof odometer which will register up to 999,999 miles or kilometres.

The Rivals

ROLLS-ROYCE SILVER SHADOW II £22,809

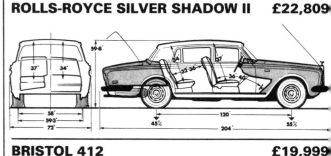

BRISTOL 412 £19,999

CADILLAC SEVILLE £14,888

VANDEN PLAS DAIMLER £13,629

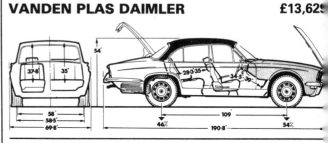

MERCEDES 450SEL 6.9 £22,999

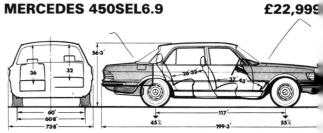

PANTHER DE VILLE £33,755

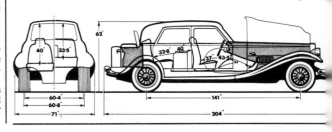

112

Press-on drivers will prefer the Mercedes 6.9, exhibitionists the stunning Panther De Ville, while those of lesser pocket could be more than satisfied by the Daimler/Jaguar XJ12

Comparisons

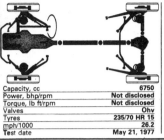

Capacity, cc	6750
Power, bhp/rpm	Not disclosed
Torque, lb ft/rpm	Not disclosed
Valves	Ohv
Tyres	235/70 HR 15
mph/1000	26.2
Test date	May 21, 1977

Numerous worthwhile changes are embodied in the MKII version of Rolls-Royce's best-seller. More responsive steering and improved suspension geometry have improved the handling and adhesion. Fully automatic air-conditioning consolidates already unmatched luxury. In short, the recipe is the same but the execution better than ever. There is no other car quite like it in the world.

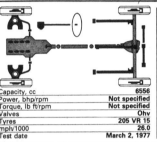

Capacity, cc	6556
Power, bhp/rpm	Not specified
Torque, lb ft/rpm	Not specified
Valves	Ohv
Tyres	205 VR 15
mph/1000	26.0
Test date	March 2, 1977

Aimed at the sporting owner/driver, the bespoke Bristol's key offerings are an excellence of manufacture and exclusivity — only 6000 have been built in 30 years. With its American V8 engine the 412 is a unique amalgam of stirring performance (achieved on two-star fuel) and the ultimate luxury of wood and leather that only the British understand. There is no shortage of customers for this rare car.

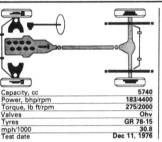

Capacity, cc	5740
Power, bhp/rpm	183/4400
Torque, lb ft/rpm	275/2000
Valves	Ohv
Tyres	GR 78-15
mph/1000	30.8
Test date	Dec 11, 1976

Small by American standards, large by European ones the Cadillac is very much a horse for an American course, comfort, quietness and smoothness being second to none on smooth to straight roads. It does not like corners, though, or indifferent surfaces. With its self-dipping lights, and chime to remind you to fasten seat belts, the Cadillac provides a novel experience for British drivers.

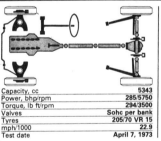

Capacity, cc	5343
Power, bhp/rpm	285/5750
Torque, lb ft/rpm	294/3500
Valves	Sohc per bank
Tyres	205/70 VR 15
mph/1000	22.9
Test date	April 7, 1973

The most luxurious version of the V12 Jaguar/Daimler series, the Vanden Plas offers opulence at an unbeatable price. Though lacking the craftsmanship and quality of its dearer rivals it sets the standard when it comes to performance and refinement. Excellent roadholding and handling, though we don't like the feel-less featherweight power steering. For those who can't afford bespoke motoring a V12 Jaguar or Daimler is the answer.

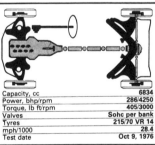

Capacity, cc	6834
Power, bhp/rpm	286/4250
Torque, lb ft/rpm	405/3000
Valves	Sohc per bank
Tyres	215/70 VR 14
mph/1000	28.4
Test date	Oct 9, 1976

Mercedes are the only company in their class that combine luxury motoring with truly sporting handling. With its superb power steering, exhilarating performance and uncannily roll-free, sure-footed handling the Mercedes doubles as a sports car and limousine. It does not cosset its occupants as well as a Rolls-Royce but whisks them along in style, comfort and utter safety. The car for the press-on driver.

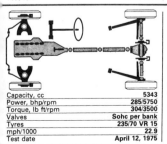

Capacity, cc	5343
Power, bhp/rpm	285/5750
Torque, lb ft/rpm	304/3500
Valves	Sohc per bank
Tyres	235/70 VR 15
mph/1000	22.9
Test date	April 12, 1975

A latterday, coach-built car styled loosely on the Bugatti Royale, the Panther causes a sensation wherever it goes. However, contemporary components provide the De Ville with performance, handling, and ride the Bugatti never knew and few modern rivals will ever match. When you buy a Panther you buy craftsmanship, opulence and above all exclusivity. That is why each V12 De Ville costs £33,755, or more!

PERFORMANCE

	Rolls-Royce	Bristol	Cadillac	Daimler	Mercedes	Panther
Max speed, mph	112.9	140.0†	103.2	135.7	131.0	‡
Max in 2nd	74	99	94	103	96	103
1st	40	58	56	62	60	62
0-60 mph, secs	10.7	7.4	12.6	7.4	9.3	9.5
30-50 mph in kickdown, secs	4.1	2.8	4.3	2.6	4.0	3.3
50-70 mph in kickdown, secs	6.8	4.3	8.4	3.7	5.0	5.2
Weight, cwt	42.6	33.8	39.3	34.8	34.6	38.3
Turning circle, ft*	35.1	‡	41.7	39.1	37.4	‡
50ft circle, turns	1.0	‡	1.1	1.0	1.0	‡
Boot capacity, cu.ft.	14.5	‡	8.0	11.8	15.0	‡

*mean of left and right †estimated ‡not measured

COSTS

	Rolls-Royce	Bristol	Cadillac	Daimler	Mercedes	Panther
Price, inc VAT & tax, £	22809	19999	14888	13629	22999	33755
Insurance group	7	7	7	7	7	7
Overall mpg	11.1	‡	12.4	11.5	13.4	12.1
Touring mpg	12.5	‡	‡	13.5	‡	‡
Fuel grade (stars)	4	2	2	4	4	4
Tank capacity, gals	23.5	18.0	16.0	20.0	21.1	22.0
Service interval, miles	6000	10000	6000	6000	5000	6000
Front brake pads £*	20.89	11.88	27.80	19.43	10.57	19.43
Oil filter, £*	4.32	4.51	6.87	4.32	5.16	2.48
Starter Motor, £*	82.08	71.55	151.83	55.16	229.27	62.58
Windscreen, £*	81.77**	194.40**	231.84**	47.52**	130.14**	124.64**

*inc VAT **Laminated ‡not measured

EQUIPMENT

	Rolls-Royce	Bristol	Cadillac	Daimler	Mercedes	Panther
Adjustable steering			●	●	●	●
Carpets	●	●	●	●	●	●
Central locking	●		●	●	●	●
Cigar lighter	●	●	●	●	●	●
Clock	●	●	●	●	●	●
Leather trim	●	●	●	●	●	●
Dipping mirror	●	●	●	●	●	●
Dual circuit brakes	●	●	●	●	●	●
Electric windows	●	●	●	●	●	●
Fresh air vents	●	●	●	●	●	●
Hazard flashers	●		●	●	●	●
Headlamp washers					●	
Head restraints					●	
Heated rear window	●	●	●	●	●	●
Laminated screen	●	●	●	●	●	●
Locker	●	●	●	●	●	●
Outside mirror	●	●	●	●	●	●
Petrol filler lock		●				
Radio	●		●	●	●	●
Rear central armrest	●		●	●	●	●
Rear wash/wipe						
Rev counter		●		●	●	●
Seat belts — front	●	●	●	●	●	●
— rear			●			
Seat recline						
Sliding roof						
Tinted glass	●		●	●	●	●
Windscreen wash/wipe	●	●		●	●	●
Wiper delay	●		●	●	●	●

CONCLUSIONS

TO MOST people, a Rolls-Royce is the ultimate status symbol: you can aspire to nothing higher. Until you drive or even ride in one it is difficult to appreciate quite how much lies behind the name. There are quieter cars, there are quicker cars but none cossets its occupants in quite the same manner, none insulates them from the outside world quite so effectively, to the point where rush-hour traffic jams, inclement weather and the other irritations of day-to-day living no longer seem to matter. On top of all this is the immeasurable pleasure offered by sheer unadulterated luxury of the sort that stems only from the use of the very best materials by the very best craftsmen. This is what the Silver Shadow is about.

In its latest form the Shadow, though no sports car, is much more enjoyable to drive than it used to be. The improvements to the steering and suspension have helped to make driving more fun and passengering more comfortable. The new automatic, split-level air conditioning system is a masterpiece, unrivalled by anything anyone else makes. Rolls-Royce say it took eight years to develop. Their time was not wasted. Inevitably, there are areas in which the Shadow can be (and is) surpassed by other cars, some of them much cheaper. But there is no other car in the world that combines all its qualities, or that could still be in its prime after 100,000 miles.

The scene is assorted individuals, three glorious days in June, two of the most expensive cars ever to come into our possession at the same time, and a stately home. Mike McCarthy played the millionaire and wrote the scenario, Peter Burn recorded the events for posterity onto celluloid, and Messrs Rolls-Royce and Panther provided the props.

Continued after colour

RICH THE TREASURE...
Continued

CROSSROADS has been kept going for weeks on thinner story lines. Somewhere there has to be a sparkling, witty, urbane piece about the fact that we have, in our possession, here and now, two cars worth between them roughly £53,000, give or take a pound or two. Books have been written about Rolls-Royces yet here we have not just a Rolls-Royce but a genuine honest-to-goodness Panther De Ville as well.

Obviously it would have to be about someone really, *really* stinkingly rich: the sort of person I admire. Say an Arab sheik (but I'm no good on desert scenes) or an industrial tycoon. That sounds more feasible: I could throw in words like 'Cartier', and 'Gucchi'.

The setting would be the annual meeting of the Amalgamated Global Finance Corporation, and its associated companies the Instant Damp-coarse Co Ltd, the Bessarabian-Poonah Far East Bank and the Shell-shock-a-go-go Niteclub, Harlesden.

The Chairman stands.

Crowd scenes of much booing and hissing. Cries of "Resign".

"Before I get on to the financial statements and audited accounts (more cries of "boo") I feel I must say a few words about the scurrilous and misleading attacks that have appeared in the media recently about me and my relationship with this — your, *our* — company, attacks compounded by insidious and libellous (snort from the company secretary) scandal from, of all people, my fellow directors who are even at this moment in time seeking my resignation.

"It is true that in the last few months the company has purchased two motor cars which to you may seem extravagant. But let us ask ourselves this: what is the purpose of a motor car? Is it simply to get from A to B? In this context the answer must be a resounding NO (more boos). Think, ladies and gentlemen. Without a Rolls-Royce, could I possibly hold my head up when I go to meet Sid Jones, the works convener, who drives a Jaguar? Let him put one over on me and *who knows where it may end?* If only our Prime Minister would use a Rolls-Royce the country would not be the hot-bed of seething left-wing anarchy it is today, and Wedgy Benn would still be a lord!"

"Many of you too are no doubt querying the necessity to provide a Panther de Ville for Miss Daisy De'l Eiffel, our Publicity Director and star of the Shell-shock-a-go-go Niteclub. What better publicity can there be, as she pointed out, than to arrive in something so splendid that bystanders fall off the pavement to see what the car is, and *who is in it?* And may I point out that it is in fact the economy model? We could have chosen the telephone and television as options, essential to this day and age for such a high-powered executive as her."

No, no, that line of thinking doesn't seem right. Sounds far too sleazy, and couldn't possible happen in real life. Perhaps something more in the Barbara Cartland-Virginia Holt style. Romantic, like.

"Basil stretched across, his Yves St Laurent silk jacket sleeve immaculate. He pressed in the cassette cartridge, and the full Pioneer quadraphonic sound boomed around the interior of the gliding Rolls-Royce, uninterrupted by even the merest whisper of engine, wind or road sounds. He gazed ahead, through the windscreen, across the patrician bonnet, past the proudly flying mascot and classic radiator, his steely grey eyes piercing the gloom of a Cote d'Azur evening. Patience, long, leggy and beautiful, cocooned beside him in Conolly hide, slipped her Charles Jordan shoes off and let her dainty toes luxuriate in the deep pile of the carpet, colour-keyed to the interior. Basil adjusted the upper control of the magnificent air-conditioning system, giving a trace more cooling to their faces, but leaving their nether extremities warm..."

No, no, this will never do. It sounds too much like a commercial for a cheap deodorant or something. Think, man, think.

This is getting more difficult than it seems. Dammit, the two cars are as different as chalk and cheese, or Flanagan and Allen, to name but four. And the sort of person who drives a Rolls-Royce would probably not be seen dead in a Panther, or vice quite likely versa. I mean, any story would have to be based around the individual attractions of each. How about something along the lines of millionaire's daughter with tyrannical daddy meets Punk Rock quick-fortune star?

Let's get back to that last line of thinking, about considering the cars individually. Consider the Rolls-Royce. Magnificent, of course, but more: subtle. The Rolls-Royce is definitely a *subtle* car. It may have sprouted a beard at the front, but is otherwise undated, and looks like the thoroughly engineered car it is: Mercedes and Jaguars have the same feeling about them. Solid — there's a good word, it quite definitely *is* solid — but delicate, if you know what I mean. Take the way the switches work: smooth, light, precise, yet definite. Dammit, even they've been engineered to perfection. Oh yes, and of course luxurious. Absolutely luxurious without being ostentatious. No flash or pizzas here. Seats that move — no, glide — this way and that, just by the merest movement of a little joystick on the centre console. And armrests adjustable for height, not to mention the individual ashtrays with their own lighters in the rear.

And in the back, of course, is where our heroine would have to sit. Just looking superior, which isn't difficult when you sit in the back of a Rolls-Royce. Occasionally she would flick her long blonde hair over the leather of the seat back, now and then glancing into the mirrors at each side to make sure — as if she needs reassurance — that she is as lovely as ever. She may even occasionally rest her arm against the highly polished veneer of the door cappings...

And wouldn't the dreaded up-tight Daddy (who would only ever buy a Rolls-Royce) flip his lid over Rick Dick Superstar's Panther Deville? But no, I'm being cruel to the owner of a Panther, because it is certainly not tasteless, which Punk Rock is. Bizarre, maybe, ostentatious of course, but it is in its own way as magnificent. Picasso versus Rubens, perhaps. So he would have to be someone who seeks attention since there is no way you can escape the stares, gawps, and astonishment of

Walnut, leather and deep pile carpets — both the Panther (above) and the Rolls-Royce (below) exude the richness and elegance of traditional British quality and craftmanship at its very best

Twenty cylinders shared between two cars should be more than adequate for the average sybarite. Twelve of them are a tight fit under the Panther's long and narro bonnet (above), while the eight of the Rolls-Royce are surrounded by ancilliaries..

ut seriously, though...

ANTHER DE VILLE £33,755

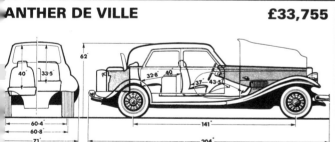
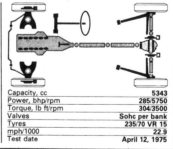

Capacity, cc	5343
Power, bhp/rpm	285/5750
Torque, lb ft/rpm	304/3500
Valves	Sohc per bank
Tyres	235/70 VR 15
mph/1000	22.9
Test date	April 12, 1975

A latterday, coach-built car styled loosely on the Bugatti Royale, the Panther causes a sensation wherever it goes. However, contemporary components provide the De Ville with performance, handling, and ride that the Bugatti never knew and few modern rivals will ever match. When you buy a Panther you buy craftsmanship, opulence and above all exclusivity. That is why each V12 De Ville costs £30,000, or more!

OLLS-ROYCE SILVER SHADOW II £22,809

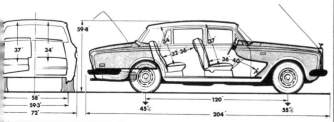

Capacity, cc	6750
Power, bhp/rpm	Not disclosed
Torque, lb ft/rpm	Not disclosed
Valves	Ohv
Tyres	235/70 HR 15
mph/1000	26.2
Test date	May 21, 1977

Numerous worthwhile changes are embodied in the MKII version of Rolls-Royce's best-seller. More responsive steering and improved suspension geometry have improved the handling and adhesion. Fully automatic air-conditioning consolidates already unmatched luxury. In short, the recipe is the same but the execution better than ever. There is no other car quite like it in the world.

PERFORMANCE	Rolls-Royce	Panther
Max speed, mph	112.9	‡
Max in 2nd	74	103
1st	40	62
0-60 mph, secs	10.7	9.5
30-50 mph in kickdown, secs	4.1	3.3
50-70 mph in kickdown, secs	6.8	5.2
Weight, cwt	42.6	38.3
Turning circle, ft*	35.1	‡
50ft circle, turns	1.0	‡
Boot capacity, cu.ft.	14.5	‡

*mean of left and right †estimated ‡not measured

COSTS	Rolls-Royce	Panther
Price, inc VAT & tax, £	22,809	33,755
Insurance group	7	7
Overall mpg	11.1	12.1
Touring mpg	12.5	‡
Fuel grade (stars)	4	4
Tank capacity, gals	23.5	22.0
Service interval, miles	6000	6000
Front brake pads £*	20.89	19.43
Oil filter, £*	4.32	2.48
Starter Motor, £*	82.08	62.58
Windscreen, £*	81.77**	124.64**

*inc VAT **Laminated ‡not measured

EQUIPMENT	Rolls-Royce	Panther
Adjustable steering		●
Carpets	●	●
Central locking	●	●
Cigar lighter	●	●
Clock	●	●
Leather trim	●	●
Dipping mirror	●	●
Dual circuit brakes	●	●
Electric windows	●	●
Fresh air vents	●	●
Hazard flashers	●	●
Headlamp washers		●
Head restraints		●
Heated rear window	●	●
Laminated screen	●	●
Locker	●	●
Outside mirror	●	●
Petrol filler lock	●	●
Radio	●	●
Rear central armrest	●	●
Rear wash/wipe		●
Rev counter		●
Seat belts — front	●	●
— rear		
Seat recline		●
Sliding roof		
Tinted glass	●	●
Windscreen wash/wipe	●	●
Wiper delay	●	

passers-by and yet, hopefully, also likes quality of the very highest order. If there's any justice in the world he would also be an enthusiastic driver who would make the most of the Panther's excellent Jaguar-derived handling and roadholding.

Hey! That's it! We'll make him a famous racing driver, with long blonde hair and blue eyes, known for being a bit of a scallywag and very much a ladies' man. *He* should appreciate good road manners, and the superb mechanical refinement which the Jaguar bits again endow this splendid machine. We'll also make him an ex-public school boy who would accept the leather upholstery, the fine wood veneers, the immaculate paintwork, and all the other little detailed touches that is his due as a superior being. He would be super trendy, of course, wearing tee-shirts to black-tie dinners and so forth, but this would be simply a sign of his up-to-the-minute thinking.

Right, we're getting there, slowly but surely. He would waft up beside her at a set of traffic lights. The front bumpers of the two cars would be level, but his long horseshoe-radiator tipped bonnet would place him fairly and squarely beside her as she gazed out of the rear window, wrapped in opulence and her own thoughts. He would catch her eye, and smile that famous disarming grin. She would simply raise an eyebrow. They would both know, instantly, that they were on the same financial plane but poles apart otherwise — and opposites attract.

Right: you take it on from there.

The magnificent house in the background on both the cover of this week's issue and in the central colour section is Clandon Park. It is one of the outstanding Palladian country houses in Britain, and was built about 1733 for the 2nd Lord Onslow by Giacomo Leoni. Among its very fine features are the white Marble Hall, the Palladio Room, the famous Gubbay Collection of porcelain, furniture and needlework, the Onslow family pictures and furniture and a unique collection of Chinese porcelain birds.

Clandon Park is on the A247 near Guildford, is open from the 2nd of April to the 16th of October daily (except Mondays and Fridays) from 2 pm to 6 pm. Admission is 70p, children and parties less. It is owned by the National Trust.

IN THE SHADOW OF ROLLS-ROYCE

I HAVE just enjoyed a week with a Rolls-Royce Silver Shadow II. It was indeed a very pleasant experience. Nevertheless, I do not intend to be drawn into discussing whether or not, in my opinion, this fine £24,248 6.7-litre V8 motor-car from Crewe is the finest car in the World. Either you think so, or you buy a Mercedes or a moped.

Nor do I propose to inflict on you a full road-test report on this Rolls-Royce, for the very good reason that it is not a new model. It represents a worthwhile development of the original Silver Shadow, in the manner in which Rolls-Royce engineers like to incorporate improvements as and when these are thought desirable; and I dealt rather comprehensively with that Rolls-Royce in MOTOR SPORT for May, 1968. Also, the Assistant Editor told you about the Shadow II in the issue of MOTOR SPORT dated April, 1977.

Writing about the Shadow I, I headed my report "Not So Much a Motor-Car, More A Way of Life". That impression remains, after experiencing this model in its latest form. It is so different from anything one normally drives, dignified, luxurious, yet reasonably unpretentious. It hurries, up to around 120 m.p.h., if you extend it. It will out-pace a lot of quite quick motor-cars, in the traffic-lights drag-race, if you drive it that way. But primarily it is a car to enjoy, rather than to fully open-up, a car to be driven with restraint and certainly not one to be kidded round corners. That way, it still melts those miles that can seem so tedious in lesser cars, because great hush, smooth-operating steering and controls, an automatic gearbox that alters the ratios scarcely perceptible, an air of top-quality all about you, and the sense of security that bulk and lofty seating conveys, combine to do just that.

There still seems a small impression remaining at R-R that the back-compartment occupants should be even more comfortable than those in front, who might just, after all, be chauffeur and

The Editor was naturally impressed by the Silver Shadow II, which bears a familiar number plate, a legacy from previous Rolls-Royce demonstration cars.

The Shadow II is distinguished from the rear by its substantial bumper bar, twin tail pipes and the rather cock-eyed model designation badge.

valet. This, though, only applies if you are very tall, or are exceptionally long in the legs, and it isn't really valid once the electrical multi-range adjustment of the front seats, each with its own inner arm-rest, and sliding outer arm-rests, has been mastered. Then only the width of the rather flat seat backs can be criticised, as they do not always quite retain small people on fast corners, – if you must drive a Shadow in this fashion. Otherwise, criticism is encouraged only because this is such a splendid and prestigious car. Some of those who travelled in it thought the highly-polished facia and veneered door surrounds did not look quite like genuine tree-wood because of its shiny surface.

I was more interested in the power-steering. From having a far-too-finger-light action, just like that of the XJ12 Jaguar, this now operates with a much nicer feel, while still calling for only a fore-finger and thumb on the thin rim of the now smaller-diameter (15½ in.) steering wheel. I was sorry to find a slight tremor, one could scarcely call it a vibration, at the steering wheel rim over some surfaces – the Dunlop tyres were fairly new and probably required re-balancing. The power disc brakes were most powerful and pleasantly progressive. The automatic choke made the engine run too fast for a time, after cold-starts. I was pleased to find an oil-gauge incorporated in the new four-needle communal-dial (this oil-gauge, and fuel-contents, engine-heat, and ammeter), even if the needle symmetry is spoilt because the oil-pressure one rises reassurringly higher than its opposite number, at speed. At idle the big engine really *is* inaudible, so that a tachometer would be useful. The important point to concentrate on is that there is little more sound at 100 m.p.h. than at, say, 60 m.p.h. – never mind the clock.

I had been so very impressed with the ride of the first Silver Shadow, achieved by self-levelling

the complex arrangements, that I think this has deteriorated very slightly on the Shadow II. There was some mild float over some road surfaces, you see. The car rolls quite a lot, too, if you corner it ambitiously but far off the limit – yet I must record that a passenger who felt car-sick in a fast-driven Rover 3500 was quite happy in the Rolls. Whether this was because I drove the latter more sedately than the tenaciously road-clinging car from Solihull, or is it a tribute to Rolls-Royce seats and suspension, I cannot say.

I have called the car a Rolls. But there are people, the late Laurence Pomeroy was one, who maintain that the real credit for these great British motor-cars is due to the Engineer and not to the Racing Driver, and thus refer to them as Royces. By the same token I suppose we should call the Shadow I a Grylls, the Silver Shadow II a John Hollings. . . .

I used none of those names when I went by firm-appointment to collect the test car from Hythe Road (the famous R-R Service Depot) one very wet Wednesday night and found that it wasn't available! On reflection, it was interesting to know that even the makers of "The Best Car in the World" sometimes slip-up. . . .

The next morning all was well. I was asked, quite rightly, to identify myself, in view of the "considerable value of the car you are taking away". I have been surprised, on the thousand or more occasions in the past when I have collected Press cars, how seldom this precaution is taken. But in this case, as I was leaving behind a brand-new Rover 3500 Editoral car (British at last!), I felt that 2/3rds. identity should have sufficed. . . .

Anyway, we were away at last and soon discovered that to all the well-known features which put a Hollings in a class of its own, can be added the new high and low air-conditioning system, controlled by outside sensors, the clever speed-hold control, the deliberate time-lag on the interior lamps after a door has been closed (giving the Duchess time to adjust her skirt), the moveable foot-rests in the back, just like those found in Edwardian drawing-rooms, the outside-temperature gauge and ice-warning lamp, and so on. Surprises were that there is but one key, that the doors lack a central external locking arrangement (although they can be thus-secured from within), that the horn-button is still in the steering-wheel centre, and that the fuel-filler flap cannot be released by the facia button unless the ignition switch is "fully-off", although the electric windows function without the ignition switch being "on". The quality of the stereo, for which only a test-cassette was provided, was approved of, as was the smooth electrical-selection of the ratios in the automatic gearbox and, later, the

The radiator mascot on this Rolls-Royce, above left, belonging to the Harding-Rolls family, relatives of the Hon. C. S. Rolls, is based on a piece of masonry on the parapet of the Hendre, the house where Rolls' parents, Lord and Lady Llangattock, lived, above right. The statue of Sir Henry Royce in its new location on the river bank in Derby. Rolls-Royce enthusiasts Club members assembled at the Hendre, below.

power of the 270/75 four-lamp lighting set, foot-dipped, by the way.

There might well be more fore-and-aft adjustment on the front seats and the sill-buttons for locking the back doors are so far back, as to be difficult to reach from the front seats. The luggage boot has plenty of capacity and all the locks function commendably smoothly.

As the RREC was having a rally to Monmouth to unveil a plaque on the town-square statue of the Hon. Charles Stewart Rolls, BA, FRGS AMIME, on the centenary of the birth of the famous motoring and aviating aristocrat, that is where I directed the Silver Shadow on August Bank Holiday Saturday. At the assembly on the playing fields of Monmouth Secondary Modern School the present-day Grylls and Hollings were segregated from the pre-war Royces. I found myself next to Dennis Miller-Williams, the popular R-R PRO, who was engaged in washing down (oh; these chauffeur-less days!) the Camargue in which he had driven from London with his wife and daughters. A friend borrowed his bucket, to rinse the Shadow. A stately drive with Police escort took us non-stop past the statue, through the grounds of Rockfield House residence of Lt. Col. Harding-Rolls, MC, DL, JP and on to the Hendre, where Charlie Rolls wa

120

brought up (although born in London) this being the mansion of his parents, Lord and Lady Llangattock.

Here we enjoyed an excellent lunch, and visited a most interesting museum of pictures, documents and other items relating to Rolls, Rolls-Royce and the Hendre (some of the items on display are normally housed in Monmouth Museum, which those interested would do well to visit). After watching an unsuccessful attempt, in the gusting wind, to launch a hot-air balloon, (intended to commemorate Rolls' great activity in this field, before he went on to aeroplanes and was killed at the 1910 Bournemouth Flying Meeting; the intrepid balloonists received applause for their brave efforts, although their craft remained tethered to a tree) we excused ourselves from a wine-party and left, being due at the VSCC race meeting at Cadwell Park on the following day. Incidentally, one of the exhibits in the aforesaid museum is a notebook kept by Rolls of work done on his aeroplane and it shows that about the only item not checked before the fatal Bournemouth flight was the tail-plane, which collapsed. With their present love of plaqueities, the RREC has discovered the spot where Rolls fell and has duly marked it for posterity....

The usual splendid cavalcade of Rolls-Royces attended this rally but not, alas, the replica of the Rolls balloon-carrying 40/50.

My wife and I had arranged to break our long journey half-way, in Derby. We found that, in spite of much traffic on the M50 and M5, so that the overtaking-lane was transformed into a commuter-stream, we had reached our destination in a couple of hours. But then, 70 m.p.h. in the silent majesty of a Silver Shadow tends to make the driver wonder why everyone in front of him is crawling.... At the Midland Hotel by the Station, in Derby, formerly the Rolls-Royce headquarters and also where the Auto Union team, including Nuvolari, stayed for the 1938 Donington G.P., we had to park the Rolls in an open yard, in company with but one other car, resident W. E. Harker's Alfasud. At dinner he regaled us with tales of his Austin Ulster and his V8 Harker Special and the next morning he came with us to the site of the statue to Sir Henry Royce, remarking that, although a R-R apprentice in the 1920s, it was years since he had ridden in a Rolls-Royce. This statue to the great Engineer was originally unveiled at the Arboretum in Derby, in 1923 on the occasion of a VIP visit to the Rolls-Royce factory. It was removed to this present, pleasant site on the river bank in 1972, being re-unveiled, as it were, by The Worshipful The Mayor Councillor, J. Carty.

Once upon a time a Rolls-Royce would have presumably been sacrosanct in Derby. Not so, today. We had parked the Silver Shadow in the deserted 'bus-station, while we photographed the Royce Memorial. Returning some five minutes later, we found an enormous red 'bus parked tight up against the Silver Shadow's boot, to hem us in, and a little man in a brown dust-coat leaping up and down at our audacity in invading the premises of his public-service vehicles. Brushing off his suggestion that we pay a £1 parking-fee, we returned Harker to the Midland and set off on our long Sunday morning haul to Cadwell Park. In the attractive town of Louth I again filled the petrol tank with 4-star, enabling a fuel-consumption check to be made. The traffic conditions had prevented me from driving hard, and the figure came out at exactly 15 m.p.g. Previously I had looked at the sump oil-level, using the very long, rapier-like dip-stick, to discover that Mr. Grylls'

The hot air balloon.

engine certainly doesn't suffer a craving for lubricant. You can get a rough idea of sump-level on a dashboard button, which reminds me that the switch-panel, ignition-key, dashboard-switches and so on, are reminiscent of those on former Rolls-Royce and Derby Bentley cars.

And that a young, Ford-owning Son-in-Law was impressed by the detail finish, the under-bonnet accessibility (except perhaps for some of the sparking-plugs), the under-sealing, and the overall air of quality about the Silver Shadow II – which *Punch* would, I suppose, have dismissed as a blinding glimpse of the obvious, yet is something which few writers about the Rolls-Royce can leave out.

The statue to the Hon. C. S. Rolls in Monmouth.
(Photo by C. W. Hughes)

R-REC Plaqueities. –This plaque will be found at Monmouth. There are others at Le Canadel and at Bournemouth.
(Photo by C. W. Hughes)

All in all, then, it was no hardship to drive back to Wales from Lincolnshire that evening, after my wife had presented the MOTOR SPORT Brooklands Memorial Trophy to Bernard Kain. We did it *sans* Motorways, via Shrewsbury, in five hours, without hurrying. That we were nearly back in "dry" Wales became evident when, diverted along remote country lanes near Clun, due to a bridge being under repair, we came upon a lighted village pub, with many many inmates drinking therein. "Let's have one," I said, not realising that it was a little after closing-time. The publican said a firm "no," refusing us even non-alcoholic drinks, which the driver had intended to have anyway. And he hadn't even seen the Rolls. On Bank Holiday evening, going out to dinner with friends, it was necessary to drive down some exceedingly narrow Radnorshire lanes, in which the big car felt as at home as it had on the Motorways. That about sums it up! Getting out of a Silver Shadow for the first time after a fast

The R-REC Parade past Rolls' memorial-statue on August Bank Holiday Saturday. Note the crowds. The car in the foreground is A. H. Biddiscombe's 1933 Schutter & Van Bakel 20/25 saloon. It is followed by B. R. O. Harris' 1926 Twenty Hooper landaulette. (Photo by C. W. Hughes)

journey its size surprises you and you marvel that you pushed it at speed through such narrow gaps – and that surely is the mark of a good big car, whether 1920 Hispano-Suiza or 1977 Rolls-Royce? The cubby hold lid shuts nicely on the Rolls, too (and it *will* hold a Rolleiflex!).

I must confess that immediately after I had returned this Shadow and got into the Rover 3500, the latter felt dreadful, *for the first few miles* – actually, when I have done a decent mileage in it, I propose to tell you why it is really a very good car – the Car of the Year, I am informed. –W.B.

> THIS TABLET WAS UNVEILED BY
> COLONEL E.R. HILL D.S.O. THE LORD LIEUTENANT OF GWENT
> ON THE 27TH AUGUST 1977 WHEN A LARGE NUMBER OF CARS BELONGING TO
> THE TWENTY GHOST CLUB AND THE ROLLS – ROYCE ENTHUSIASTS CLUB
> DROVE PAST THIS STATUE IN COMMEMORATION OF
> THE CENTENARY OF THE BIRTH OF CHARLES STEWART ROLLS

WEST HEAD IS ONLY A SILVER SHADOW 11 AWAY

The gold printed invitation heralded a new Rolls-Royce...
who better to send along than our resident chauffeur?

SIR, I'M not really sure that you would have enjoyed today. Perhaps it is just as well the yacht developed that engine trouble and you couldn't make it in time. However, this report should fill in enough detail for you to make a decision about the car.

As you know John H. Craig — he's the chief executive of Rolls-Royce Motors International, Pacific Market — invited some of us and, I believe, a contingent of journalists — at least that's what I think they were because they carried notebooks and there were some television cameras in evidence — to a special viewing at the Elanora Golf and Country Club, just outside Sydney.

The invitation didn't say why, but I knew it was for the Silver Shadow II that had only just been released in the United Kingdom. Well, they would hardly have required such an event for the Camargue, would they Sir?

The country club is quite nice really. A little middle-class if you don't mind me saying so, Sir, and I'd rather you didn't join. I mean $405 a year is hardly going to keep the riff-raff out, is it, and the car park was full of those dreadful little Japanese cars. Afterall the club was only established about 50 years ago; still, it does have an excellent view of the Pacific and I suppose it's about the best you can expect in a colony.

There were four new Silver Shadow IIs on hand but only three were registered and could be driven and those nasty journalists rather hogged them. I suppose you can't blame them. I earnestly believe most of them have never experienced Rolls-Royce motoring before. It showed in the quite shocking way in which they hurried out of the club and rushed towards the cars after we had been shown a quite soul-stirring film which in retrospect, I feel none of them appreciated. It took me back, Sir, to our days together in London when the Rolls-Royce-powered Spitfires knocked those terrible Germans out of the sky. It proved British is still best and that the Silver Shadow is the best of the best, if you get my meaning?

123

SILVER SHADOW 11

I've saved the literature they handed out and the rather impressive suede satchel, although I did notice with some disgust, that it was made in the United States. That wouldn't have happened in Sir Henry's day.

The drive the Rolls-Royce folk had arranged took us from Elanora to Commodore Heights, at West Head, in the Ku-ring-gai Chase. You might remember we took the Duke and Duchess there to show them the view down Pittwater a few years ago. You can almost see the roof of the Palm Beach summer house from there. Well, Sir, it struck me as hardly the kind of road suited to a Silver Shadow if you recollect that the dogs were sick in the back of the old Shadow that day. But, I suppose the Rolls-Royce people thought of that because they didn't have any dogs along.

Sir, I'm not sure I approve of some of the changes to the new model. There's something called an "air dam" under the front bumper bar. I'm told it improves high speed stability and reduces wander in side winds but it doesn't look right. They have also stuck a large and undignified badge on the boot lid that is clearly out of character with the subtlety of the rest of the car. Well, apart from that damn air dam — excuse me. They are both blantly crude in the extreme.

I was also surprised to notice that the bumper bars no longer have overriders but are rubber-tipped. It makes me question their entire attitude to we chauffeurs. Don't they think we can drive carefully any more?

But I must say the dashboard looks neater, although there are still plenty of controls and switches and buttons. Sir, it probably would be better if you let me explain how things work before you take one for a drive.

The steering wheel is smaller in diameter and I approve of its black plastic rim. None of this soft-rimmed nonsense for a Rolls-Royce. Actually, I had to spend some time reading the handbook to work out how to operate the air-conditioning system. It's the same system as in the Camargue and works automatically once you comprehend the controls. It really is magnificent, Sir.

You can select any temperature between 63 and 91 degrees F and there are two temperature controls, one for the upper level and another for the lower level, and it maintains the required blend no matter how the outside temperature might change. I'm told it has the cooling capacity of 30 refrigerators and enough heating ability for a three bedroom house.

I was rather surprised, at the time, at the attitude of the writers but I now realise they just don't know any better.

SILVER SHADOW 11

I find this hard to believe but I really think they consider the Silver Shadow is just a car. Well, Sir, when it became apparent that they were going to monopolise the cars for the entire afternoon my only course of action was to quietly find myself a position in the back seat during one of the runs to West Head.

At first I thought everything was going to be conducted in a proper fashion. The discussion in the car centered around whether or not the steering really was heavier and had more road feel. I might add that I didn't feel required to become engaged in the conversation even though I could have confirmed their opinion on the steering. I even began to suspect that one or more of them may have driven the old model.

Anyway this rather pompous youth — well, he can't have been more than 35 — carried on at length about how the steering definitely was more accurate and precise. He even knew there had been a change from recirculating ball to power-assisted rack and pinion steering.

I had to wonder, though, at the standard's of today's young people. They didn't seem at all over-awed by the car. They claim to have appreciated its smoothness and quietness and the opulence but there wasn't the respect that surely belongs to a Rolls-Royce. As I said, I believed everything was going to be tolerable. Until one of them, the pompous fellow, started telling the entire car that this was his regular way home and that he intended seeing what the Silver Shadow II was all about. As if it was a Mercedes-Benz or something equally as commercial.

I was scared, and don't mind admitting it, after he claimed to have run out of brakes in a Ford on the same road and you know how it twists and turns and snakes its way to West Head.

I just cringed and grasped one of the handles and wondered what I had done to deserve such a fate. Oh, it was horrible Sir. The fool must have thought he was Mike Hawthorn or Stirling Moss. I'm sure he actually gave the car full power from a standing start because I've never heard a Rolls-Royce make so much engine noise. He flung it around corners, commenting all the way about how it still felt big and heavy, even if the body roll had been reduced, and that it still required a delicate touch if you weren't to get the car swaying, even lurching from side to side. Then he had the temerity to compare it unfavorably with a Jaguar XJ12 which he said was quieter and easier to drive quickly.

Mind you, he did praise the brakes but only after filling the car with the most repugnant smell. Then the other chaps decided it was their turn for which I was grateful, and I must add that I was pleased the national park authorites had been notified that we were coming through for we didn't have to stop at the toll. The wave was suitably discreet. Just as well, for I'm sure none of those people would have had any money.

That, I thought, is that. The wild man was now in the front passenger's seat. Except that on the way back he decided he wanted some photographs of the Silver Shadow going around a particularly tight corner. I need hardly add that I, and I might say the other occupants, very quickly left the car.

Sir, you would have considered it reprehensible. Wheel spin in a Rolls-Royce and tyre squeal. In all honesty I must admit that it didn't produce the kind of tyre scrub and moaning noises that the old Silver Shadow did when you tried to avoid that truck a few years back. It actually looked very stable, but the sight of a new $63,500 Silver Shadow II being driven — I think the term used by one of the other observers was "opposite lock" — in this manner quite appalled me.

Thankfully, he finished this madness without crashing the car, which is a tribute to British engineering. The other people were content to fiddle with the speed control while that uncouth youth made snide remarks about the fact that the new Silver Shadow has a German Blaupunkt radio and Japanese Pioneer quadraphonic cartridge system. I must admit he has a point.

Despite this they all agreed that the Rolls-Royce was still superbly built and cossetted them in rich and comfortable surroundings. But please, Sir, the next time such an invitation comes along couldn't we send the junior chauffeur? I don't think my sensibilities could take another battering.

And the sooner we get a new Silver Shadow II the better, just so that I can prove to myself that it really is a dignified form of transport. *

road test
by John Bolster

There are other cars which are more exciting, but they are rich men's toys of which one soon tires.

Incomparable engineering

When good King Edward VII was on the throne, the Rolls-Royce was first called 'The Best Car in the World'. That seemed a bold claim when the Delaunay-Belleville was there to contest it, followed by the Hispano-Suiza and other wonderful cars. Yet the name soon entered the language, or rather the languages of many countries, as the symbol of superexcellence. Almost incredibly, in a world of changing values, the magic of the name is just as potent as it was 70 years ago; but as for the Delaunay, the Hispano, and all the others . . . *où sont les neiges d'antan?*

Like the snows of yester-year, they have vanished, not because they lacked quality but because their manufacturers failed to keep up with public taste. Henry Royce built cars to a standard of engineering that was incomparable, but today even that would not be enough. The Silver Shadow II is made with the same meticulous care as Royce lavished upon his Silver Ghost, but it is because its character fits in so well with the requirements of the modern owner, that it has become by far the best-seller of all the Rolls-Royce models ever made.

There are other cars which are much more exciting, but they are rich men's toys of which one soon tires. Even after only a few day's 'ownership', the Shadow becomes so much a part of life that it is hard to live without it.

Though the Silver Shadow is more compact than most of the Rolls-Royces that preceded it, a car weighing two tons that is 17ft long can scarcely be described as small. Much of its attraction comes from its lack of sporting pretensions; a two-ton car can be a magnificent luxury carriage but if it tries to ape competition machinery it becomes absurd. Personally, I prefer the four-door model to the Corniche and the Camargue and if I were buying a Royce, I'd have the Silver Wraith II, which is the long-chassis version; when one owns a magic carpet, I feel that it is best enjoyed in the company of one's friends.

The introduction of the Silver Shadow in 1965 marked a break with tradition in several respects, for it was the first Rolls-Royce with a monocoque body shell, instead of having a separate chassis frame with the coachwork resting upon it. This, and the deletion of the live rear axle, permitted the construction of a much lower car with increased interior space. It was at a time when more and more Rolls-Royce owners were finding it convenient to take the wheel themselves, and the Shadow proved to be a far better owner-driver's car than the stately carriages that preceded it.

The many subsequent improvements to the car have been largely aimed towards increasing the driver's enjoyment. In introducing the Silver Shadow II, the manufacturers have gone much further in this direction, which the new rack and pinion steering emphasizes with its considerably smaller wheel. The air dam beneath the radiator, for improved high-speed stability, suggests that the happy owner will habitually motor in a manner that 'James' was not encouraged to emulate in the chauffeur-driven era.

Similarly, the suspension geometry has been modi-

The finish is, of course, superb . . .

fied to give increased cornering power and less understeer and the many controls and switches have been laid out for easier identification and operation. You may still send 'your man' to the Rolls-Royce school on a course which takes a fortnight, but you can now jump straight into the car and drive off, without having to acquire any of the special skills that were needed to operate a Ghost or Phantom. Indeed, the Shadow II is probably the easiest car on the road to drive, after a fashion, though it is infinitely rewarding to the accomplished driver who has hands, to use an equestrian phrase.

The car has many refinements which no other vehicle possesses, such as a heating and aircondition-

road test

ing system that took eight years to perfect, of which more later. These things are certainly of great complexity to the layman, but the basic engineering of the machine is straightforward, as has been the case with all Rolls-Royce designs. The belt-and-braces philosophy of Henry Royce has been adopted, which renders any serious failure virtually impossible, but the main components are actually simpler than their predecessors. For example, the big V8 engine would cost less to overhaul than the the previous six-cylinder unit.

Some people have suggested that the Silver Shadow ought to have a V12 engine, in the Phantom III tradition. Such engineering would be attractive to the purist but, at the present stage of the art, a big V8 with lots of low-speed torque is more suitable for use with automatic transmission than a higher-revving and less torquey V12. Even when the power unit and transmission are perfectly matched, the engine that will pull the highest gear and will need the fewest gearchanges is the one which will prove the least tiring to the passengers on a long journey.

Behind the wheel

Entering the Shadow, you notice the lightness of the door, for any part that may be moved by the owner — including the boot lid and the bonnet — answers to a touch. The seats, Connolly leather of course, may not give the location of a rally-type bucket, but they are very easy to slip into or leave, without effort. Breathes there the man with soul so dead, that the view down the long bonnet, with the silver lady upon her Palladian perch, does not cause an anticipatory thrill? On a less poetic plane, though the forward vision is unrivalled it's difficult to judge the exact length of the boot when reversing.

The engine starts instantly on the key and though the starter pinion is more audible than the chain-drive of the 1920s, one can scarcely hear it idling. The selector lever of the transmission is an electric switch, calling for no effort, and the car moves off without hesitation, cold or hot. Whereas I tended to override the earlier transmission manually, the present arrangement is so effective that I never did so, the acceleration with automatic operation proving just as rapid as with the hand lever. The changes, up or down, are scarcely perceptible.

Most cars have some lost motion in the accelerator pedal but the Shadow has none, the merest touch sending the big machine into rapid action, for the acceleration can be quite fierce. After a few minutes, one learns to graduate the pressure of the foot and the car will then glide away in the manner of the best chauffeurs, but that astonishing acceleration is always in reserve, called into play by a little more pressure of the right foot.

The transmission is completely silent and the engine
By far the best-seller of all the Rolls-Royce models ever made.

is very quiet, while such sound as it makes is at a slow tempo, for it does most of its work at comparatively low revs. Though it is possible to reach about 75mph in second gear, top gear is engaged at a much lower speed in ordinary driving. One seldom knows which gear the car is in, but the changes are so smooth that it is a matter of indifference; even the kick-down is smooth, though the acceleration is very rapid indeed thereafter.

This Rolls-Royce seems particularly lively in the 70 to 100mph range, and as there is virtually no increase of sound at the higher speed, it behoves one to glance frequently at the speedometer. If it is desired to stick rigidly to the speed limit, it is perhaps best to make use of the automatic speed hold, the controls for which are combined with the gearlever.

A remarkably quick getaway can be made with very little wheelspin, thanks to the excellent traction of the huge Avon tyres on wet roads or dry. They also allow full use to be made of the immensely powerful brakes and provide cornering power which is somewhat unexpected with such a heavy vehicle. Even Avon have not completely solved the problem of road noise, always more difficult with very large tyres, but the Silver Shadow II is noticeably better in this respect than its predecessors. Though the steering is light, it contrives to give the driver some feel of the road and the new rack and pinion offers quicker action with less wheel winding; the Silver Shadow has always had a much smaller turning circle than earlier Rolls-Royces.

The soft suspension has a very long travel and this is not reduced when the car is heavily laden, thanks to self-levelling under Citroën patents. The ride is always very comfortable and though there is a floating sensation at speed, this does not affect the roadholding, the tyres remaining glued to the road. In order to exploit the higher cornering power of the new tyres, some changes have been made to the suspension geometry, noticeably reducing the angle of roll during fast cornering, as well as giving less understeer.

In one respect, the Silver Shadow II is so far ahead of all other cars that no comparison is possible. The air conditioning system is extremely sophisticated, yet it could not be simpler to operate; one simply preselects the required temperatures in the upper and lower parts of the car, refrigerating and heating simultaneously if desired, and a computer does the rest automatically. Whatever the required temperature, the incoming air is first brought down to freezing temperature to remove humidity, to avoid mist on screen or windows. On first switching on, there is a delay of 13 seconds before the fans start, to avoid unfrozen air entering the ducts, and the switching of the heated rear window also occurs automatically when misting is possible.

Once the system is set, it is entirely independent of ambient temperature and totally automatic. A dial on the instrument panel does tell the driver what is going on outside, however, reinforced by a warning lamp if

The exquisitely trimmed luggage boot, and the engine bay with its 6.7 V8 buried deep down.

freezing conditions are experienced. There is a whole battery of well identified warning lights and, in the unthinkable occurrence of overheating or low oil pressure, not only instrument dials and warning lights, but also an audible buzzer, would go into action.

The finish is, of course, superb and I think that the quality of the wood veneer of the test car was even better than on the many previous Rolls-Royces I have driven. It might be thought that the high price cannot be justified, but this is about the only car that one can be sure of selling at a profit, so it must be a good investment. Don't forget that a Rolls-Royce with 100,000 miles on the clock is considered to be a low-mileage example, the speedometer recording up to 999,999 miles.

Regarded as a status symbol by some, the Rolls-Royce car is far more than that. It provides a standard of comfort for all its occupants that no other wheeled vehicle can approach and a performance which, like its secret power output, is 'sufficient'.

Specification and performance data

Car tested: Rolls-Royce Silver Shadow II 4-door saloon, price £26,740.
Engine: V8 104.1×99.1mm (6750cc). Compression ratio 8 to 1. Pushrod-operated overhead valves with hydraulic adjustment. Opus electronic ignition. Twin SU HIF7 carburettors.
Transmission: Fluid torque converter and 3-speed automatic gearbox, ratios: 1.0, 1.5, and 2.5 to 1. Hypoid level final drive, ratio 3.08 to 1.
Chassis: Combined steel body and chassis with aluminium panels for boot, bonnet, and doors. Independent front suspension by wishbones, coil springs, and anti-roll bar. Power-assisted rack and pinion steering. Independent rear suspension by semi-trailing arms, coil springs with automatic self-levelling, and anti-roll bar. Disc brakes all round, ventilated in front, with two twin-cylinder calipers for each front wheel and one four-cylinder caliper for each rear wheel, operated by dual circuits, each with its own high-pressure pumps and hydraulic accumulator. Bolt-on steel wheels, fitted Avon 235/70 HR 15 tyres.
Equipment: 12-volt lighting and starting with front and rear fog lights and reversing lights. Speedometer. Ammeter. Fuel and oil level, oil pressure, and water temperature gauges. Outside air temperature gauge. Clock. Two-level air conditioning system with automatic operation. Electrically-operated door windows. Automatic speed control. Electrically-adjusted seats. 2-speed and intermittent windscreen wipers and washers. Central door locking. Flashing direction indicators with hazard warning. Quadrophonic radio/tape player. Headlamp washers and wipers.
Dimensions: Wheelbase 10ft 0½in. Track 5ft 0in/4ft 11¼in. Overall length 17ft 0½in. Width 5ft 11¼in. Weight 2 tons 2½cwt.
Performance: Maximum speed 117mph. Standing quarter-mile 17.3s. Acceleration: 0-30mph 3.3s, 0-50mph 7.5s, 0-60mph 10.3s, 0-80mph 19.5s, 0-100mph 35.5s.
Fuel Consumption: 11 to 15mpg.

Not the peerless performer some imagine, the
Rolls-Royce Shadow II is still . . .

THE STANDARD

Rolls-Royces have been described as everything from
"the best cars in the world" to "exorbitantly dreary".
WHEELS, after driving the Silver Shadow II and
watching how it is built, feels that the wide discrepancy
in criticism is because the critics tend to judge what the
car *stands for,* not what it *is.* We say the Shadow II is
one of the best cars in the world.

THE NEXT person who buys a Rolls-Royce Silver Shadow II and expects to get $80,000 worth of hard-nosed value for money, will probably be the first.

Admittedly the exchange *includes* a large luxury car, faintly old-fashioned in style and layout but superbly trimmed, assembled and equipped, a car which many people call The Best Car In The World. But there is a lot more to it than that. The Rolls owner also acquires automatic licence to be seen as rich, discerning and at least a bit aristocratic. He might even be famous. Certainly he is a member of a 74-year line of people which the community has come to imagine is pre-eminent.

The detached observer would say that the second part of the deal is what attracts most buyers.

Consider this. If an equivalent of the Rolls Shadow appeared in modern, visible show-rooms in our capital cities today — a car equal in quality of manufacture, opulence, size, performance and price but bearing the name Wartburg — would it sell? No chance. Rolls owners would pretend not to notice or never to have heard of the interloper. Jaguar and Mercedes-Benz drivers would sneer as they swept by with a low growl of V12 or V8, Volvo owners would lecture one another about the excesses of the new car and Holden/Toyota owners would rapidly hatch and spread the rumor that the "Red lead sleds" tended to break in half at 10,000 kilometres.

In short order, examples of our Wartburg Pinko Shadow II would be found on the second-to-front row of Kevin Dennis or John □ lineups with "$25,000 OFF" slashed across their screens. Rolls sales would not be af-fected one jot.

The bare bones definition of a Rolls-Royce Silver Shadow II is a large, expensive, fairly old-fashioned luxury car. If a Mini van is a small, cheap, basic delivery vehicle then a Rolls is "a large, expensive fairly old-fashioned luxury car". The Shadow was the car WHEELS used recently to travel to Crewe, north of London, to visit the Rolls assembly works and to meet the company's managing director, David Plastow.

Though a current model, it wasn't a new Rolls or even a near-new one. The bronze-colored car we drove had covered nearly 50,000 miles (80,000 kilometres) and it was at least 18 months old. It was produced for the launch of the Shadow II and according to the Rolls man it had become something of a hack. "Usually we don't keep them this long," he told us, "but a few stay on and get used by everyone. We call them trials cars."

It is not often that motoring journalists are given 80,000 kilometre-old cars for road test. Most cars that "have been driven by everyone" are breathless and saggy-kneed by 15,000 kilometres. The Rolls had some scratches on the door kickplates and some wear on the front carpets as evidence of its many drivers, but apart from that there was simply nothing wrong with it. Not that we could see.

Our "Roller" — even the PR men call them that — was strictly the base model. The Bel-mont. Admittedly it was as completely equipped as any of them, but nothing changes the fact that the car we were driv-ing is only about half as expensive as the top car Camargue (price here $147,000) and that's not discussing the Phantom VI limou-sine, one of which they recently gave the Queen.

The models go like this: Silver Shadow II,

Silver Wraith (a Shadow with an extra four inches of body inserted behind the front seats), the Corniche (the two-door which resembles the Shadow front and rear and comes in either saloon or convertible form); then comes the Camargue which owes little in appearance to the lower models. That is the two-door "personal car" which is extremely luxurious but is known mostly for its extraordinarily high price tag. The limousine, Phantom VI is a "traditional" car made for processions with dignitaries aboard. It's a "Queen's chariot" as the Rolls man said. You don't even speculate about its price.

Cheap it may be, but the Shadow is Rolls-Royce's most successful car ever. It appeared first in 1965 and appears likely to be around for a few more years yet. While we were at Crewe — in fact as we drove in the front gate quite early on a summer morning — we were passed by a silver-painted prototype of a car which we judged to be the Shadow replacement. It is a lower, sleeker car than the upright Shadow II but our information is that it is still two to three years away. By that time there will have been close to 30,000 Shadows built.

Rolls-Royce describes the Shadow as a compact car — and alongside the biggest American aircraft carriers it is — but the Rolls is still a bulky car by our standards. It is 10 centimetres longer, 270 kg heavier and about a mile higher than a Mercedes Benz 450 SEL which itself is a little longer, heavier and higher than the Jaguar/Daimler 5.3. The Rolls is narrower than the Benz, but not the Jag.

The body is a superbly crafted all steel integral structure. It is built, believe it or not, by a division of British Leyland but checked minutely for quality flaws when it gets to Crewe. Before the Shadow, Rolls-Royces had separate chassis — and even this car has a hefty subframe to carry the front suspension and engine/gearbox mountings. If you see one of the frames lying on the shop floor you could be forgiven for thinking that it came from a sophisticated truck. Underbody components are huge.

The Rolls' power is an aluminium alloy V8 engine which has its valves operated by pushrods and breathes through two large SU carburettors. The engine's capacity is 6.75 litres — the biggest British made car engine there is. The V8 is not intended for spectacular power; R-R still describes its engines' outputs as "adequate", but reliable estimates put it at about 170 kW. Torque is not specified either, but the peak figure is probably around 400 Nm. The engine is utterly routine in specification — apart from the fact that it's all-alloy — but if you lift the lid and look at it the sight is wonderful. There is gleaming black stove enamel contrasting with polished castings that have R-R badges glowing out of them. The piping and wiring has been recently tidied up and now runs in a most orderly way around the engine bay. The compartment is very full of components, but it's not the bewildering layout of, say, a V12 Jaguar. It's a shame that the sight will be lost on most owners. They will never open the bonnet.

The big engine puts its power through a three-speed GM400 gearbox, the same box

that has recently been adopted for the Jaguar 5.3-litre V12 engine. The box is built specifically for big engines and big torque ratings. In the Rolls, the box excels itself. Its changes are unobtrusive, smooth yet fast. Its performance is not as "mechanical" as that of the box which goes behind the Mercedes Benz 4.5 and 6.9-litre V8s. Those boxes are marvellous at regulating revs, ratios and slippage — juggling each for maximum smoothness — but somehow you can hear their smoothness. The Rolls' box is simply unobtrusive.

Adding to the set-and-forget nature of the Shadow transmission is an electric selector mechanism. The driver can select gearbox mode with his little finger. An initial problem for Australian drivers is the fact that the selector lever sprouts from the right hand side of the steering column and it is possible to absent-mindedly select neutral when intending to signal a left turn. The turn indicator lever, which doubles as screen washer control and headlight flasher, is on the left.

The Rolls has a most traditional dash and fascia layout. Dominating everything is the walnut veneer which is decidedly real and selected, matched and fitted by artisans at Crewe who spend their days doing nothing else.

In description the R-R dash is simple enough — two large dials, one for speedo and odo/trip meters and the other a composite arrangement of small gauges for oil pressure, fuel, water temperature and amperage. Each of the big dials has easily legible white figures on black. On the right of the dials is a block of eight warning lights to back up the instruments. They warn of the failure of either one of the two brake circuits, that the parking brake is on, that fuel or engine coolant or washer fluid or brake fluid are low, that the stop lights aren't working or that there is ice in the engine's cooling system. In centre dash there is an interior thermometer and an electric clock with traditional face and hands (none of your digital nonsense).

The arrangement of the dash was altered for the introduction of the Shadow II and you'd have to say that the part of it that supplies information works well. It's the placement of some of the controls we have argument with. The light switch is on the dash in a confusing little roundel that also houses the ignition key and, incongruously, warning lights for oil pump and alternator malfunction. The wiper switch (intermittent, slow, fast) is on the lower dash between a switch which puts sump oil level on the fuel gauge and a fog light switch. When you've reached for it a dozen times in a mile it gets annoying.

Incredibly, the high beam switch is on the floor. There is a headlight flasher lever (it doubles as the turn indicator, remember) yet you must grope with your toe to keep the lights up.

On the other hand the air conditioning controls are easy to follow and convenient. There is a simple twist control (off, low, auto, high) and there are separate controls for upper and lower temperature. These switches are grouped near the steering column instead of on or above the centre console as they are in most other cars.

Ventilation comes from numerous outlets scattered about the Rolls' front and rear cabin. The visible vents are two eyeballs in the fascia and a rectangular one between radio and tape player in the console. Fresh air ventilation is probably available, though it is assumed that the air will be used all the time. Rolls-Royce now regards its air system (the cooling components are American Delco, though the switchgear is R-R's own) as better than that of any other car. It really is excellent, easy to use, unobtrusive and capable of a very wide range of settings. It's difficult to think of a way it could be better.

The Rolls' controls are a mix of good and poor. Every switch or lever works with satisfying precision, but that foot-operated dipswitch and on-the-dash wiper switch let the layout down.

The rest of the R-R interior is very comfortable. There are two huge leather bucket seats which practically envelop their occupants. They are so big that they are almost reckless in their use of interior space. A little "joystick" controls movement of each seat by electric motor. A seat can go up/down, back/forward at the touch of one of these levers. This facility is one of those things which seem superfluous in the description, but once used to them, would make it hard to go back to the old underseat grope for a lever release and a series of lunging jerks to put the seat in its rear position.

The front occupant of a Roller is utterly comfortable, though he needs a fairly long trip and an opportunity to relax to appreciate it fully. He sits high amid hide, walnut and the finest of door and headlining trim. The rear passengers have a surprisingly small amount of rear knee room because the front buckets are so bulky. Not that there's insufficient room, but tall people won't find that there is too much. A refinement is a rear footrest under each front seat for rear passengers. Naturally there are conveniently placed and well made lights, ashtrays, map pockets and air conditioning controls which make it easy for rear passengers to wallow in the opulence of it all.

The Rolls Shadow is very quiet, particularly in the rear, though we'd be surprised if it were as silent as a 5.3 Jaguar. Mind you, these arguments over degree are interesting only to enthusiasts; the users of Rolls-Royces probably won't even think about it.

Fuel economy is something which we did not explore; nor would you if you owned a Rolls. The only reason you might like to know that a Shadow sedan will return between 3.6 and 4.6 km/l (10 and 13 mpg) depending on usage is because that enables you to calculate the car's range with the standard 107-litre tank is between 380 and 480 kilometres, run to the last drop of fuel. When cruising fast, we can't imagine ourselves trusting the Shadow to cover more than the 380 between fuel stops.

Which brings us to performance. The

Above: Not the last word in modern styling, the Rolls is nevertheless clean and elegant in its lines. We find that people who dislike what it stands for are the ones with most to criticise about the way it looks.

Above: Shadow's fascia layout works for the most part, though confusing wiper and high beam switches need urgent improvement. Assembly is simply the best.

Right: Rolls-Royce 6.7-litre V8 is fed by twin SU carbs at present, but fuel injection is likely to appear in coming model.

Below right: Rolls is a big car, and its boot is cavernous. Trimming is first class. That's the toolkit in centre front. It contains tools which are as substantial as the car itself.

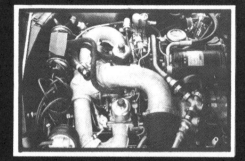

Shadow has exactly the right amount. It is not a fast car, but an effortless one. In European tune the cars are said to be able to get through a standing 400m in less than 18 seconds and to have a top speed of better than 175 km/h. Australian versions, which use what the world calls a "California 1973" engine can be expected to be slower, though according to R-R engineers not a lot slower.

The British spec car we drove could cruise at an honest 150 km/h with enough in reserve to get to 160-165 without much delay. The cruise was quiet and stable on motorways. Yet if the test car was typical the Rolls is not so powerful as to underline any steering or chassis shortcomings or to encourage the kind of balls-out driving which would emphasise the great weight of the car. "Adequate" was a better description of the car's power than we ever realised.

The engine and transmission are not intended to be analysed or even thought about by the driver. They are there simply to provide the go and to power the air conditioning and steering. They are good enough at being unobtrusive that a road tester bent on making comments on engine noise or drive train antics can only conclude that there is a lack of either.

It is certain that Rolls-Royce never intended the transmission to be changed manually. Oh, you *can* do it if you insist, but there's no noticeable benefit. It's like staying at the Waldorf and insisting on cleaning your shoes.

Until you drive the Shadow you tend to be critical of R-R for not keeping up with Jaguar (who manufacture that fabulous 5.3-litre V12) and Mercedes Benz (whose range of SOHC V8s up to 6.9 litres aren't too bad, either). But if you accept that the 6.7-litre alloy Shadow engine provides the right amount of power and you've given up in your search for noise/harshness quirks in the power train, then you have no arguments left. Perhaps the V8 could do with fuel injection, but then we understand that's coming anyhow.

The driver sits, as mentioned earlier, high in the car with a commanding view of the long bonnet. The Rolls itself is a high car, so it is possible for the driver to see right over the roof of some sedans. Though visibility is commanding, it doesn't impress the driver of modern cars with its big glass areas.

The Shadow's steering wheel is fairly small for a car of this size and weight and now has only 3.2 turns from lock to lock. Mind you, the Shadow has had three reductions in steering box ratio and three reductions in steering wheel diameter since its release in October, 1965. R-R started with a specification which it thought would please the Americans and retreated progressively from that.

The latest reductions came with the Shadow II release. The cars also were given rack and pinion steering (powered, of course) instead of the old recirculating ball system, changes in front-end geometry for the same reason as GMH made its recently RTS modifications to Holdens — to keep the outside wheel "more upright" during cornering.

The car now works pretty well. It still rolls its body under hard cornering and its tyres will still scrub if it is hauled into tight bends too quickly, but for its size and

WHY THE SKY IS BLUE AT CREWE

At the mention of Rolls-Royce, the man in the street is still apt to ask: "They went bust, didn't they?" But at Rolls-Royce Motors, they're setting records for sales and profits.

Rolls-Royce managing director David Plastow: "I am a passionate supporter of the capitalist system".

weight it deserves to be called a predictable, neat handler. That wasn't always the case. Even English weekly motoring writers — more given to forelock tugging than most — have run now-it-can-be-told stories of massive understeer, prodigious body roll, frightening tyre scrub, coarse steering and constant moans from maltreated tyre walls. That was in the old car.

A Rolls is not intended to be driven with verve. It is supposed to be driven to take advantage of the effortlessness of its chassis and power train. Any examination of a Rolls-Royce or a trip around any city block will soon educate you about the priorities its designers worked to.

Thanks to the chassis and steering changes it is now capable of travelling very quickly to the point where effortlessness passes into white-knuckling, where real driving skill needs to be exercised to make the car go faster with safety. The Rolls is certainly going quickly enough to swallow miles, to make the well-heeled traveller (at least in Europe) think twice before taking the plane.

As for Rolls-Royce's claims to build the best car in the world, it can't be true if the dynamics of the car are the prime consideration. There are plenty of cars faster, plenty of cars with better roadholding and there is a car or two which is quieter. Here, the "best" claim does not stand up. But if it is honesty and excellence of manufacture plus long life that we're discussing then the Rolls might just be best.

However, a complex combination of qualities called desirability that makes people buy cars and the Rolls must have much more than its quota of this. A WHEELS staff man who has previously never regarded a Rolls as a fit member of his more favored dozen cars has elevated the Shadow II to second or third in the lineup. He says it's a bastion of hand-built solidity in an ocean of the plastic, the frail and the short-lived. If that kind of reasoning appeals to you, you are just the person who should buy a Rolls. And after you trade in that Datsun 180B there's only about $75,000 to find . . .*

IT COMES as a mild surprise to the visitor to discover that the halls of Rolls-Royce are not of the classical hallowed kind.

The Crewe car-building concern from which 3500 to 4000 cars will roll this year (the exact figure depends on the state of labor relations) is a rather unimposing collection of brick buildings which were opened just before the outbreak of the 1939 war to build Rolls-Royce Merlin aero engines.

Even the wide, well carpeted foyer of the administration building, from which a wide gradual staircase leads to the executives' offices above, doesn't get a chance to impress the visitor. The main doors were closed permanently when terrorist bombings were threatened a year or so back. Everyone who wants entry to Rolls-Royce must now enter through the factory gate past a couple of venerable but quick-eyed men in uniform.

The appearance of Rolls-Royce Crewe is one of established solidarity. There is no hint that this is a company which has recently been reborn, though this is a term which Rolls applies to itself, even in its own publicity material. Before 1971 there was one Rolls-Royce; now there are two.

Before '71, Crewe was just one of a much larger Rolls group's divisions. Car building made reliable profits though fairly small ones considering the amount of capital expenditure that had been required, especially since the Silver Shadow had been introduced only five or six years before. In 1971 the Rolls group got into financial difficulty "associated", as the handouts say, with production of the RB211 aero jet engine. The company crashed on the stock exchange causing pandemonium. Old ladies lost their lives' savings.

Eventually the ailing areas of Rolls were taken over by the good old British Government. The two healthiest divisions, the car building business and the diesel (Rolls calls it the oil) engine division, were floated on the stock exchange as Rolls-Royce Motors Ltd, a company completely separate from Rolls-Royce Ltd.

Rolls-Royce Motors now consists of the Crewe works which employs 5700 people, the industrial engine business at Shrewsbury which employs another 3000 people and a coachbuilding business, Mulliner Park Ward which makes bodies for the more special Rolls cars and is in London. About 1200 people work there.

The only substantial connection between Rolls-Royce Ltd and Rolls-Royce Motors is now the name and trademarks. These are owned by the Government's Rolls and leased to the publicly-owned Rolls on condition that its ownership stays in Britain. It would still be possible for American or Arab interests to take over the car building business at Crewe, but they'd have to think of another name for the cars.

The man in charge is David Arnold Stuart Plastow, a 20-year employee of the car building company who has held the top job since about five weeks before the crash of the big company. Even though that happened seven years ago, Plastow still has the manner of a new broom.

A genial, informal but dedicated man of 46 years, Plastow has presided over a great improvement in the fortunes of Rolls-Royce Motors. In his words, when he took over "the place was a mess, it wasn't making any money". That isn't quite true — it was making some money but not enough to satisfy its ambitious new management.

In 1971 Rolls-Royce was building around 2000 cars a year and returned a pre-tax profit of about $2.3m on car sales valued at more than $30m. Since then, profits have always improved, the value of exports has increased by 450 percent and Rolls-Royce is now extremely healthy.

In 1977, the latest year for which results are available, Rolls-Royce Motors made nearly six times its 1971 profit (before tax) on sales which were not quite four times greater. And that was in a year when coachbuilders were on strike for 16 weeks.

Plastow calls himself "a passionate supporter of the capitalist system" and he sees neither the reason nor the circumstances which will slow the growth of Rolls-Royce. Expanding profits and volume, he says, are needed to generate the funds to maintain the company's high standards. "We are needed at the top," he says. "People have always needed something extra to strive for — a fourth bedroom, a second color TV, a swimming pool . . . or a Rolls-Royce."

Plastow is something of a phenomenon among traditional managements. He is neither old nor stuff-shirted. He greets you informally, coat off, and you sit in armchairs away from the big desk.

It is clear early in any "shop" conversation with Plastow that he is an enthusiast of Rolls-Royce cars. He will hear you out gravely as you opine that the Jaguar 5.3 V12 is a faster, quieter and better riding car than the Rolls-Shadow in which you've just come 300 kilometres to see him — and it's a lot cheaper if not in the same class for assembly. But you can see his defence mechanisms coiling for the spring.

"Look," he says, "we're not cocky here. We don't believe in taking on the V12 makers. Their approach is totally different from what we're trying to do. The Jaguar is quicker, I grant you, but I'd fight you tooth and nail on ride. And as for noise — it's not a black and white subject. You don't call a car 'noisy' or 'not noisy'; there's a lot more to it than that. Look, we're just doing things a different way." At this stage David Plastow is leaning forward out of his armchair, willing you to reach out and grasp his point of view. Practically doing a Billy Graham on you.

"We're producing good value," he goes on. "Our cars keep their value. Ask the man who buys one — it doesn't cost him a thing. He can sell it after 10 years and get what he paid or perhaps a lot more. A Rolls is clearly the best for a man who is always in his car, who works from his car. End of conversation."

But Plastow does not take critical questions personally. His answers reflect considerable self-confidence. In fact, if there is a reason for his excellent relations with the press and for the considerable staff loyalty he has generated it is the fact that he can leave you with the feeling that he is fundamentally a nice bloke, whether he's lecturing you on the necessity of $80,000 cars, or replying to serious criticisms of present models — or (doubtless) when dressing down some minion for a job badly done. As you shake hands with Plastow before he heads off to another meeting he grins at you and says to the hovering PR man, "Don't bring me any more rough-arsed buggers like him".

Plastow doesn't protest when you put the proposition that the present badge engineering of Bentley cars — the Bentley T2 is a Shadow II except for the grille — demeans both Rolls and Bentley, although he points out that the Crewe company "invented" badge engineering before the war.

But they're doing something about it. In a year or two there will be a "more individual" Bentley, based on the Camargue. It will be much more its own car. Coming too is a Shadow replacement which is lower, lighter but occupying a similar amount of road to the present cars. Rolls-Royce will say little about this car but if a prototype we chanced to see while driving in the front gate and a model on Plastow's desk (which we weren't allowed to photograph) are indicators the new Rolls will be a sleek, chiselled, modern car with detailed attention to aerodynamics, and whose styling appears to break ties with Rolls-Royces of the past.

It is almost certain to use a fuel-injected version of the present aluminium V8 and to utilise suspension and running gear componentry which has been well-sorted in the Shadow II. It will need better controls and more glass, but it could almost be said that the Shadow II is a rolling test bed for the new car's under-body mechanicals.

The managing director is adamant that there will be no small Rolls; no V6-engined, plastic-bodied lightweight. For one thing, they regard the Shadow as a compact car at Crewe and for another Rolls buyers evidently want nothing less than a "real" five seater which "you don't have to be an athlete to get in and out of".

Fuel crises of the future hold concern but no real threats for Rolls-Royce, according to Plastow. It appears that the US laws which were going to require all cars to exceed a high fuel consumption figure have been softened to the extent that those which do not meet it will incur a fine of $1500. "I think the fine is right", says Plastow, "but once it is paid then it becomes socially right to drive a Rolls-Royce".

But if there is a problem ahead for Crewe it is in getting weight out of the cars and trying to improve fuel consumption.

If Rolls' main problem is an engineering one — however "agonising" — then it must be said that the company is lucky. Engineering problems are eventually solved. If the problem were finding a market for unproven cars, or trying to find a way of producing cars down to a highly competitive price or counteracting a generally-held poor opinion of the product — problems which many other car builders have — then it could be said that the future of Rolls-Royce was not secure. But as things stand, it is hard to think of a way that it could be more so.   *

HOW THEY BUILD A ROLLS

There can be no doubt that there are several cars which go, stop, handle, ride and are quieter than a Rolls-Royce Silver Shadow. It is the extremely high standards of materials and workmanship in the cars which keep them in touch with the "best in the world" title. Here, in pictures, is how the cars go together.

1. This is as close a Rolls-Royce comes to having a production line. Cars do move in an organised line and assembly operations are performed at particular stations, but most of it is done by hand.
2. The Rolls-Royce 6.7-litre aluminium V8 is made to unusually high standards. It is dynamically balanced during assembly and each engine is run for several hours before being installed in a car.
3. Rolls-Royce Crewe has a great deal of machining equipment for an operation of its size. The company employs outside suppliers for its castings but does almost all its own machining — for brake discs, differential components, engine blocks and components, axles, suspension components
4. Rolls Shadows wait outside the plant for the fitment of their brightwork and dispatch to markets. Picture shows Crewe factory is more functional than beautiful.
5. Rolls-Royce still has artisans who spend their day selecting and matching walnut veneers to see that the dash grain matches the doors and that both match the woodwork on the inside of the rear pillar.
6. A Camargue in the process of interior fitting out. At this stage the engine is in, but the interior is still bare. Camargue is Rolls-Royce's most expensive generally available car (only the Phantom limousine is dearer).
7. This front subframe has been the subject of much re-engineering in the Silver Shadow Rolls lifetime. Components are extremely robust.
8. The dash of a Shadow before the dash trimming is done. Great care is taken in the assembly of small electrical components — a surprising number of bits are made at Crewe to ensure good quality control.
9. The Rolls radiator, not as tall as it used to be but still imposing, is built entirely by hand. Sides and edges look straight but are convex to give the impression of solidity.
10. Body shells get to Crewe looking like this. They are made by Pressed Metal, a division of Leyland at Cowley, and are checked minutely when they reach the Rolls plant.
11. Rolls-Royces in the garage awaiting road test. Tests are run without grille and bumpers to minimise the chances of minor damage. Each Rolls does about 240 kilometres of testing before being passed fit for sale.

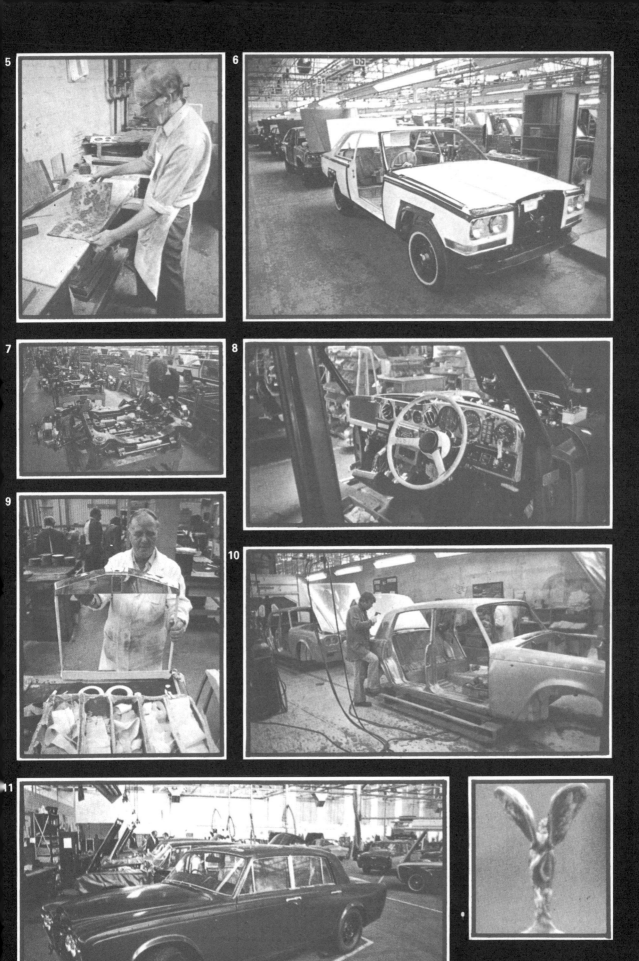

T2 to Le Mans

1,000 miles of effortless motoring in a current Bentley

By Warren Allport

Right: The Bentley at rest on a typical French tree-lined road. The black bumpers identify it as a T2 model. Above right: The high seating position gives excellent visibility and helps the driver to maintain high average speeds. The superb air conditioning system helps to reduce fatigue

A LONG JOURNEY is the best way to appreciate the latest Bentley, for it is in the effortless way in which this big car covers the miles that its greatest virtue lies. Le Mans was our destination to photograph and drive the 1928 4½ litre Bentley which is the subject of a colour article on pages 26-28. An overnight ferry crossing from Newhaven found us leaving Dieppe at about 3.30 a.m. French time, yet despite a coffee stop and a detour, we still reached Le Mans some 250 miles further on before 8 a.m. in time for breakfast before a full day's work. From Dieppe our route lay via N27 to Rouen and then on N154 by way of Evreux and Dreux to Chartres to pick up the recently opened *autoroute* as far as Le Mans.

Ride, comfort and noise levels

I find motorway driving tells me little about a car except noise levels at various speeds. Main road driving is far better in this respect, while the roads through French towns are usually as good a suspension test as any. The T2 Bentley, of course, has all-independent suspension with wishbones and coil springs at the front and semi-trailing arms and coil springs at the rear. High pressure hydraulics provide for self-levelling of the suspension irrespective of the number of passengers carried. Though undoubtedly the suspension works well, soaking up quite large bumps with ease, it is a little on the soft side with a tendency to give a slightly floating feeling on some surfaces. As far as rear seat passengers are concerned the ride is very good and there is hardly any sensation of speed, so little in fact that 100 mph seems like about 60 mph.

Putting the tyre pressures up to the high speed pressures of 28/32 psi gives a firmer ride and also improves the feel of the power-assisted rack and pinion steering which must now rate as one of the best on any big car. Initial body roll when cornering, which can upset passengers at first if the car is driven hard, is also reduced somewhat. Raising the tyre pressures unfortunately increases the bump-thump from the large 235/70HR-15in. tyres. At one time it was claimed that the loudest noise at 60 mph was the ticking of the clock. The clock is now electric, with quite a loud click, but at speeds up to 80 mph the loudest noise is tyre roar. Above 80 mph some engine noise becomes apparent, while at speeds over 100 mph the engine can be both heard and felt working when accelerated hard. Even so at 100 mph one can still listen to speech on the radio. With a maximum of around 120 mph — we actually saw 130 mph on the car's speedometer (uncorrected) on a downhill section of *autoroute* — there is more acceleration than most people expect available at 100 mph and a speedometer speed of about 115 is quite quickly and easily reached in normal driving on a fast road.

Driving controls

For a big car — this Bentley is 17ft 0.5in. long, 6ft wide and has a 10ft wheelbase — it is surprisingly easy to drive. Of course, the three-speed GM 400 automatic gearbox means there is no gearchanging to do — though the electric gear selector on the steering column is a delight to use for manual changes up or down — and the four-wheel 11in. ventilated disc brakes with a high power hydraulic servo system take al

Left and below: Differences between the Bentley and Rolls-Royce are now cosmetic only. The distinctive radiator is an obvious difference, but the facia is now identical and even has RR on the instruments, while the wheel trims are different in name only

the effort out of stopping, yet provide plenty of feel. All the driver has left to keep him from going to sleep is control of the throttle (the ingenious electronic speed control can be set to take care of that function too) and steering the car, which thanks to more power assistance is virtually effortless. Placing the car on the road exactly where one wants it is easy and is greatly helped by the commanding driving position — one looks over most modern cars — which allows both front wings to be seen. The winged B on the bonnet serves as an excellent aiming mark. Driven hard the Bentley understeers as might be expected, but the roadholding is remarkably good and most drivers are unlikely to come anywhere near the limits of adhesion of the tyres. Rearward visibility is much better than one might expect — in any case one is seldom overtaken by lesser cars — and even reversing and parking present few problems for the Bentley owner. Power-assisted steering with 3.2 turns from lock to lock and a 39ft turning circle mean the Bentley is not nearly so difficult to manoeuvre in confined spaces as its outward size might suggest.

Automatic Air Conditioning

Inside the car one of the best features — appreciated by driver and passengers alike — is the fully automatic air conditioning system. Designed and built by Rolls-Royce, this allows for different temperatures to be set for face-level and feet. Sensors inside and outside the car feed information on temperatures to the electronics of the air conditioning unit, which then maintains the temperature set regardless of what the weather is doing outside. The system has been so arranged that in the event of the refrigeration compressor under the bonnet ceasing to work, then hot air is still available by leaving the air conditioning switch in the Auto position. If the fault occurs in hot weather, the unit can be switched off, which cuts off all air supply into the car, and a window opened as a temporary expedient. Until one has experienced the delights of a fully automatic air conditioning system, it is difficult to appreciate how much more relaxed it makes driving. Gone is all that fiddling with controls to obtain the right temperature and then altering them again when the outside temperature changes. With the Bentley system you just set it and forget it.

Driving the car in the wet reveals a minor shortcoming. The wipers are not as good as they might be, and certainly not up to the standard of those on the more expensive Rolls-Royce Camargue. Switching on the wipers results in only about three quarters of the wiper arc being swept on the driver's side on the first wipe. There is also an area of screen, again on the driver's side, which is not wiped by the blade effectively because of the curvature of the screen.

Interior appointments

The seats are, of course, trimmed in Connolly hide and quite comfortable but somewhat slippery and lacking in sideways location. This is not so bad for the driver who has the steering wheel to hold on to, but passengers can find themselves sliding when the car is cornered hard. In this respect the built-in armrests on the front door and in each front seat help matters a little; they also help the driver steer a straighter course. The electrically-operated seat adjustment for the front seats should meet most requirements and is very easy to work. As well as the usual fore and aft movement, the seats can be raised or lowered and tilted at the back or front.

Interior finish, as one might expect from a car costing £32,000, is of a high standard with leather and walnut veneer much in evidence. A Bosch Blaupunkt stereo radio with electric aerial and a Pioneer quadraphonic cartridge player, both using a speaker in each of the four doors, are standard equipment as are electric windows and centralized door locking. Rear seat passengers are provided with stowage space in the backs of the front seats, a centre armrest and individual vanity units and reading lamps.

Anyone who can afford the price of this Bentley will not need to worry about the fuel consumption — even with petrol at £1.50 a gallon as it was in France on our trip to Le Mans — but for the record in almost 1,000 miles driving across France and back with the speedometer often around the 100 mark the T2 averaged 13.7 mpg. Not bad for a big car weighing almost 2¼ tons driven fairly hard. In more gentle useage we know from previous experience that 15 mpg is easily attainable, with 17 mpg possible with very restrained driving.

There is no doubt that the T2 represents an effortless means of travel and makes it possible to cover long distances with the minimum of fatigue. It is, of course, equally at home among town traffic where the automatic gearbox and power-assisted steering relieve the driver of the hard work. The aspect of the car that disappoints most is that it does not really live up to the sporting Bentley image. It is not entirely a matter of performance, more a matter of being almost too refined. The two-door Corniche with its extra performance is rather more in the Bentley mould, but it would be good to see a different car that was not sold as a Rolls-Royce marketed under the Bentley name — a mixture of more acceleration, a two-door body perhaps with a hatchback, and a somewhat firmer ride. Such a car could stand in its own right as a Bentley, rather than as a badge-engineered version of a Rolls-Royce. □

Under the bonnet the V8 engine of 6,750 c.c. produces over 200 bhp but is to the same state of tune as the Rolls-Royce, whose name now appears on the rocker covers

Bentley T2
Behind a different radiator

NOT EVEN strictly speaking, this Bentley T2 Road Test may be read without any reservations as that of the current Rolls-Royce Silver Shadow II. The test car bears a Bentley radiator — which, to some eyes suits the Shadow shape better than the Royce one does, thanks to way its more rounded profile blends with that of the car — it has a winged-B mascot, and other minor changes.

So, although the last *Autocar* Road Test of a Bentley was as long ago as 1960 (Bentley Continental S2 by James Young), the last time we tested this car was effectively only three years ago, in the Rolls-Royce Silver Shadow test of 1 May 1976.

Bentley T2

Rolls-Royce's classic Silver Shadow II with a differently labelled radiator, headlamp surrounds, mascot, nave plates and tail marking becomes a Bentley — a lovely motor car, if a misuse of a once great name.

Rolls-Royce Motors Limited
Crewe
Cheshire

There was, it is true, the Silver Wraith II Road Test of a year ago (21 October 1978), but that car has 4in more wheelbase and is not therefore exactly the same mechanically, although it did incorporate the considerable Shadow Series II changes.

Since the Series II modifications, there have been other detail ones, not all of them found on the test Bentley — from a Road Test point of view, a commendably high mileage car — because some came after it was built. The engine-driven live hydraulics for power braking and self-levelling now use mineral oil, Citroen-style, rather than brake fluid; an extra switch allows the driver getting into the car on a hot day to direct refrigerated air direct to his face via the facia eyeball vents; the electric aerial switch is deleted, in favour of automatic radio switching; and brush-type wash wipe has been added to the paired headlamps.

Performance
Effortless obedience

Dealing with the actual performance figures first, the current Series II car weighs a shade under 43 cwt, 57lb (a negligible amount in 4,809lb) more than the 1976 Road Test Shadow I, which had the same capacity 6,750 c.c. engine with the lower same-as-today (8.0-to-1) compression ratio but the bigger HD8 SUs, and 149lb more than the first Shadow test car of 1967, which had the original shorter-stroke engine size of 6,230 c.c., a 9-to-1 compression, and HD8 carburettors. The current Bentley (and Silver Shadow) weight is given as 99lb less than the Silver Wraith II without division between driving compartment and rear.

For many years, Rolls-Royce have been famous for not stating engine outputs. The power figures are probably not by any means the highest for the engine size class, as the company have obviously sought quietness as well as enough performance. There is no rev-limiting device other than the driver's nerve and hydraulic tappet pump-up; in our case it was the former which pumped up first, at 49 and 82 mph in the intermediate gears, which corresponds to 4,650 rpm assuming no torque converter slip. The engine is clearly past its peak then, so that one would guess that that peak is not much more than the 4,000 rpm which is the average for similiar-sized vee-8's of American and European (Daimler-Benz) manufacturers. That being so, the car is almost certainly a mite undergeared, for the sake of good top gear acceleration; its maximum speed of 119 mph — the highest figure we have seen from any Shadow — is equivalent to a calculated 4,550 rpm, with a best one way of only 1 mph more (120 mph and 4,600 rpm) in pretty good testing weather.

In spite of its greater weight, as well as having the highest maximum, the test Bentley is the most accelerative Silver Shadow of the four we have now tested; for the sake of anyone interested, previous tests were published in the *Autocars* of 30 March 1967, 16 November 1972, and 1 May 1976. The test day weather may have helped a little, with 10 mph less wind than for any of the other cars, but there is no doubt that this car performs well. The Silver Wraith II had been the fastest Shadow variant previously tried, yet it is now displaced; here are the full set of standing start figures, done leaving the excellent GM400 box to its own devices— we could not beat it by changing up higher than the set points at 44 mph (4,150 rpm) and 72 (4,050) — which are strong hints that the power peak is close to 4,100 rpm; 0 to 30 mph in 2.9sec (3.6sec for the Wraith), 40 in 4.6 (5.3), 50 in 6.6 (7.4), 60 in 9.4 (10.1), 70 in 12.8 (13.3), 80 in 16.9 (17.7), 90 in 23.2 (23.7), 100 in 30.7 (32.4) and 110 in 44.6 (47.5).

The transmission can be over-ridden with Rolls-Royce's superb steering column electric selector, still unapproached by any other maker. In Low, there is no safety change up, nor is there any sort of freewheel, so that it may be used for engine braking at slow speeds. The box will not drop into Low (first gear) from too high a speed however. In Intermediate, the same applies; at higher speeds of course. Maximum kickdown speeds are 27 mph from 2 to 1 and 68 mph from 3 to 2. Changes are made beautifully smoothly, which together with an exemplary light yet ideally progressive throttle linkage makes fast sweet movement of the car remarkably easy. Big cars should be so, because their mass absorbs jerks, yet the Bentley is remarkably good all the same in these respects. The discerning driver derives enormous quiet satisfaction from the way one can hurry this large saloon along any sort of road, restricted only by width of course. When the engine is warm, the only transmission jerk noticed is an unavoidable, slight but perceptible one when selecting Intermediate on the move — it changes up again almost imperceptibly. Kick down changes are done very well indeed. It will be good when Crewe adopt fuel injection, as one small drawback of the present carburettor cold start device is the high engine speed from cold, which inevitably means a jerk on engaging Drive.

The engine is not however as inaudible as one might hope and expect. Amusingly, several drivers agreed that the noise it makes are unacceptable in a Rolls-Royce, but about right in a Bentley. It is beautifully quiet at town speeds when one is not using the acceleration too much; the way the car drifts along at 30 to 40 mph is one of the greatest delights of Rolls-Royce motoring. But quite a few cheaper cars can do the same thing notably Jaguar (see our interior noise comparison tests in *Autocar* of 15 September) — and on the open road, when the driver puts his foot down, even to maintain speed up a motorway gradient at 80, there is a subdued growl from in front, not unpleasing but definitely there. It grows louder on full acceleration.

Noise
You can hear the clock ticking

. . . But only at up to about 30 mph, partly no doubt because modern car clocks are much quieter than they used to be. There are other sources of noise besides the engine, which

Apart from the radiator, only badging distinguishes the Bentley T2 from its Rolls-Royce sister. Black moulded bumpers are a distinguishing feature of the T2. Rear quarters are not as blind as they appear from the side

The T2 outside the automobile museum at Clères in Normandy. Note the spoiler beneath the front bumper

together mean, as on much lesser cars, adjusting the radio volume according to speed — and road surface. The car is still disappointingly sensitive to coarse road dressings, to a surprising extent until you consider how quiet some other usual noise makers are. The steering power assistance is inaudible, which is most unusual. There are no transmission noises. But that road noise is there, and it is the car's only real let down; bump-thump is there, and if you hit a particularly sharp bump, it is made slightly worse by a hint of boom resonance in the body. Brakes on the test car tended to make graunchy noises towards the end of a gentler slowing down, and the Avon tyres squeal too readily at low cornering speeds.

Economy
Surprisingly good for the size

If you make nearly two and a quarter tons of large, quite tall motor car accelerate from 0 to 60 mph in under 10 seconds, you are going to burn a fair bit of petrol. It is remarkable then with such performance, even if one

Right: Under the bonnet there have been numerous changes. Series 2 improvements include two fans, mechanical and electric, smaller SU HIF7 carburettors.

uses it only occasionally, that up to 15 mpg is possible at times, and, given gentle driving, that 16 mpg is fairly easily obtainable. Town driving takes a heavy toll, reducing the figure to around 13 mpg. The car's overall test consumption of 13.6 mpg is equal to the previous best figure obtained from the 1976 Silver Shadow test car. The 23½ gallons tank means a useful range of between 275 and 350 miles, and time is not wasted at petrol stops by that tiresome modern disease of too many cars, fuel fillers that blow back and are slow to brim; the Bentley one takes full delivery very nearly all the way to the last quarter-gallon. A neat feature is the remote button which you press to release the fuel filler flap. As a safety measure, it only works when the ignition is switched off.

Above: Our test car did not have headlamp wash/wipe but we tried another car which did and found it more than adequate. There is a washer jet for each headlamp. Wash/wipe operates when the headlamps are on and the windscreen washer is used

Road Behaviour
Superb direction

The change to power-assisted rack and pinion steering is perhaps the best thing that has happened to the design in its 14-year evolution. It puts to headlong flight the notion that rack and pinion is not suitable for very large cars, since the system combines the best of that simple mechanism's virtues — simplicity through a far less complicated linkage, directness which gives real feel of what the front tyres are doing, accuracy and response — with the safeguards of more old fashioned steering gear, adequate insulation from road shock.

That last point may sound contradictory; it is not in this case, because the system doesn't kick back at all alarmingly, yet it does kick just enough to give the steering life and true feel without it becoming anywhere near out of hand, even in the most gentle driver's charge. The fact that it kicks at all on this sort of car is astonishing enough; and that the gearing is so delightfully high, at only 3½ turns — the same as for a really very handy-for-the-size-of-car 37ft turning circle. It is just the right weight, no longer tip-of-little-finger as on the old recirculating ball system, but light enough to take all real effort out of steering, regardless of what you are demanding of it (some power systems can be caught out, when manoeuvring for example, when they begin to lose their power temporarily). You find yourself sitting in the car, controlling it with wrist and fingertips, elbows supported by the adjustable door armrest and its mate which swings down from the left side of the seat squab. It self-centres well too, making town driving in tight places a pleasure.

The car handles therefore pretty well for its size and class.

There is of course still some understeer when you start cornering at all quickly, but not too much. One cannot help suspecting that the construction of the Avon R-R Turbo Steep Speed 70 radials is biased towards ride rather than handling, since even when we pumped them up for high speed banking tests, they rolled over too readily on to their shoulders and the tops of their sidewalls as one approached the car's cornering limit, reducing the effect of the improved front suspension geometry on the uprightness of the loaded outside wheel. There is then quite a lot of roll, which doesn't help. The ultimate limit is set by break away in front if you keep your foot down, or at the back if you decelerate in the corner, when the camber change due to weight transfer and the semi-trailing arm rear-end geometry takes immediate but not unacceptably sudden effect. The quickness of the steering makes opposite lock correction of what can be a very undignified, wide slide relatively easy.

One must naturally be careful with the throttle in the wet; provided there is enough room on the road, leaving right-angle junctions then makes one wonder whether a limited slip differential might not be a good thing. Harder than standard tyre pressures make response and accuracy of steering that much more pleasing, but if one takes such changes too far, one realises how much of the car's remarkable low-speed, sharp-cornered-bump ride is owed to tye deflection. The Bentley is wonderfully soft in ride, to an extent few if any other cars attain, without an unacceptable cost in steering response. Big bumps it rides well too, except that there is a suggestion of

under-damping on the far side of the bump, the car giving a long extra pitch — really a very slow bounce as the nose drops. It may, on some other cars so equipped, be a fraction of the Citroen-design self-levelling which is otherwise very much appreciated in the way it keeps the Bentley level.

Brakes are beautifully weighted, with a virtually linear response curve — retardation goes up almost exactly as pedal effort does — and an extremely enough 1g maximum stop for 50 lb pedal — slightly under the norm for big cars but not too much so. The all-disc system, ventilated in front, copes impressively well with our fade test. Stopping ten times in immediate succession at ½g from the car's standing-quarter-mile speed (79 mph) at quarter-mile intervals without anything worse than pad smoke and some shudder during the last two stops is good going when there is so much weight to arrest. The handbrake takes advantage of the Bentley's not too front heavy 52.6/47.4 weight distribution by returning an excellent 0.4g stop from 30 mph; you have to pull very hard on the umbrella handle, as you do for it to hold, just, on 1 in 3.

Right: Facia is the traditional walnut veneer and incorporates an impressive warning lamps panel on the right; speedometer is electronic. The facia eyeball air conditioning switch is just to the right of the centre console beside the upper and lower air conditioning temperature controls
Far right: Standard installation is a Bosch Frankfurt AM/FM radio plus a Pioneer stereo cassette player. Our wash/wipe car shown here had a combined Pioneer radio/cassette player available to special order at extra cost

Far left: Rear legroom is good, as one would expect, and each door has its own cigar lighter and ashtray. Vanity mirrors and reading lamps are positioned in the rear quarters. Left: Upholstery is in best Connolly hide and door armrests are adjustable. Seat multi-way electric adjustment is worked from just behind central ashtray. Each front door has a switch for the central locking

Behind the wheel
That valuable thing, height

Rolls-Royces and Bentley's have one advantage that puts them usefully above all other cars — the height of the driver's eye level. You sit only a little above other cars, your eyes roughly level with your neighbour's roof — but it is enough to give one a commanding extra view around one.

The driving position is in any case good, and although paradoxically the car is available only to a minority, thanks to the electrically moved fore-and-aft and vertically adjustable seat, it will suit the majority of drivers. The armrests on each side have been mentioned; taller drivers may wish that the door one had another half-inch of adjustment range. The centre armrest provides, crudely, a rough sort of sideways location that is sadly lacking in the seat itself. The seat's rake adjustment is ideal, in that it can be moved fast to roughly where you want it, by pulling the usual lever, then adjusted finely with the rotary knob; the combination is a perfect answer.

Generally controls are done very well. We like the retention of a horn button in the middle of the steering wheel, which is still the ideal place for it, since in any emergency, whatever the driver is doing with the steering, that is the easiest place in which to find and work it quickly. It is high time on the other hand that Rolls-Royce put the wipers control on a stalk (easier to work quickly) instead of on an old-fashioned facia switch — and that they modernised the wiper action; at present it still insists on making that hesitant start, making three of four half strokes before settling into its stride with the full cycle. The 82 wipe cycle per minute fast setting is excellent, but we would like a variable rate control for the interruption wipe. The Vintage style switch panel to the right of the comprehensive warning lamp one looks well amid the walnut of such a car, and its switches — the lamps and ignition key ones — like all others on the car work very pleasingly.

The driver's door carries both switches for all four electric windows and the central locking system. The windows open and shut exceptionally quickly as well as quietly, taking just over 2sec to wind up and 1.3sec down — the norm for most other cars so equipped is around at least twice those times. The central locking control locks all doors and the boot, but leaves the boot locked when used to unlock the doors; the boot remote lock release button is hidden inside the (key-lockable) glove compartment.

Visibility is generally good. The rear quarter panel is quite large, but too far behind one for its size to make anything more than a small blind spot. The car's pronounced shoulders at the front make it surprisingly easy to "place".

The highest compliment one can pay the automatic air conditioning-cum-heater system is to say that having selected the temperature which suits you, you forget about it thereafter. It works very well indeed, and is an enormous yet unnoticed contributor to the way the car relaxes one for the entire length of a long journey. The only other car we know to be the Bentley's equal in this respect is the Jaguar XJ family, which has the very slight advantage of being quieter in operation when first working to bring the temperature under control.

Living with the Bentley T2

The first of the many pleasures of such a life is the smell of leather on first getting in; it isn't overpowering, but just enough. Another is to find a foot dipswitch again, ideal on a two-pedal controlled car since whilst the hands may be busy when one needs to dip, there is always the left foot available. The hundreds of little details found in the car never cease to please; window switches which work when the ignition's off (handy for passengers left in the car); the oil level test switch which, although the handbook says you should treat as a rough indicator only, and then (by implication) only when ticking over at rest, in fact works quite well enough at steady straight-running speed; the outside air temperature gauge (which certainly only works accurately with the car moving); the ideally placed bonnet release, under the steering column, which instead of the stiff yet fragile-feeling action of so many other cars, works smoothly and firmly; the lensed map light which peers over the passenger's should on pressing a switch in front of him; the provision of an ashtray on the driver's right.

For back seat passengers, there is as there should be plenty of knee and leg room, with ideal support for the feet on inclined toe boards under each front seat. Behind, on top of the seat, there is a padded roll for a headrest. In front are elasticated pockets of decent size in the front seat backs. Roof grab handles are provided, with what some will use as coat hooks. There is enough headroom for a 6ft passenger, and just the right amount of visibility.

Not every detail pleases. We

Above: Boot is fully carpeted and houses the battery and fitted small tools as well as the jack. Spare wheel lives under the floor and winds down

twice cut ourselves quite messily on the unduly sharp-cornered catch for the petrol filler flap, when undoing the handsome screw cap; it is perfectly placed to skin the knuckle of the middle finger. The boot floor gets warm at its front six to eight inches when laden; a minor detail, of no importance until the day one left say, a box of chocolates there. The spaces between the cartridge holder in between the front seats and the seats themselves are just high enough to lose a pencil in, which is very difficult to get out again. Few drivers will encounter the tendency for the front window frames to start vibrating quite badly as one approaches maximum speed, but it happened on the test car nevertheless. The same probably applies to the way's one right knee, searching for the sideways location missing from the seat when cornering very fast, works the driver's switch for the door window.

The Bentley range

The car tested is the cheapest Bentley, at £36,652.42 — the same price as the equivalent Rolls-Royce Silver Shadow II. Two other models flying the winged B mascot are available; the more powerful Corniche and its very dashing convertible variant (£53,322 and £56,636 respectively).

141

HOW THE BENTLEY T2 PERFORMS

ACCELERATION

FROM REST

True mph	Time (sec)	Speedo mph
30	2.9	32
40	4.6	42
50	6.6	52
60	9.4	63
70	12.8	74
80	16.9	85
90	23.2	97
100	30.7	97
110	44.6	119
120	—	129

Standing ¼-mile: 17.7 sec, 79 mph
Standing km: 32.3 sec, 101 mph

IN EACH GEAR

mph	Top	2nd	1st
0-20	—	—	1.8
10-30	—	—	2.4
20-40	—	—	3.3
30-50	—	4.2	—
40-60	—	5.2	—
50-70	—	6.4	—
60-80	7.7	—	—
70-90	10.2	—	—
80-100	13.6	—	—
90-110	22.0	—	—

TEST CONDITIONS
Figures taken at 7,500 miles by our own staff at the Motor Industry Research Association proving ground at Nuneaton, and on the Continent.
All Autocar test results are subject to world copyright and may not be reproduced in whole or part without the Editor's written permission.

Wind: 0-12 mph
Temperature: 11 deg C (52 deg F)
Barometer: 29.3 in. Hg (993 mbar)
Humidity: 70 per cent
Surface: Dry asphalt and concrete
Test distance: 1,947 miles

MAXIMUM SPEEDS

Gear	mph	kph	rpm
Top (mean)	119	192	4,550
(best)	120	193	4,600
2nd	82	132	4,650
1st	49	79	4,650

FUEL CONSUMPTION

Overall mpg: **13.6** (12.0 litres/100km)

Constant speed:

mph	mpg	mph	mpg
30	22.0	70	17.1
40	21.1	80	15.0
50	20.1	90	12.8
60	18.8	100	10.3

Autocar formula: Hard 12.2 mpg
Driving Average 15.0 mpg
and conditions Gentle 17.7 mpg

Grade of fuel: Premium, 4-star (98 RM)
Fuel tank: 23.5 Imp. galls (107 litres)
Mileage recorder reads 1.0 per cent long

Official fuel consumption figures
(ECE laboratory test conditions; not necessarily related to Autocar figures)

Urban cycle: 11.1 mpg
Steady 56 mph: 19.5 mpg
Steady 75 mph: 15.9 mpg

OIL CONSUMPTION
(SAE 20W/50) 850 miles/pint

BRAKING

Fade (from mph in neutral)
Pedal load for 0.5g stops in lb

	Start/end		Start/end
1	25-20	6	25-35
2	25-25	7	25-38
3	25-25	8	25-35
4	20-20	9	25-30
5	25-30	10	25-30

Response (from 30 mph in neutral)

Load	g	Distance
10lb	0.21	143ft
20lb	0.41	73ft
30lb	0.60	50ft
40lb	0.78	39ft
50lb	0.98	30.7ft
Handbrake	0.40	75ft

Max. gradient: 1 in 3

WEIGHT

Kerb, 42.9 cwt / 4,809 lb / 2,181 kg
(Distribution F/R, 52.6 / 47.4)
Test, 46.5 cwt / 5,209 lb / 2,362 kg
Max. payload 1,063lb/482kg

PRICES

Basic	£29,420.00
Special Car Tax	£2,451.67
VAT	£4,780.75
Total (in GB)	**£36,652.42**
Seat Belts	Standard
Licence	£50.00
Delivery charge (London)	£70.00
Number plates (approx)	£20.00
Total on the Road	**£36,792.42**
(exc. insurance)	

EXTRAS (inc. VAT) — selection only)
Mascot anti-theft alarm	£130.19
Chubb fire extinguisher	£37.06
Rear armrest compartment	£84.10
Front seat head restraints	£59.49
Picnic tables, per pair	£265.98
Non-standard paint	£495.22

*Fitted to test car

TOTAL AS TESTED ON THE ROAD £36,792.42

Insurance: Group 7

SERVICE & PARTS

Interval

Change	6,000	12,000	24,000
Engine oil	Yes	Yes	Yes
Oil filter	Yes	Yes	Yes
Gearbox oil	No	Yes	Yes
Spark plugs	No	Yes	Yes
Air cleaner	No	Yes	Yes
Total cost	**£94.76**	**£143.18**	**£163.18**

(Assuming labour at £8.00/hour)

PARTS COST (including VAT)
Brake pads (2 wheels) — front	£28.75
Brake pads (2 wheels) — rear	£14.95
Exhaust complete (stainless steel)	£851.00
*Tyre — each (typical)	£102.53
Windscreen (laminated, tinted)	£143.75
Headlamp unit	£19.55
Front wing	£213.90
Rear bumper	£230.00

WARRANTY
Mechanical: 36 months / 50,000 miles
Body: 12 months

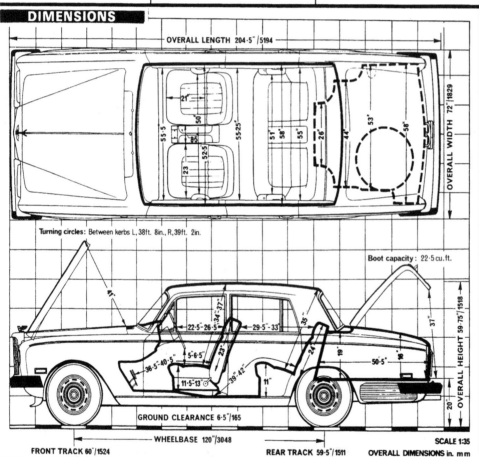

SPECIFICATION

ENGINE
	Front, rear-wheel drive
Head/block	Aluminium alloy
Cylinders	8, in 90 deg vee
Main bearings	5
Cooling	Water
Fan	Viscous fan drive, plus electric fan.
Bore, mm (in.)	104.1 (4.098)
Stroke, mm (in.)	99.1 (3.902)
Capacity cc (in³)	6,748 (412)
Valve gear	Ohv pushrods, hydraulic tappets
Camshaft drive	Gear
Compression ratio	8.0-to-1
Ignition	Breakerless, transistorised
Carburettor	2 SU HIF7
Max power	Not stated
Max torque	Not stated

TRANSMISSION
Type	GM400 3-speed epicyclic automatic with torque converter
Clutch	
Gear	Ratio mph/1000rpm
Top	1.0-2.0 26.2
2nd	1.48-2.96 17.70
1st	2.48-4.96 10.56
Final drive gear	Hypoid bevel
Ratio	3.08-to-1

SUSPENSION
Front—location	Independent, double wishbone
springs	Coil, self-levelling
dampers	Telescopic
anti-roll bar	Yes
Rear—location	Independent, semi-trailing arm
springs	Coil, self-levelling
dampers	Telescopic
anti-roll bar	Yes

STEERING
Type	Rack and pinion
Power assistance	Yes
Wheel diameter	15.2 in.
Turns lock to lock	3½

BRAKES
Circuits	Two, split front/rear
Front	11.0 in. dia. disc
Rear	11.0 in. dia. disc
Servo	Hydraulic, engine-driven
Handbrake	Umbrella handle, rear disc

WHEELS
Type	Steel disc
Rim Width	6.0 in.
Tyres—make	Avon (on test car) or Dunlop
—type	R-R Turbo Steel 70 radial tubeless
—size	235/70HR-15
—pressures	F24 R28 psi (normal driving)

EQUIPMENT
Battery	12V 68 Ah
Alternator	75A
Headlamps	4 lamp system 115/195W
Reversing lamp	Standard
Hazard warning	Standard
Screen wipers	2-speed and intermittent
Screen washer	Electric
Heater and air conditioning	Auto temperature control
Interior trim	Leather seats, cloth headlining
Floor covering	Carpet
Jack	Screw
Jacking points	One each side
Windscreen	Laminated
Underbody protection	Galvanised, phosphated, bitumastic

HOW THE BENTLEY T2 COMPARES

Bentley T2 (A) £36,652
Front engine, rear drive
Capacity 6,748 c.c.
Power Not stated
Weight 4,809lb / 2,181kg
Autotest Rolls-Royce Silver Shadow 1 May 1976; Silver Wraith II 21 October 1978

BMW 733i (A) £15,012
Front engine, rear drive
Capacity 3,210 c.c.
Power 197 bhp (DIN) at 5,500 rpm
Weight 3,585lb / 1,627kg
Autotest 6 August 1977

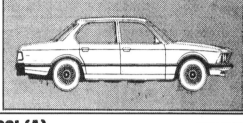

Ferrari 400I (A) £31,809
Front engine, rear drive
Capacity 4,832 c.c.
Power 340 bhp (DIN) at 6,500 rpm
Weight 4,145lb / 1,880kg
Autotest Ferrari 365 GT4 2+2 4 October 1975

Daimler Vanden Plas 5.3 (A) £22,672
Front engine, rear drive
Capacity 5,343 c.c.
Power 285 bhp (DIN) at 5,750 rpm
Weight 4,310lb / 1,955kg
Autotest Jaguar XJ12 5.3 9 September 1978

Mercedes-Benz 450 SEL (A) £19,161
Front engine, rear drive
Capacity 4,520 c.c.
Power 225 bhp (DIN) at 5,000 rpm
Weight 3,904lb / 1,772kg
Autotest 4 May 1974

Mercedes-Benz 450 SEL 6.9 (A) £30,476
Front engine, rear drive
Capacity 6,834 c.c.
Power 286 bhp (DIN) at 4,250 rpm
Weight 4,060lb / 1,841kg
Autotest 24 March 1979

MPH & MPG

Maximum speed (mph)

Ferrari 400I*	150*
Daimler Vanden Plas 5.3**	147**
Mercedes 450SEL 6.9	140
Mercedes 450SEL	134
BMW 733i***	122***
Bentley T2	119

Acceleration 0-60 (sec)

Ferrari 400I	7.1*
Mercedes 450SEL 6.9	7.3
Daimler Vanden Plas 5.3**	7.8**
BMW 733i***	8.9***
Mercedes 450SEL	9.0
Bentley T2	9.4

Overall mpg

BMW 733i***	19.4***
Mercedes 450SEL	14.1
Bentley T2	13.6
Mercedes 450SEL 6.9	13.6
Daimler Vanden Plas 5.3**	13.2**
Ferrari 400I*	11.0*

*Figures for 365GT4 2+2 manual gearbox (1.8 per cent lower power to weight ratio).
**Figures for 1.8 per cent lighter Jaguar XJ5.3.
***Figures for manual gearbox BMW 733i.

Even before one considers the price and quality of the Bentley, it has to be admitted that there are precious few cars made which are completely comparable with it, for all sorts of reasons. Big luxury saloons are, if not a dying breed — Rolls-Royce sell every car they make up to at least two years before it rolls out of the factory — a rare one today. Of the cars variously obvious or not which are available we have not tested the Lagonda (£39,931), Maserati Quattroporte (available here in left hand drive manual box form only at £28,900), de Tomaso Deauville (£24,418), or any of the odd Americans "legally" available here, the most obvious of which must be the Cadillac Fleetwood (£15,739). With the exception of the Ferrari, which though the biggest car Modena produces is only a 2+2, the cars chosen here are after something like the same market; as the figures show, the Bentley pays for its ride and quietness, to some extent achieved with sheer mass — it is 11½ per cent heavier than the next heaviest rival, the Daimler — in performance. It is only fair to point out that the less roomy Daimler does better than the Bentley in quietness and performance, thanks to less frontal area and weight, probably better shape, and a more powerful, smoother V12 engine, whilst the BMW wins easily on economy.

ON THE ROAD

Best all-rounder here is certainly the Daimler, with its superb ride, handling and roadholding, and its unrivalled quietness; its road manners are impeccable. For pure saloon car pedigree and entertainment, the Ferrari is of course tops — and you don't buy a Ferrari for quietness or the best of ride. Interestingly these two cars are the only ones with the near-ideal of double wishbone geometry front and rear, avoiding too much camber change with changes of attitude. The other four are all semi-trailing rear-ended, which sets a certain limit on their ultimate roadholding and behaviour (due to camber change at the back); but the Bentley must be singled out for its superb ride and, since the adoption of a rack and pinion steering set-up, its really magnificent direction, which is one area where Crewe still does build the Best (Big) Car in the World; the two Mercedes better it with their exemplarily high gearing, particularly in the case of the 6.9 (2.6 turns lock to lock for a 36ft turning circle), but haven't got quite such remarkable feel.

SIZE & SPACE

Legroom front/rear (in.)
(seats fully back)

BMW 733i	42/40
Mercedes 450SEL 6.9	43/39
Mercedes 450SEL	43/39
Bentley T2	40/41
Daimler Vanden Plas	39/36½
Ferrari 400I	41/29

Legroom dimensions are a little misleading here, since they take no account of headroom, where the Bentley and BMW are paramount; all of the four-door cars are in fact more than roomy enough, whilst the Ferrari does not pretend to be anything greater than a 2+2. In interior layout the German cars win, particularly the BMW with its very good clarity of instrument layout, though they have the least informative range of instruments. Best equipped in that respect is the Ferrari as you would expect; it has an oil temperature gauge which arguable necessity the others lack.

VERDICT

Final choice in this peculiar case boils down to a case of horses for courses; the man who wants something like the Ferrari won't look at any of the others for a start, though he might consider an Aston Martin Vantage (£30,878) not listed here. Amongst the five four-door cars, things are a little more even, with the Daimler winning on refinement and performance, value for money, the BMW followed by the 450SEL doing something like the same things the least thirstily, and the 6.9 offering something marvellously different in Mercedes motoring; sadly, it must be remembered that it is now extinct, with the coming of the new S-class — which will undoubtedly mean its financial canonisation as an appreciating classic of the 1970s. The Bentley is of course merely a differently labelled Rolls-Royce Silver Shadow; in passing we must say that we find such a cavalier misuse of a great name a mistake, about which rumour suggests the name's owners are eventually to do something. Names aside, it is nevertheless very impressive, and it undeniably offers a sort of progress about the place, nothing to do with motoring snobbery, which thanks to that height, that shape, that refinement and finish and that taste, is still unapproached by other cars made here or, even more so, abroad.

THE SHEER VOLUME probably wouldn't excite General Motors very much, but about 25,000 Rolls-Royce Silver Shadows have been built since 1968, making it by far the most successful model ever built by the company since it was founded 75 years ago. During the last 11 years the car has been much modified, and the Shadow II was introduced in 1977, so when Rolls-Royce called up and offered us a test car we felt that it was time to reacquaint ourselves with the marque. Anyway, the opportunity of driving a Rolls-Royce for a few days is a bit like being invited to dinner at Maxim's in Paris; it's the sort of offer you can't really refuse.

The Rolls-Royce Silver Shadow is not like other cars and people who know the cost of everything and the value of nothing should forget it. In addition, if performance in the broadest sense is your basis for judgment, then it won't measure up to a Mercedes-Benz 6.9 or a Jaguar XJ12L but it will cost you a lot more. But if you are looking for pure, unadulterated luxury which cossets you and insulates you from the annoyances of the world, then there is nothing to beat a Rolls-Royce.

Since we last tested the Shadow in September 1976, various refinements have been incorporated in the car. Among the major ones are a switch from recirculating-ball to rack-and-pinion steering, combined with some modifications to the steering geometry to improve handling and decrease tire wear. Sports car handling has never been a Rolls-Royce virtue, but the steering and suspension changes have improved the handling and significantly reduced the amount of body roll during hard cornering conditions.

Inside the car the instrumentation and switches have been improved in the interest of ergonomics, there is a smaller-diameter steering wheel, although our test car had a non-standard wheel, and the very sophisticated air conditioning system introduced on the Camargue is now standard on the Shadow. It is a two-level, fully automatic system, with separate thermostatic controls for the upper and lower levels.

The transmission is the General Motors Turbo Hydra-matic 400, the unit which has the distinction of being probably the best and probably the cheapest automatic transmission for big cars in the world today. Coupled to the 6750-cc V-8 in the Rolls, it works like a charm. Selection is by a column-mounted electric selector switch, which requires only fingertip selection. Once in drive, shifting is barely perceptible and it is virtually impossible to tell which range the transmission is in when driving around town. Full throttle upshifts occur at 40 and 75 mph and the shift into 3rd is barely perceptible. An interesting feature is an electronic override that locks the transmission whenever the ignition key is removed, regardless of the selector position. This is not just a safety feature as it also makes stealing the car a two-man job.

Our test car was finished in two-tone red and garnet, and it was one of the most impressive paint jobs we have seen in a long time. On more than one occasion when we had parked the car, a stranger came up and said, "What a beautiful car you have," which must make life seem worthwhile to the Rolls owner.

The interior is tremendously impressive with a warmth that can really only be approached by the Jaguar XJ series. One has the impression of being enveloped in leather and lamb's wool, with walnut veneer to delight your eyes and everything possible for your comfort and convenience within your reach. The fleece floor mats make you want to drive the car in your bare feet, and the smell of the interior is delicious. The seating position is really high, so you can look down on lesser mortals and, of course, it is adjustable six ways. Visibility is excellent all around and the silver lady at the end of the hood is always there to remind you that you are not driving just any car.

The cold starting is impeccable and the engine settles down immediately into a silent idle. Moving off, one notices that the steering is considerably improved compared to the old recirculating-ball system, having much more feel to it. On the road the car is extremely polite and handles all road conditions except California freeway hop in impeccable fashion with the silence broken occasionally by a muffled thump from the tires. At speed the suspension has a peculiar combination of softness and firmness so the car seems to float along, but without any of the alarming feeling that it is floating. It soaks up road irregularities amazingly well and it doesn't lose its dignity or concern itself at all over culverts.

When pressed hard into a corner, it tends to oversteer and particularly if you lift off the throttle, which is characteristic of

ROLLS-ROYCE SILVER SHADOW II
Looking for a comfortable family sedan?

the semi-trailing arm independent rear suspension. It has a lot of anti-dive built into the front suspension, but it lifts its nose very majestically on hard acceleration.

We hustled the car over our mountain test route and to those who were following in more mundane vehicles, it was deceptively fast from one point to another, belying its 5000 lb, while the lack of effort was much appreciated by its driver. The various people who drove the car all remarked on how deceptive it is as far as speed is concerned, because you can easily find yourself driving down the freeway at 80 mph thinking you are doing only 60 mph. ➤

PHOTOS BY JOE RUSZ & JOHN LAMM

AT A GLANCE

	Rolls-Royce Silver Shadow II	Mercedes-Benz 6.9	Aston Martin V-8
List price	$69,900	$51,829	$53,925
Curb weight, lb	4970	4270	4070
Engine	V-8	V-8	V-8
Transmission	3-sp A	3-sp A	5-sp M
0–60 mph, sec	11.3	8.2	7.4
Standing ¼ mi, sec	18.4	16.4	15.7
Speed at end of ¼ mi, mph	75.5	90.0	90.5
Stopping distance from 60 mph, ft	159	171	158
Interior noise at 50 mph, dBA	65	65	72
Lateral acceleration, g	na	0.693	na
Slalom speed, mph	52.6	53.7	na
Fuel economy, mpg	9.5	13.0	9.5

This is primarily because of the extremely low noise level and it may also have something to do with the upright and very high seating position.

The brakes take a little getting used to when driving hard because the operation is rather similar to Citroën's full hydraulic system, and there is a tendency toward light pulsations at the pedal, which makes smooth brake modulation difficult until you become accustomed to the system. As with almost everything Rolls-Royce, the brakes are exceptionally complicated with redundant twin circuits pressurized to 2500 psi by pumps driven from the camshaft, backed up by a direct-acting system in case both pressure systems should fail.

In general, the brakes were very good, but with a tendency for rear locking with only the driver aboard, although the control was very good with no tendency to slew. The stopping distances were 159 ft from 60 mph and 281 ft from 80 mph, which are commendable figures for what is today a very heavy car.

The detail work on the car in general was not without flaws, which surprised us. Nothing much more than uneven stitching here and there or signs of a file mark on the exterior sheet metal, but nonetheless something one would not expect to find on a $70,000 car. However, in mitigation one can say that our test car had 7000 miles on it and had been used for extensive testing.

As far as the Rolls-Royce reputation for silence is concerned, the Shadow II carries on the tradition, comparing well with other very quiet cars such as the Cadillac Seville, the Buick Riviera S and the Jaguar XJ6L.

Putting a Rolls-Royce through our test program seemed a little sacrilegious. Standing on the line before our acceleration runs with all the wiring coming out to our 5th wheel, she looked a little bit like a duchess in intensive care, but she managed to chirp her tires merrily coming off the line, much to the delight of the onlookers, and ran the quarter-mile at a rather sprightly and very stately 75.5 mph in 18.4 seconds.

Driving around in a Rolls-Royce is an illuminating experience and one quickly finds that the car draws a lot of attention and also has its detractors, which doesn't matter at all because none of them could afford one anyway and would probably be the first to buy one if they could. Whether or not a Rolls is the best car in the world is entirely up to you because, when you buy one, you buy a way of life as much as an automobile. What you actually get is a surprising feeling of well being, contentment and security, totally removed from all the mean and horrid things that are going on outside your rich and luxurious surroundings, and all that combined with one of the most refined pieces of automobile engineering in the world.

PRICE
- List price, all POE $69,900
- Price as tested $70,300
- Price as tested includes standard equipment (air cond, AM/FM stereo/tape deck, leather upholstery, self-leveling suspension), leather dash & console ($400)

GENERAL
- Curb weight, lb/kg ....4970..............2256
- Test weight ....5140..............2334
- Weight dist (with driver), f/r, % 52/48
- Wheelbase, in./mm ....124.1..............3152
- Track, front/rear ....57.5/57.5....1461/1461
- Length ....211.5..............5372
- Width ....71.8..............1824
- Height ....59.8..............1519
- Trunk space, cu ft/liters ....19.4..............549
- Fuel capacity, U.S. gal./liters ......22.5..............85

ENGINE
- Type ....ohv V-8
- Bore x stroke, in./mm ....4.10 x 3.90...104.0 x 99.1
- Displacement, cu in./cc ....412..............6750
- Compression ratio 7.3:1
- Bhp @ rpm, SAE net/kW ......est 190/142 @ 4000
- Torque @ rpm, lb-ft/Nm ....est 290/393 @ 2500
- Carburetion ....two SU (1V)
- Fuel requirement ....unleaded, 91-oct

DRIVETRAIN
- Transmission..........automatic; torque converter with 3-sp planetary gearbox
- Gear ratios: 3rd (1.00) 3.08:1
- 2nd (1.50) 4.62:1
- 1st (2.50) 7.70:1
- 1st (2.50 x 2.00) 15.40:1
- Final drive ratio 3.08:1

CHASSIS & BODY
- Layout ....front engine/rear drive
- Body/frame ....unit steel
- Brake system ....11.0-in. (280-mm) vented discs front, 11.0-in. (280-mm) discs rear; actuated by dual high-pressure hydraulic system
- Wheels ....steel disc, 15 x 6JK
- Tires ....Michelin Wide X, 235/70HR-15
- Steering type..........rack & pinion, power assisted
- Turns, lock-to-lock 3.2
- Suspension, front/rear: unequal-length A-arms, coil springs, tube shocks, anti-roll bar/semi-trailing arms, torque arm, coil springs, tube shocks, anti-roll bar; self-leveling

CALCULATED DATA
- Lb/bhp (test weight) .... est 27.1
- Mph/1000 rpm (3rd gear) 25.5
- Engine revs/mi (60 mph) 2350
- R&T steering index 1.28
- Brake swept area, sq in./ton 183

ROAD TEST RESULTS

ACCELERATION
Time to distance, sec:
- 0-100 ft 3.5
- 0-500 ft 9.8
- 0-1320 ft (¼ mi) 18.4
- Speed at end of ¼ mi, mph 75.5

Time to speed, sec:
- 0-30 mph 3.3
- 0-50 mph 8.1
- 0-60 mph 11.3
- 0-70 mph 15.4
- 0-80 mph 21.3
- 0-90 mph 30.3

SPEEDS IN GEARS
- 3rd gear (4250 rpm) 114
- 2nd (4000) 95
- 1st (4000) 59

FUEL ECONOMY
- Normal driving, mpg 9.5

BRAKES
Minimum stopping distances, ft:
- From 60 mph 159
- From 80 mph 281
- Control in panic stop..........very good
- Pedal effort for 0.5g stop, lb 18
- Fade: percent increase in pedal effort to maintain 0.5g deceleration in 6 stops from 60 mph 39
- Overall brake rating............very good

HANDLING
- Lateral accel, 100-ft radius, g ...... na
- Speed thru 700-ft slalom, mph ... 52.6

INTERIOR NOISE
- Constant 30 mph, dBA 59
- 50 mph 65
- 70 mph 69

SPEEDOMETER ERROR
- 30 mph indicated is actually........27.5
- 60 mph 59.0

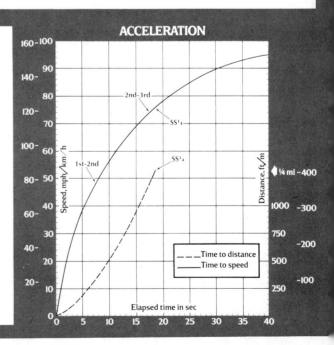

146

R-R's SILVER SHADOW II

Old-fashioned? Yes. But, what opulence! What luxury! And so quiet now, you can't even hear the clock.

by BILL QUINN

THE HEADY AROMA of unblemished, natural leather hides covering opulent seats that embrace you in a near-sensuous manner. Wilton carpeting covered by lambswool rugs so soft you want to luxuriate in them with bare feet. Walnut veneer from Milan that sets off the instruments worthy of a VanCleef & Arpels display. A discreet silence that once caused a Rolls-Royce ad copywriter to pen, "The loudest thing you hear at 60 mph is the clock."

That's the Rolls-Royce Silver Shadow II. A car that Beautiful People are well into. A car that's no stranger to Beverly Hills or the Hamptons, West Palm Beach or Nob Hill, Palm Springs or Las Vegas. Yet, a car that driven almost anywhere continues to draw admiring glances.

Still, here's a car whose basic design is 13 years old. As the original Silver Shadow, it was in production for 11 years with more than 2000 improvements made in it before it became the SS II. A Rolls-Royce seems to be—must be—the epitome to many. It's the apex. The pinnacle. Owning the "best car in the world" makes *you* the best in the world.

It's a car in which only the best materials are used. A car built to exacting standards of finish and craftsmanship. A car that carries a "history book" as it's being produced—detailing the materials used, processes undergone and tests made—which many of the craftsmen who attend to it during its genesis sign with pride.

Under the Hood

The relatively-large (by today's standards, at least) 6.75-liter/412ci V8 fills up almost the entire compartment. Still, there is a modicum of room for access to the necessary maintenance items—if you, as an owner, are interested.

What an owner would be interested in is the fact that the all-aluminum engine is a masterpiece of design, built to standards usually reserved for racing engines. It's perfectly balanced for long life. For example, each cast-iron cylinder liner has its own matched piston, one of a weighed, matched pair. Connecting rods are also weighed and

matched, then crack-tested for imperfections as well. The crankshaft is machined to a tolerance of only one ten-thousandth of an inch and is individually matched to its main bearings.

The completed engine is carefully run-in on a test bench, using natural gas, for an equivalent of 150 miles, minimum. One out of a hundred engines is also run for eight hours on gasoline. Then, it is stripped down and minutely examined piece-by-piece. Every tolerance is checked against that which was originally specified.

Coupled to the engine is a General Motors-supplied Turbo Hydramatic (automatic with torque converter). The engine/transmission combination is such that upshifts are both positive and virtually unnoticed, even at full throttle.

Behind the Wheel

To say that slipping behind the wheel of the Rolls-Royce for the first time is a thrill is putting it somewhat mildly. It's not only the fact that you're in a Rolls, but that you soon realize this car is *engineered*, engineered as "a command post for the driver."

Instruments are laid out in logical groups: directly in front is the electronic speedometer and flanking it are two dials with the various operating gauges; to the right is an outside temperature gauge and clock; on the lower panel, the air conditioner/heat/vent controls; above the console the separate radio and cassette tape player.

Why an electronic speedo? To eliminate "the last possible source of mechanical noise in the instrument system." Coupled to it is an optimistic odometer that can read up to 999,999.9 miles!

Warnings are handled both visibly and audibly. In the "unlikely event" that the engine overheats, you hear a buzzer as well as seeing a yellow "overheat" light. You can then check coolant and oil levels without leaving your seat. And to make sure the lights are working, there's a manual check of them by pushbutton.

Both front seats are infinitely adjustable fore and aft, up and down and through several degrees of tilt through use of a small "joystick" at the side of each seat.

Passenger/Luggage Space

Given the size of the Rolls-Royce Silver Shadow II, it's not so remarkable that entry and exit are excellent. The SS II has a body somewhat shorter than either the Cadillac 4D DeVille or Lincoln Continental Mark VI, but it's 3-4 ins taller and 5-6 ins narrower. Yet, it has adequate interior room inside for full passenger comfort.

Adding to the comfort in the rear are individual footrests, vanity mirrors, cigarette

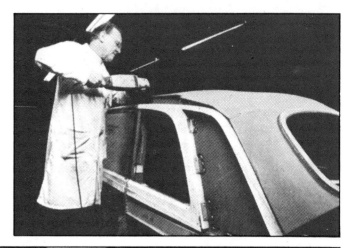

lighters and ashtrays. All that's missing are lap robes.

The spacious trunk (19 cu ft) is fully-carpeted, with the spare in a separate compartment beneath the floor. It's not necessary to remove luggage to remove the wheel, either. A comprehensive tool kit is also provided.

Driving the Rolls

All controls operate with a smooth and precise action. R-R attention to detail in this respect is evident in many ways. For one, the horn button relays are made of real gold. Another, the glovebox door is fitted with a Yale lock and opens and shuts like a bank vault.

Starts from cold were always instantaneous, though "cold" in this respect was never below freezing. We never encountered any stumble or hesitation under any condition.

Compared to other Shadows, the most immediate and gratifying change is in the steering. The new, power-assisted rack-and-pinion system is much more responsive and positive than the former recirculating ball setup. The new system is slightly heavier in operation, gives a better feel of the car and the road, and has minimal kickback.

Compared to previous Rolls-Royces, the driver tends to slightly overcorrect, but within a few miles this inclination disappears. The car is stable at all speeds and gusty sidewinds such as we encountered on a trip from Los Angeles to Las Vegas have little effect on directional stability. Severe dips are taken in stride, with little evidence of bottoming out.

This, of course, it owes to the suspension, which is independent all-around. In front by lower wishbones, stabilized upper levers, coils and telescopic shocks. In the rear by trailing arms, coils and teleshocks. Anti-roll bars are used front and rear. Automatic height control keeps it level during all but the most violent maneuvers.

Visibility in all directions from the unusually high seating position is excellent. Oddly, the sun visors are fixed, not allowing their use in deflecting glare from the side.

During full throttle acceleration, some air induction roar is evident, but unlike pre-1977 models, almost no engine fan noise. The new fan is geared lower, and also slips at a lower rpm than previously. This is through the use of an auxiliary electric temperature-sensitive fan mounted in front of the radiator.

At speeds of 70 mph, the engine is all but inaudible. The decibel level is only 68 at this speed, which puts it on a par with the Cad Seville and Jaguar XJ-S.

Cruising at high speeds was a distinct pleasure, with one important malfunction evident in our test car, though. The cruise control was incapable of maintaining preselected speeds on winding, uphill mountain roads. Speed would vary by 7-8 mph and additional throttle pressure was needed to keep it steady. Downhill, it was fine, with not more than ½ mile difference being noted.

How It Performs

By no stretch of the imagination is the Rolls considered a performance car. The luxurious feel of the car tends to discourage brisk driving. Yet, after a few miles at the wheel, the good handling, coupled with the big-bore V8 and 4-wheel discs, begin to get to you. In no time—on the open road, that

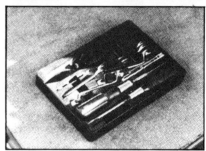

is—you begin to swallow up the miles at surprisingly high speeds.

What's so deceptive is its quietness, at 90 mph, and its stability, at 110 mph and over.

During our mountain driving, the brakes were used fairly hard, but there was never any indication of fade, pulling to one side or tendency to lock up. The good balance of the car, the front/rear proportioning valve and Michelin wide-profile radials all share in the credit.

What About Economy?

Come now, do you really expect to get good fuel economy with a Rolls? If it wasn't for the U.S. government decreeing it, you probably wouldn't get even the 10 mpg the EPA rates it at in city driving. Nor would you care, particularly.

Out of curiosity, we carefully monitored fuel consumption on a roundtrip to Las Vegas and observed the following: open road driving at steady 60-65 mph, air conditioner on half the time—13.4 mpg; open road at steady 55 mph, air conditioner off—16.1 mpg; extremely high speeds (holding at 110 and touching 115 mph) coupled with "normal" highway speeds—8.0 mpg; and, an average of 12.5, which is slightly higher than can be expected in the normal mix of city/highway driving.

So far, Rolls-Royce has exceeded both the stringent automotive safety and exhaust emission requirements set down by the federal government. But, says David Plastow, group managing director and CEO of Rolls-Royce Motors, "improving fuel consumption is a taller mountain to climb."

As part of the "great technical effort into getting better mileage," R-R engineers are working on several new engine designs and plan to reduce the weight of their cars.

Making a point of the fact that Rolls-Royce production is so low—something over 3000 cars annually—and that the cars last so long, Plastow said last September,

"Think of all the energy we are saving by making cars to last 50 years instead of seven or eight. Plus, the difference between the fleet average requirements—and we have no fleet—and our current performance amounts to about 30 barrels of oil a day out of 5 million barrels a day used by all cars in this country."

Last year, 1109 Rolls-Royces were sold in the U.S.

Summing Up

How much does all this luxury, craftsmanship and attention to detail cost? And is it worth it?

Would you believe that the Silver Shadow II is the bottom-of-the-line Rolls, at a mere $77,600? (You can pay up to $140,000 for the Corniche Convertible, a totally-crafted car that takes 5 months to build.)

As to whether it's worth it or not, that depends on the ability to pay for it. For those who can, the satisfaction comes from knowing that they have a car built with patience and responsibility, with the utmost quality and the best technology available. These are the evolutionary influences that make a Rolls-Royce what it is—a Rolls-Royce

1980 ROLLS-ROYCE SILVER SHADOW II

BODY STYLE, 4D, 5-seat sedan.
PRICE. Base, As Tested, $77,600 (completely equipped).
ENGINE, OHV 90-degree aluminum block V8 with cast iron wet liners, mounted up front. **Bore/Stroke,** 4.1/3.9 ins. **Displacement,** 6750cc/412ci. **Comp Ratio,** 7.3:1. **Carburetion,** twin SU carbs. **Ignition,** Electronic. **Max BHP/Torque,** N.A. **Fuel Req,** Unleaded.
DRIVETRAIN, 3-speed automatic with torque converter, single-piece prop shaft, hypoid bevel final (rear) drive. **Ratios,** 2.5, 1.5, 1.0, **Final Drive,** 3.08:1.
CHASSIS & BODY. Integral, steel. **Suspension, Front,** independent by lower wishbones, stabilized upper levers, coil springs and tube shocks. **Rear,** independent by trailing arms, coils, tube shocks. Anti-sway bars front and rear. **Wheels & Tires,** 15-in steel discs with HR70 HR15 or 235/70HR15 low-profile radials. **Steering,** Power-assisted rack-and-pinion, 17.5:1 ratio, 3.2 turns lock-to-lock, 38.5 ft turning circle curb-to-curb. **Brakes,** 11-in discs, all wheels.
DIMENSIONS & CAPACITIES. Wheelbase, 120.1 ins. **Length,** 207.5. **Width,** 71.8. **Height,** 59.8. **Headroom,** F/R, 36/36.5. **Curb Weight,** N.A. **Fuel Tank,** 22.5 gals. **Luggage,** 19 cu ft.
FUEL ECONOMY. EPA MPG, City, 10. **MPG Range/Avg on test,** 8.0 - 16.2/10.1. **Tank Range,** 180 - 365 miles.

High-Roller Ragtops

Wherein four auto editors abscond with a quarter-million dollars' worth of crème de la crème convertibles.

• We've all heard the oft-told tales of sure-fire investment schemes gone bust, oil wells that dried up, and house-limit rolls at Vegas that came up craps. But ponder, for a moment, a far more serious calamity, the tragedy of *gaining* unexpected riches overnight. The 7-Eleven clerk who inherits a bank full of money and the securities secretary who bought Belridge Oil before merger face a tougher problem than any of us realize: how to *spend* their money. Today's gold prices are tracking 1928 Dow Jones trends point-by-point, the Susan B. Anthony dollar has no more cachet than a

PHOTOGRAPHY BY AARON KILEY

quarter, and millions' worth of California real estate slides into the ocean every time it rains. Which brings us to this portfolio of three convertible assets that could safely and conveniently soak up a full quarter-million dollars of your excess cash, and simultaneously entertain you with inestimable pleasure.

Just in case you wake up filthy rich some day next week, you should be aware of the fine art of spending big money on cars. The first rule of high-roller etiquette is never, ever come right out and *ask* the price. We'll save you that trouble. The Mercedes-Benz 450SL is the bargain buy in this league at a mere $37,526. Of course, you should expect to pay a little extra for a back seat; the two-plus-two Aston Martin Volante is justifiably a bit more costly, at

$79,650. And for you really conspicuous consumers, we have the most expensive car on the face of the earth, the Rolls-Royce Corniche convertible, at a cool $140,925. Cash, check, and Manhattan office buildings are acceptable forms of payment.

The purpose of gathering these three blue-chip commodities together was not to name the one, true Smart Buy, or even the World's Best Ragtop, but rather to try and find out how much of this $258,101 mix is myth and how much is genuine coachbuilder's magic. If we had a good time in the process, well then, that's just one of the pratfalls of working at *Car and Driver.* So we started with some of our usual test procedures, carried on with several afternoons of photography in and around various high-

Yeah, but What'll She Do on Rodeo Drive?

• Money talks. I've always known that, but from where I'm usually sitting, the conversation is too far away to make out the words. Then again, I'm not usually sitting in a Roller or an Aston either. When I'm trolling around in these gilt-edged convertibles, the dollar dialogue just about drowns out the stereo.

When money talks in such dulcet tones, I'm inclined to listen. And when asked to rate the status of these symbols, I'm inclined to vote along strict money lines.

Let's face it. Nobody buys one of these cars merely to make sure he gets to work every morning. Rather, they are bought to make a statement, to show that, even when you're trapped in the everyday amber of traffic, you're not just another fly. And that's why the Mercedes goes to the bottom of this exalted heap. Too many people have M-Bs already. Hell, the cleaning ladies in Bev Hills show up in them. They've discovered that the resale value holds up better than on their Cadillacs. As far as they're concerned, a Mercedes is just your basic good deal, and nothing has less status than that.

An Aston Martin will trump an M-B any day. An Aston is a gambler's car. You unroll 80 large for this convertible and you don't even know if the company is going to be in business next week to sell you spare parts. That takes guts. Drive an As-

ton Martin and people are going to know that the kind of trivia you worry about would be life-and-death stuff to them.

Still, 80 large is not what it used to be. Any halfway-decent welfare scam artist could pile up that much by Thanksgiving. So, for the kind of status based on pure unattainability—which is the only sort that really holds up in the cut and thrust of Rodeo Drive—it pretty well comes down to the Corniche convertible, at 141 large, or nothing. You find only the finest class of pretenders behind this wheel.

Moreover, the Rolls also happens to be a regal motorcar in every detail. It's hard to find anything lacking in a car that protects its carpets with wall-to-wall floor mats of dyed lambskin. No matter how closely I looked, there were no flaws. The inlaid veneer was magnificent. The chrome was mirror-bright. The emerald-green paint was polished to a jewel-like luster. And somehow, unexplainably, perhaps even miraculously, this car seemed to resist dirt. It was always shiny and unsullied.

Perhaps it was for this reason that finding a sprinkling of California winter gnats splattered on the gleaming grille momentarily took me aback. But then I realized that bugs, like people, probably figure a Corniche convertible is the only way to go. —*Patrick Bedard*

roller enclaves, and spent the remaining hours racking up miles behind the wheel. Sure, it's tough to go back to Subarus and Toyotas after Aston Martins, Sunday-afternoon soaring, and pâté de foie gras, but we manage.

Our high-rolling resulted in three editors and one photog wearing winter tans back to the Michigan office. Every belt buckle was adjusted a notch wider because of the aforementioned high-cal sojourns. And, oh yes, we did learn a thing or two about expensive cars, some of which was surprising. Money may not buy you love, but, well spent, it'll definitely buy plenty of attention. A week of stirring up a wake of twisted necks, lifted eyebrows, and more oohs than boos gave us a good feeling for the level of respect each of these high-rollers can command.

The Rolls was far and away the winner here; it never failed to elicit the full red-carpet treatment, complete with fawning valets and gracious maître d's. The Parthenon-perfect radiator shell, sculptured mascot, and interlocked Rs that make way for the Corniche are more recognizable than the presidential seal. The Rolls-Royce that follows commands at least as much esteem. It's still the benchmark of class in cars, and quite secure in its status.

Unfortunately, California was the wrong venue for the Mercedes 450SL. This was *the* in-car a few years back, but now that upwardly mobile attorneys and physicians drive SLs in droves, its awe-factor has fallen. In Beverly Hills, the rich folks' Disneyland, it's not at all unusual to stop for a light amid a three-car-wide silver streak, with little more than different top configurations to distinguish one SL from another: stylish ladies prefer the chic-looking soft top; men go for the steel roof no matter what the weather; while only exposure-hungry starlets seem to select topless running. Blond hair and bare shoulders will certainly turn heads, but the other two SLs invariably glide by without a nod.

Likewise, the Aston Martin Volante flew way over these blasé Californians. For most of them, 80 Gs' worth of fine British craftsmanship registers the same

Continued

	acceleration, sec 0–60 mph	¼-mile	top speed, mph	braking 70–0 mph, ft	idle, dBA	interior sound level full-throttle acceleration, dBA	70-mph cruising, dBA	70-mph coasting, dBA	EPA estimated fuel economy, mpg
ASTON MARTIN VOLANTE	7.8	15.9 @ 92 mph	128	225	59	81	75	75	10
MERCEDES-BENZ 450SL	11.4	18.1 @ 79 mph	108	199	49	76	73	72	16
ROLLS-ROYCE CORNICHE	12.5	18.6 @ 73 mph	106	205	52	76	72	72	10

impact as a nice, clean '65 Mustang. The few interested enough to search out the nameplate inevitably ask, "Say, mister, how much do those Austin Martians go for, anyway?"

That's the risk you take in flaunting this sort of wealth before the proles. Fortunately, there are compensations. You've got to limit your exposure if you want to travel about like a true high-roller. These convertibles are best savored privately, by the owner and a close circle of friends en route to the week's special event.

Both the Aston Martin and the Mercedes-Benz are drivers' cars, while the Rolls is most appropriately enjoyed in the hands of a trusty chauffeur. Even though the 450SL's chassis lacks the up-to-the-minute refinement of Mercedes' big sedans, and in spite of a debilitating loss of power for 1980 (twenty horsepower sacrificed to tighter emissions controls), it still feels good when you're in a hurry. The stout control efforts, the open-arms way your hands lie on the steering wheel, and the rock-solid sensations that register through the bodywork, seats, and suspension settings give the SL an integrity never seriously challenged by hard running or bad roads. The best part is the unique-to-Mercedes shock damping that keeps each foot firmly planted, with just enough harshness telegraphed back to keep the driver assured that all systems are under control.

The Aston Martin takes this business-like attitude toward motoring and adds the finest hardwoods, leather, and aluminum-bodywork artistry money can buy. The instant you fire up the four-carburetor, four-camshaft V-8, the Volante starts speaking in distinctly male tones. A basso rumble to the pipes and heavy efforts at every control leave the undeniable impression that this is a car for men of stature. Fine-limbed ladies are welcome as passengers, but they have no business worrying about driving the Volante the only way it should be driven: hard and fast. Around town, this car is sluggish and cantankerous; it's too heavy and tall-geared to launch itself from a light with authority. A Sci-

	base price	price as tested	engine	wheelbase, in	curb weight, lbs	EPA interior-volume index passenger compartment, cu ft	trunk space, cu ft
ASTON MARTIN VOLANTE	$78,650	$79,650	DOHC V-8, 4x2-bbl carburetors	102.8	4110	85.5	5.7
MERCEDES-BENZ 450SL	35,839	37,526	SOHC V-8, mechanical fuel injection	96.9	3740	62.8	6.6
ROLLS-ROYCE CORNICHE	140,000	140,925	V-8, 2x1-bbl carburetors	119.5	5180	81.0	10.0

rocco will beat it in maneuverability, if only because the VW has rear visibility where the Aston has a folding top. But on the road when the motor's on-cam and the bends are 80-mph or better, the direct steering, hard brake pedal, and built-in forthrightness are just right for the mission: speed with grace.

Meanwhile, the Rolls conveys just the opposite message: "Why hurry?" Toss it about like one of these sports jobs, and the 2.6-ton Corniche whips its tail like a trailer. The Rolls-Royce craftsmen have toiled diligently to achieve a "ball of silk" feel throughout, from the electrically operated shift selector to the swift-but-silent power top, and they'd appreciate due respect. One doesn't bat a ball of silk around with a tennis racket, after all. The Corniche is overassisted and underdamped by today's standards. Unless, of course, you're ready to take life at a more leisurely pace, in which case the world that flickers and flashes back at you from the depths of the Brewster Green paint seems entirely acceptable. Set the auto-temp, crank up some Eagles on the Blaupunkt, wrap the top under its leather cover, and the Corniche will take you to all the right places. Wise owners make a point of reserving the Rolls strictly for special occasions, while striving to keep the annual odometer accumulation less than 10,000 mellifluous miles. With such a policy, the sheer envy of one's peers will appreciate this investment even more than the Corniche's devoted owner.

No doubt you'll rise above such piddling concerns as market value once you achieve full high-roller status. Which is exactly why we won't trouble you here with mundane gas-mileage, luggage-space, and operating-cost statistics. If it's necessary to pack a few things for the weekend in Vermont, just pick up the Aston's $2000 fitted luggage. Without a twinge. What could it possibly matter when you're into blue-chip convertibles? You've already learned the single most important fact of life while getting here: it's only money.

—*Don Sherman*

fuel-tank capacity, gal	steering	suspension front	rear	tires	brakes
25.8	rack-and-pinion, power-assisted, 3.0 turns lock-to-lock	ind, unequal-length control arms, coil springs, anti-sway bar	de Dion, 2 trailing arms, Watt linkage, coil springs	Avon R-R Turbo Steel 70, 235/70HR-15	all disc, power-assisted
27.2	recirculating ball, power-assisted, 3.2 turns lock-to-lock	ind, unequal-length control arms, coil springs, anti-sway bar	ind, semi-trailing arm, coil springs	Michelin XVS, 205/70HR-14	all disc, power-assisted
28.5	rack-and-pinion, power-assisted, 3.5 turns lock-to-lock	ind, unequal-length control arms, coil springs, anti-sway bar	ind, semi-trailing arm, coil springs	Michelin Wide X Radial, HR70-15	all disc, power-assisted

Just another Roller

Survivor of the Shadow series, defender of the faith and king of the overweights, the £66,000 Rolls-Royce Corniche convertible is now better than it used to be. But is that good enough? By Mel Nichols

I SAW A YOUNG MAN DRIVING A Rolls-Royce Corniche convertible. He was dark, tanned, handsome, his shirt open deep at the neck and his sleeves rolled well back from the wrists. He was in a hurry, gunning the Rolls hard as he went past me so that its nose lifted high and its radio telephone antenna whipped urgently. I looked down to watch him disappear in the mirror on the door of the Corniche I was driving, and wondered. What did *he* make of the Corniche? What had he expected when he began driving it?

It is, of course, a mistake to think that Rolls-Royces are the best cars in the world. If you're in touch with the broader standards of automotive engineering and have driven Silver Shadows, Corniches and Camargues before you will know that they are not – although, to maintain a decent reserve, the new Silver Spirit announced on October 1 may be a different kettle of fish. It is true that no other production car is as superbly detailed and beautifully finished as a Rolls-Royce; but if the claim is to be made that they are the best cars in the world then surely it's appropriate to expect that they are also the best engineered, so setting the standard for a combination of ride, roadholding and handling? A Rolls-Royce is seen by most people to be the pinnacle of motoring achievement; pardon the dogma, but I think it really should be, especially if, as in the case of the Corniche convertible, it costs £66,366.

Even at that price, the Corniche convertible is not the most expensive Rolls-Royce. The Camargue costs a phenomenal £76,120. But the Corniche is considered to be the ultimate

> 'The Corniche convertible has now been brought into line with its new brethren, not simply left as a relic'

personal Rolls-Royce and is thus the most glamorous car in the range. With the Camargue, it is also the only car of the old Shadow-derived series that survives the arrival of the Spirit and its sisters, the long wheelbase Silver Spur that replaces the Silver Wraith limousine and the Bentley Mulsanne, which takes over from the T-series. The Corniche coupe dies, right now, with the Shadow.

In its way, the Corniche convertible has been brought into line with its new brethren rather than being left simply as a relic of the previous generation. Last year, Rolls-Royce quietly but significantly altered the Camargue and the Corniche convertible by replacing their Shadow-type rear suspension with that of the new Spirit, which is hydropneumatically damped and self-levelled and has less unsprung weight. This wasn't only to set up the Corniche (now in production for almost 10 years) and Camargue for their run into the '80s but was in accordance with the Rolls-Royce tradition of introducing major new mechanical components first into the two-door models. By this method, Rolls are able to produce the new components in small numbers and have them in proper service long before full-scale production begins. In this case, the installation of the new suspension was a very well-kept secret.

Under its new body, the Spirit isn't all that different from the outgoing Shadow II. Its front track is half an inch wider than the Shadow's but otherwise it uses the rack and pinion steering and revised front suspension that Rolls introduced when they updated the Shadow in 1977. 'Having sorted out the front end,' says Rolls' chief engineer J H ('Mac') MacCraith-Fisher, 'we thought it was about time we did some work on the rear suspension. We'd had criticism from some of our customers and certain elements of the press, you know.' Over the phone, you can almost hear him grinning as he speaks.

Rolls-Royce's targets for the new suspension were a better ride, better suppression of road noise, less body roll and keener handling. Previously, the trailing arms were anchored to one cross-member and the axle assembly to another. Now, the trailing arms have been inclined closer together (in plan), altering the geometry to provide *more* camber change – more 'swing axle' effect. This makes the wheels remain more upright, thus reducing tyre scrub when the car is cornering. At the same time, the roll centre is raised by 2in, reducing body roll and further improving the handling. Rolls have now, through a series of stiff links, tied the leading cross-member to the after one, so that

the entire rear suspension is carried o a very rigid assembly. The new, and newly-located, mounts that hold it to th body permit minute fore and aft movement of the suspension frame bu have terrific resistance to lateral movement. Thus lateral movement of the rear wheels is much better controlled, and, through the better mounting material, transference of roadnoise is significantly reduced.

The Corniche's weight of 5200lb is carried by mini-block coil springs (of t Opel Senator type). But the ride qualit is provided by new hydropneumatic struts, which pick up from points 4in behind the wheels' centre line. Becau these struts – which also provide the car's self-levelling facility – have very good rising rate capability, Rolls-Royc were able to set the suspension more

meant that now it's a pleasure to touch the big, thin-rimmed wheel and to turn it; the weighting is excellent and the smoothness of the motion quite delicious. And the response of the car to the steering is now laudable. Then you notice that, at low speeds over knobbly city streets, the Corniche is riding very smoothly and quietly; the old traces of lumpiness and tyre patter

'In typical RR style, he doesn't mind admitting it: "We're following in the footsteps of Citroen"'

have been all but eliminated. Moreover, the car feels more stable and better-controlled, and is thus even easier to whip through city streets. The Corniche has a surprisingly tight turning circle, is conveniently narrow for its length and high enough to give the driver very good vision in traffic. While the ultimate performance is good rather than breathtaking (120mph and 0-100mph in around 30sec), the bagfuls of torque from the 220bhp 6.9litre V8, multiplied by the big GM automatic transmission, mean that there is real oomph away from standstill. Thus in town the

softly over all and achieve, at the same time, better load carrying ability while maintaining a much more stable condition from lightly laden to fully-laden. The effect is very much like that of the Citroen GS and CX suspension, and Mac – in typically self-effacing Rolls-Royce style – doesn't mind admitting it. 'We're following in Citroen's footsteps,' he says.

From the instant you begin driving this revised Corniche you notice that it has benefited a great deal from the presence of the new suspension, and from other refinement work. The steering, which used to be so awful before Rolls switched to rack and pinion with the introduction of Shadow II in 1977, is now beautiful. Initially, the rack wasn't all that impressive, but constant tuning (as Mac puts it) has

Corniche is a handy car, a marvellous car – it blends its manoeuvrability and response with quietness and real luxury so that the driver and his passengers are effectively removed from the hubbub around them. It doesn't seem to matter whether the top is up or down. And the condition is real: it isn't just that you're riding behind that famous mascot.

While the Corniche benefits discernibly from its update, its body's

'Not even Rolls-Royce can overcome the inherent difficulties of making a large, open car'

lack of rigidity and the noise allowed in by the cloth roof mean that it obviously isn't the ultimate showpiece for the new rear suspension. That will require the Silver Spirit itself. In the Corniche, the suspension itself copes very well with poor surfaces, but there are often bump-created tremors right through the Corniche's vast, open body. You can feel them in the seat, at the wheel rim and see them at the scuttle. They're not severe, but they're there; not even Rolls-Royce can overcome the inherent difficulties of making a large open car

rear suspension from the Silver Spirit and er mods have smoothed out Corniche's and sharpened handling but limitations sist: body tremor over bad bumps and oversteer at speed. Looks good, though, space apart, has exquisite cabin and trols – unlike visually poor engine bay

159

that doesn't even have the structural assistance of a rollover bar. You learn to live with these tremors.

You must also learn how to live with the Corniche when it rains. It is a quick handling car in that, with any sort of power applied, it corners neutrally with an increasing bias, as the prodigious weight takes over with speed, towards roll oversteer. In other words, the back comes around very quickly. In the wet, it will come around and let go with very little provocation. So you learn to use only the slightest trace of power when you're cornering if you wish to avoid what many might regard as the unseemly sight of a Corniche going very sideways. All this happens at very modest speeds: 20mph on a South Circular road right-hander, for instance. Although there is *more* roadholding than before, the level in the dry still isn't very high either. There's just too much weight to contend with, and building in understeer would be no answer at all.

On the open road, this means that you drive the Corniche as a stately open carriage, slipping along easily and tidily – and very comfortably – at

Corniche has limited room for four adults, though boot (right) is definitely Fortnum hamper size. Open road pottering with hood down in fine countryside is car's forte

50mph where something like a Peugeot 505, 604, Opel Senator, Mercedes or Jaguar (to name but a few of the better saloons) would be just as effortless and comfortable at 70mph. If you press on harder in the Corniche, you find yourself encountering the roll oversteer far too often for the comfort of both driver and passengers. In this area, the Corniche is an antiquated car with only the almost-as-heavy Bristol for behavioural company.

Driving within its limits is, however, a smooth and pleasant experience most of the time because the steering is so smooth and accurate and the body motion is (then) so well-controlled. Occasionally, over bends which are vertical as well as horizontal, the steering goes uncomfortably light in the driver's hands but the car as a whole retains commendable poise. In the back, so long as the driver doesn't go too fast and stays smooth and clean, you ride in an outstandingly peaceful manner. The ride itself is now magnificent and, top down, the car needs to exceed 70mph before wind buffeting becomes uncomfortable. In the front, the windshield provides superb protection when the side windows are up. Beyond 70mph, windnoise becomes noticeable to the point where it spoils the stateliness of the Corniche. The problem in the cabin, front and rear, is that, despite the car's 17ft length, there isn't enough legroom. The front seats adjust electrically and

multi-directionally (through finger pressure on small toggles on the respective doors) but when they're even right back there isn't enough legroom, or reach to the wheel, for anyone much over 6ft. In the back, unless the front seats are well forward (and that's out of the question unless the driver is short) legroom is Fiesta-class tight. The other things that are more than just disappointing about the Corniche are that although the roof is raised and lowered electrically, it has to be manually and awkwardly clipped home, and covered by a separate tonneau when it's down. The seat belts are diabolical.

But what is so nice about the car is the way things feel – the steering (a merciful recent addition in this category), the brakes, all the minor controls, the door locks, the action of the bonnet when you lift it. Rolls-Royce have made the quality of the movement of such things an art form, so beautiful to the touch that a Rolls-Royce is the most sensuous car there is – more sensuous to the fingers and the feet than any Ferrari or Lamborghini. That, of course, is the Rolls-Royce secret, the area where so much of their developmental time and money goes. From the moment you put the key in the lock in the morning you know you're in touch with fine craftsmanship. That's a good feeling; a satisfying feeling. And I can tell you that you grow to like it very much. The difficulty, as I see it, is accepting that this element, with the Corniche's sheer visual elegance and (now) its excellent ride, is both its greatness and its weakness. Can you accept that, in your £66,366 convertible, you can cruise above 80mph on a

'Can you accept in your £66,000 convertible that you cannot hurry on a by-road you wish to take?'

motorway if you don't mind a lot of windnoise but cannot really hurry on any by-road you may wish or need to take? Can you accept roadholding that, despite the new suspension, is still at a very modest level, and that at the first sign of rain you will need to be very, very circumspect? Can you accept that you may well be able to enjoy motoring in the Corniche, but not really driving? The car's mien, and still rather too much of its ability, is of another era altogether, and without an enormous reduction in its weight it's hard ever to see it being any different. Enjoy it then for its feel and its image, not its prowess; its craftsmanship and development, not its design. I wonder if that's how the young man in the Fulham Rd reacts to his Corniche?

PRACTICAL CLASSICS BUYING FEATURE

Bargain or bankruptcy?

Series One Rolls-Royce Silver Shadows are now both cheaper in real terms and more widely available than they have ever been. But has this really made ownership more practical for the majority? Peter Simpson investigates.

Bargain or bankruptcy? Bargain or bankruptcy? Bargain or bankruptcy? Bargain or bankruptcy? Bargain o

Although they might not admit it, most car enthusiasts would probably like to own a Rolls-Royce one day, but most probably also realise that for them this can be little more than a pipe-dream. But if such a person, out of curiousity, were to investigate the prices of Series One Rolls-Royce Silver Shadows, they may be surprised to find that 'up and running' cars are available from around £6,000 (the cost of a new, very down-market car), whilst for £8,000 it should be possible to find a very nice example of around 1970 vintage. £8,000 is also about the present value of a 2-3 year old Jaguar XJ6, and apart from the increased prestige, the older Rolls-Royce will be unlikely to depreciate unless it is neglected. The Jaguar, even given the very best attention, will probably lose value for several years

So, perhaps Rolls-Royce ownership is a possibility; after all, in theory, many of those who spent £6-8,000 on a car last year could have bought a Rolls-Royce.

However, before everyone reading this trots down to their nearest Rolls-Royce dealer, there is of course another side to the story. We, experience in the art of running older vehicles, know that the initial purchase price is only the tip of the iceberg. Will the negligible depreciation cancel out all the running costs? How practical will the car be for everyday use? And, above all, will a Rolls-Royce that is 10-15 years old still be 'the best car in the world'? If it has lost any of its legendary comfort, silence and reliability, there seems little point apart from sheer snob value, in even considering it.

Model history

The Silver Shadow was announced at the 1965 Earls Court Motor Show, though due to production difficulties it was not generally available until the spring of 1966. The new model had been under development for ten years, and when the car appeared it was not difficult to see why. The Shadow, along with the virtually identical Bentley T series (even the best car in the world utilises badge-engineering!) was easily the most technically advanced Rolls-Royce yet — for the first time Rolls-Royce had produced a car with unitary (chassis-less) construction. This was by now an accepted feature with 'mass market' manufacturers, but for Rolls-Royce it was very much a new departure and brought with it significant design problems, such as how to achieve an acceptable noise level. This problem was solved by mounting all the suspension components on separate sub-frames which were then secured to the body by 'Vibrashock' mountings. These consist of wire mesh that is compressed into a cylindrical pad, and proved to be more effective in eliminating larger movements than the more usual rubber mountings. Independent front and rear suspension was fitted, and with it, automatic, hydraulically operated, height controls. These consist of hydraulic rams fitted above the springs. As the car is loaded, with the engine running, hydraulic oil is admitted to the rams, restoring the correct ride-height. The same power source (a pump in the engine compartment which can pressurise the hydraulic oil to as much as 2500 psi) serves the braking system. The brakes themselves were discs (again a 'first' for Rolls-Royce, who were not prepared to fit them until they had worked on a system enough to be certain it would be an improvement on 'their' conventional drum brakes). Three systems were fitted, the suspension pump providing 31% of the total effort, another pump providing a further 47%. The remaining 22% was supplied by a conventional hydraulic master cylinder operating on the rear wheels only. As well as

Even the common 'three box' body styling applied to so many cars in the sixties and seventies looks somehow more distinguished when seen on a Rolls-Royce. Unusually, this 1970 Bentley T series has a 'year letter' registration; after spending at least six thousand pounds, many owners consider that a few hundred more to buy a non-year letter registration and disguise the cars age is worthwhile.

The 'Silver Lady' or Spirit of Ecstasy is, along with the Rolls-Royce radiator shell, a Rolls-Royce trademark. Surprisingly, although the mascot was available on all Rolls-Royces from February 6th 1911, it was as an optional extra until after WW2. Contrary to popular belief, they are not made from solid silver. The Silver Lady mascot was designed by Sir Charles Sykes, and the model was Eleanor Thornton, secretary to John Scott Montagu, later 1st Baron Montagu of Beaulieu. Once the most "stealable" feature of the Silver Shadow, most owners now 'alarm' these.

The "Flying B" Bentley mascot and radiator-top badge.

providing the entire system with 'feel', this supplied a back-up system in the unlikely event of both power systems failing.

The engine was a modified version of the V8 unit used in the Silver Cloud III, with different cylinder head castings and exhaust ports (more power was needed to drive the hydraulic pumps). The automatic transmission (which had of course been a standard Rolls-Royce feature for many years) was initially supplied on right hand drive cars by a refined version of the Silver Cloud box. This unit, which dated back to the late 1940s, was becoming out of date (it did not have a torque converter), so for left-hand drive cars a new box was selected, the intention being that this would eventually be fitted 'across the range'. After much deliberation, the General Motors GM400 box was chosen and unusually for them, Rolls-Royce bought in the complete units rather than taking out a licence to manufacture. Clearly, Rolls-Royce were satisfied

Bargain or bankruptcy? Bargain or bankruptcy? Bargain or bankruptcy? Bargain or bankruptcy? Bargain o

with the quality of GM's product. RHD cars received this box from chassis number 4483, in the summer of 1968.

Many readers will not realise that the bodyshell for the 'standard' (as opposed to the coachbuilt) Shadows and Ts was supplied by Pressed Steel (which became a BL subsidiary), with light-alloy doors, bonnet and boot lid. Rolls-Royce and Bentley shells were identical apart from the shape of of the bonnet. Despite a few teething troubles on early cars (which were very thoroughly sorted, so early cars need not now be regarded as suspect) the new car was very well received, and a lengthy waiting list soon developed. Indeed, a purchaser of a new car who had just taken delivery (after waiting years in some cases) could often, if he so desired, sell the car immediately for a substantial profit, such was the demand from people prepared to 'jump the queue' and pay heavily for the privilege of so doing.

Although the Rolls-Royce Silver Shadow and Bentley T Series look almost the same from a distance, there are numerous detail differences. Obviously, the badging has to be different, and Rolls-Royce and Bentley badges appear a lot on their respective cars. Hubcaps seem to have replaced mascots as the most common target for theft — not surprising when a new set of wheeltrims cost over £600. A locking kit is available from The Owners Club (UK) Ltd.

Between 1966 and 1970, as one might expect, there were numerous changes to the standard cars, so many in fact that there is insufficient space to list them all. The changes included a suspension modification (to improve handling) in mid 1968, and in 1969, a number of changes to meet new American safety regulations. The most noticeable of these was that the dash now had much more padding around the edges than previously. At about the same time, the self-levelling arrangement was removed (chassis number 7404) as in practice the difference in load between driver-only and fully-laden was very slight. Other changes included stainless steel exhaust (from late 1969), viscous fan for engine cooling (from late 1968) and swing-needle SU carburettors from spring 1969.

In 1970 however (from chassis number 8742) the capacity of the V8 engine was increased to 6,750cc by increasing the stroke length, the new engine being fitted to the Shadows for the 1971 model year. Rolls-Royce do not publish BHP figures, but it is likely that the increased capacity made little difference as the bigger engine was somewhat strangled by the requirements of USA emission-control laws.

Probably the most significant improvement was made in mid-1972 (chassis number 13485). From the start, there was a certain amount of concern about the roadholding, steering and stability, and various attempts to improve matters over the years had not been particularly successful. From this date, the entire suspension was reworked and the 'Vibrashock' mountings replaced. The new 'Compliant' suspension was a great improvement (though *Autocar* testers were still not 100% happy about the steering and the amount of understeer). Subsequent experience showed that tyre life was also lengthened on 'Compliant' cars. Other significant changes before 1977, when the series I Shadow was replaced by the series II, were modified frontal treatment (including shortening the radiator grille) to allow '5 mph' American-type shock-absorbing bumpers to be fitted (from autumn 1973), fatter, low-profile tyres, and subtle, almost unnoticeable extra flaring of the wheelarches along with an increase in wheelbase (from chassis number 18269, spring 1974), and in the autumn of 1975, the brake system became all high-pressure, the conventional system being discarded. The series I Shadow was superseded by the series II in 1977. All models (except the LWB Shadow, see later) were available as Rolls-Royces or Bentleys, but in the event only about 8% left the factory as Bentleys.

Coachbuilt variations

The introduction of these Rolls-Royces made the job of the traditional coachbuilder more difficult, and although the Bromley-based concern of James Young produced a two-door Shadow, this was not a great success, only 50 cars (35 RR, 15 Bentley) being produced. The James Young car was simply a two-door modification of the standard Shadow; only the doors and quarter windows were new, and beyond the longer front doors, offered no advantage over the standard Shadow. However, arrangements were made for H.J. Mulliner Park Ward (who were a Rolls-Royce subsidiary based in Willesden, North London) to be supplied with Shadow underframe/bulkhead/scuttle assemblies, onto which specialist bodywork was erected. Thus, the Mulliner Park Ward 2 door saloon

The view in front of the Silver Shadow driver. As one would expect, everything is top-quality, and expensive to restore. Interiors last well and will usually have been looked after properly.

Specifications
Engine cc 6230cc 1965-70, 6750cc 1970 onwards.
BHP Never revealed; "adequate".
Transmission Early RHD RR/GM Hydramatic 4 speed, later RHD and all LHD GM400 3-speed with torque converter.
Rear axle ratio 3.08:1 throughout.
Overall Length 16'11½".
Overall Width 6'
Overall Height 4'11¾" 4 door, 4'10¾" 2 door and convertible.
Tyre Size 8.45 x 15 crossplys non-'Compliant' suspension (1965-72)*; 205VR15 radials 'Compliant' suspension (1972-74); 235/70 x 15 or HR70HR15 radial ply tyres 1974-77.
* Radial-ply tyres not original equipment on pre-1972 cars, but many owners have fitted them and found them to be satisfactory.

was announced in March 1966. The two-door had sleeker lines than the four, with a distinct dip in the crown line just in front of the rear wheels. Despite an asking price 50% higher than the standard saloons (the two-door cost £9,849 in 1966) it enjoyed good demand. In September 1967, a convertible version was added, with a power-operated hood (that could only be moved with the handbrake on and the car in neutral) and, unusually for a convertible, almost the same rear seat space as on the saloons. The coachbuilt saloons took at least twenty weeks to complete, with the convertibles taking a little longer; the hood alone took a week!

Throughout the production run, it was

Bargain or bankruptcy?

Silver Shadow chassis numbers

For their first monocoque cars, it was appropriate that Rolls-Royce should adopt a new system of chassis numbering. The system outlined below was used right up to 1980, when the Silver Shadow II was discontinued, Rolls-Royce thereafter using VIN numbers. The Shadow numbers consisted of three letters, followed by four or five numbers. Thus, a typical number might be:

SRH 1999

The first letter, denoting the body style, could be one of the following:
S = Standard 4 door saloon
C = Two-door saloon
D = Two door convertible
L = Long Wheelbase.

* only after mid 1969 (chassis number 6646), before that all coachbuilt cars carried 'C'.
The second letter stood for the make.
R = Rolls-Royce
B = Bentley

The third letter denotes left or right hand drive, and in the case of post 1972 cars for North America, the model year.
H = Right hand drive
X = Left hand drive

A North American 1972 model year
B North American 1973 model year
C North American 1974 model year
D North American 1975 model year
E North American 1976 model year
F North American 1977 model year
G North American 1978 model year
K North American 1979 model year
L North American 1980 model year
In addition, 1980 cars for California had the letter C after the number.

The number was of course the actual number of the vehicle, all models sharing the same number sequence. Thus, whilst number 1067 was the first James Young two-door, number 1068 could be a Rolls-Royce 4 door saloon, 1069 a Bentley 4 door etc etc. The small number of Phantom VI models produced to special order during this period were given chassis numbers within the main sequence, along with the body style letter P.

Buyer's Guide

In preparing this article, and this section in particular, I would like to acknowledge the help freely given by Neville Williams and his staff at The Owners Club UK Ltd (260 Knights Hill, West Norwood, London SE27, telephone 01-761 6565). The Owners Club are specialist repairers of Rolls-Royce, and pride themselves on offering customers a good, quality personal service at competitive prices. Neville was quick to point out that proper servicing of a Shadow is vital, so when viewing a car one should insist on seeing the service history (and make sure that it does relate to the car you are looking at) and never buy a car without one. In particular, Shadows require a very comprehensive service every 48,000 miles, which involves a ruthless examination of all the hydraulics and replacement of anything that 'isn't right'. This takes three days and can cost well into four figures, so it is not surprising that some less well-off owners are tempted to give it a miss, with potentially expensive results later. Neville strongly advises professional inspection of a potential purchase; he has frequently seen cars that have been bodged up in potentionally dangerous ways (which I will not go into here, just in case it tempts anyone) and which will cost many times their purchase price to put right. The main problem, according to Neville, is that because a Shadow is so silent and smooth-running, it can hide potentially expensive defects that would be spotted immediately on a lesser car. For these reasons, the following notes are intended not as a comprehensive buying guide, but as a way of enabling the purchaser to rule out any real 'duds' for himself, before getting his expert to look at the shortlisted cars.

Many Shadows have been company owned at some time, and on balance this is probably an advantage as companies are generally

Plenty of usable space in the boot, which as one would expect is fully-carpetted. The box on the left contains the battery.

better placed to pay for the maintenance. It is also quite common for a car to change hands every 2-3 years; this again is no disadvantage as long as the maintenance record is complete. Few Shadows are chauffeur-driven.

With such a complicated car, on which there is so much that can go wrong, it makes sense to buy from a known source, such as a reputable specialist dealer, who has something to lose if the car fails to give satisfaction. By law, a dealer is obliged to offer a

general policy that the coachbuilt cars received most design changes before the standard saloons, although the padded facia and the 6¾ litre engine were introduced simultaneously. Both the two door saloons and the convertibles were replaced in March 1971 by the Corniche, though in fact the Corniche was 'only' an updated Mulliner Park Ward Shadow.

It is also appropriate to consider the Long Wheelbase Shadow here, as this was also produced by Mulliner Park Ward. These cars, available with or without a central division, had an extra 4" let into the roof, floorpan and rear doors, and also had a smaller rear window. Available from May 1969 (though a prototype was supplied to HRH Princess Margaret in 1967), the series I LWB car was only produced as a Rolls-Royce, the only model in the Shadow I family not available in Bentley guise.

Plenty of space for rear-seat passengers too and rear seat passengers have individual cigarette lighters as well as ashtrays.

guarantee, and to be worthwhile this should be valid for at least a year, and cover parts and labour. It is dangerous to be dogmatic, but I would advise against buying a Shadow from a 'general' secondhand car dealer, as even he may not be aware that the 'Roller' he has just bought at auction and put on the forecourt has serious defects.

Even Rolls-Royces rust if they are not properly maintained (I have seen a ten year

Bargain or bankruptcy?

Air-conditioning, powered by this Fridgidaire compressor was a standard fitting from chassis number 7500, though it was a popular option before that. This system can be difficult and expensive to repair, so ensure that it is working as it should. Leaks can occur, particularly on a system that is switched on after a longish period of idleness.

old Shadow with virtually non-existent sills) so the sills should be examined, along with the areas behind and around the rear wheels. Look also at the leaded panel joints; these deteriorate with age, and the first sign of problems will be cracks in the paintwork. Check the paintwork also for evidence of poorly repaired accident damage, which will show up in badly matching paint, crooked coachlines and so on. It is also worth checking that the gap around the bonnet is constant — if it isn't the shell may be twisted. It has been suggested that the brightwork on Mulliner Park Ward cars has not lasted as well as on the standard ones; a pity, as it is extremely expensive. As one would expect, the interior lasts well. Few Rolls-Royce owners allow their cars to get tatty inside, but if you are offered one that is less than pristine, restoration can be expensive if the leather is torn, though re-Connolising scruffy leather is not as dear as one might think.

On the mechanical side, it is important to check that the car rises to its normal riding height as soon as it is started. Your expert will certainly want to check the condition of the hydraulics underneath the car, for leaks and other problems. As one would expect, the engine should be good for 100,000 miles, but partly because of the very effective sound-deadening and smooth-running qualities of a Shadow, even severe faults like bearing rumble may be extremely difficult for the layman to spot. The hydraulic valve gear should be silent when the engine is warm, some noise when cold is normal. In fact, when a car that has been out of use for a day or two is first started, it may well sound like a 'bag of nails' until the oil has got round the hydraulic tappet mechanism. Carry out the 'usual' checks for an engine that is burning oil (look for signs of oil being burnt in and around the exhaust tailpipe). Silver Shadow oil consumption incidentally should be negligable, so check the oil level. If it is low, at the very least it is a sign of an uncaring owner. It could mean there are leaks, or that the engine is past its best. Both gearboxes fitted to RHD cars are reliable, the GM unit in particular will last a long time. As it is a standard GM unit repairs are not, by Rolls-Royce standards anyway, that expensive. Give a thought to the condition of the tyres too; they are very expensive, and on non-'Compliant' cars rarely last more than 15,000 miles, though the later cars are better. Particularly on cars that are not used regularly, the brake discs corrode around the edges and, again especially on a car that has had periods of idleness, it is also sensible to check all the rubber bushes for wear.

When carrying out the checks outlined here, put any idea that the costs of rectification will be in any way related to the cost of doing similar work on an ordinary car firmly out of your mind. It is also highly unlikely that anyone other than a specialist will be equipped to carry out the work; a large number of special tools are required.

Which model to buy

There are no differences between a Rolls-Royce and Bentley that make the latter in any way an inferior car, and as there is some evidence that Bentleys are cheaper than Rolls-Royces, the buyer choosing a Shadow for its excellence as a car rather than just 'image' may find a Bentley will suit him better as well as being more unusual. Don't be tempted, incidentally, to buy a Bentley and then 'change' it into a Rolls-Royce. You will have to replace a lot of very expensive parts, and there is no legal way you will be able to get the registration document changed. In addition, Rolls-Royce and Bentley models had different number series, so an expert will spot an 'undercover Bentley' immediately.

The choice between two door, four door or convertible will always be a personal one, though influenced no doubt by the coachbuilt cars being dearer (the convertibles up to 50% dearer) to buy than the four door saloons. The long wheelbase cars with divisions are much less practical than those without; not only does the division impair conversation, it also restricts front seat space, and removes almost all the increased space in the rear.

Rolls-Royce style brake-fluid reservoir. The level can be read off after the car has been running for four minutes without removing the filler caps. It is imperative that the correct type of fluid is used; ordinary fluid can damage brake and suspension systems.

What to pay

In general, the coachbuilt cars are more sought-after than the 4 door saloons, and a two-door Mulliner Park Ward saloon can be expected to be 20-25% dearer than a four door, with a convertible at least 50% dearer. Prior to 1972, age appears to have less effect on price than condition, and a 4 door saloon in worthwhile condition should be in the £6,000-£9,000 range. 1972-1976 cars seem to be dearer, £8,000-£12,000 appears to be normal. Unlike most cars we write about in *Practical Classics*, Silver Shadows and Bentley T Series seem not to be advertised very often as 'needing a little attention'; they are almost always "perfect". The buyer has to satisfy himself or herself as to the condition of the prospective purchase. The only way of getting a car in need of rebuilding is probably to buy an accident or fire-damaged one, though given the price of parts and the complexity of the cars, I would certainly not recommend anyone to do this unless they are very keen and very skilled. Overall, our frequent advice in these features to 'buy the best and forget the rest' certainly holds good.

At the wheel of a Silver Shadow

Not unnaturally, I was keen to see what a good Rolls-Royce Silver Shadow feels like to drive, but I was also a little unsure of what exactly to expect. My first surprise came on sitting in the car, I seemed to be much further off the ground than I expected. As soon as I started the engine, the whole car seemed suddenly to come to life. It seemed to lift itself a good three inches, and the steering and brakes, which had hitherto seemed totally lifeless (it was impossible to turn the steering wheel without the engine running, and the power-assistance) was suddenly 'all there'. I had been warned that the power-brakes were extremely effective, so it was no surprise that the foot-pedal felt totally different to that on anything I had ever driven before. A light pressure on it was enough to stop the car, and you jab at it at your peril! As I moved off, I was struck by how quiet and effortless the whole business was. I would not go as far as to say that all I could hear was the clock, but it certainly was the quietest car, both in terms of road and engine noise, that I have ever driven. The automatic transmission changed up and down so smoothly that it was impossible to feel it change, and although I did not feel inclined to give the performance a thorough test the car certainly had lots of 'go'. Driving through the crowded streets of South London I was also struck by how manoeuverable the car was, despite its size. On other cars I have driven that have power steering, the power assistance has sometimes been too great and deprived the steering of all its 'feel', but that was certainly not the case here. Everything on the car, even down to the heater and ventilation controls, felt to be in exactly the right place.

After I returned the car to Neville Williams' premises, and whilst I was taking a few final photographs, imagine my horror when one of the mechanics working in an adjacent garage, asked me to move the Rolls-Royce, so they could get a car out. I had no choice but to manoeuvre the Shadow in the tightly packed yard. I needn't have worried! It was a pleasure, and certainly far easier than many smaller cars I have driven, and thanks to the good all-round visibility I was able to re-park it within a couple of inches of a wall and another car at the first attempt. Overall, I think is probably obvious, I enjoyed driving the Silver Shadow immensely, more I think than any car that I have ever driven, and I do not think that was simply because I knew that I was driving a Rolls-Royce. Rather, it was because the car felt right in every department. There was nothing that I disliked about it, and the car had that difficult to pinpoint but unmistakable feel of quality about it. I accept that there are probably cars that are better than the Silver Shadow in some specify areas (like road holding) but these would almost certainly fall behind it in many others. I know of no other cars from the same period with such a good overall standard, the ultimate motoring experience.

Conclusions

I started this article with the intention of proving it was possible to afford and run a Rolls-Royce on a similar budget to a nearly new,

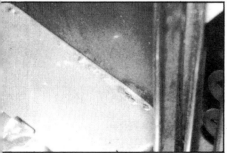

Even Rolls-Royces rust, and here's the proof! This is the front panel/valance area of a 1974/75 Shadow, and as can be seen, rust is breaking through.

Parts availability

Everything for the Silver Shadow is available, at a price. The potential owner should not be surprised to find out that Rolls-Royce parts are expensive, but as a guide, here are a few examples, all for a 1970 Rolls-Royce Silver Shadow 6230cc four door saloon.

Factory engine: £5,000 exchange (plus £3,000 surcharge on old unit)
Gasket set (engine): £83.24
Radiator grille: £850 (surround) £334.78 (grille)
Front bumper (corners): £125.02 each (centre): £129.59
Oil filter: £10.68
Refrigeration compressor: £284.82
Rear light cluster: £77.43
Front wing: £467.29
Automatic gearbox (GM type): £750 exchange (£250 surcharge)

The above prices were obtained from Appleyard Rippon in Leeds; however Silver Shadow spares should be obtainable through any official Rolls-Royce dealer.

If the price of these parts frightens you rather, it is of course possible to buy secondhand. In recent years, a number of specialist Rolls-Royce/Bentley breakers have sprung up, and many items, particularly chrome trim, body panels and mechanical spares can be obtained from these, in perfectly good condition and for worthwhile discounts. Don't expect to get (say) a set of hubcaps for the fiver you would pay for a set of Cortina ones in the breakers yard down the road though. The Rolls-Royce breakers are well informed on new prices (and if any parts are temporarily unavailable new) and charge accordingly.

The Silver Shadow is a heavy car, so not surprisingly it puts a lot of strain on the suspension. Check all suspension bushes for wear; these rear tie-bar bushes on 'Compliant' cars are a particular weak spot.

Though minor bodywork defects like this may not be that important on a more downmarket car, they certainly are important on a Rolls-Royce, because having the damage repaired to the original standard is likely to be costly. A full respray, to RR standards, will cost at least £5,000.

but more downmarket car. However, I seem to have proved that whilst it is possible to buy a reasonable Silver Shadow for a similar outlay, the running costs will be considerably higher, even if only a low annual mileage is covered. It is also important to remember that a Silver Shadow is not at present an appreciating asset, so unless you are very lucky, you are unlikely to be able to resell in a couple of years for a profit sufficient to pay for all the repairs during your ownership. On the other hand, it is unlikely to depreciate at anything like the rate of a new vehicle (those of us who spend all our time with older vehicles tend to forget that a new car can depreciate by over 20% in its first year) so at least part of the increased cost can be set against this. However there can be no doubt that a well looked after Shadow, even one that is over fifteen years old, is a better car than a nearly new 'Rustmobile Hatchback 1.6GLS, and provided you do not regularly park in dimly lit back streets (where, in any case, only a fool would leave a nearly new car) and have a garage big enough to take a 16'11½" x 6' car, the Silver Shadow is a perfectly practical proposition — particularly if you cover a lowish annual mileage, as the 12-17mpg will then be less significant. As we have seen, the Silver Shadow is a very good car, well worthy of its parentage, and what can replace the feeling of driving effortlessly and silently,

Bargain or bankruptcy? Bargain or bankruptcy? Bargain or bankruptcy? Bargain or bankruptcy? Bargain o

behind the most famous motoring mascot of them all, the Spirit of Ecstasy? Take your time choosing a car though:- remember that well over half the Silver Shadows ever made still exist, so it should be possible to find the right one!

Even the rear-view is distinguished! Note the individual reversing lights on the bootlid, these were later incorporated into the main rear light cluster.

Insurance

It is a mistake to try and insure a pre-1972 Silver Shadow through a High Street insurance broker; we were quoted (for a 26 year old journalist, living in North Kent, driving a 1970 6230cc Bentley T series, pleasure use only, clean record, maximum N.C.B.) between £328 and £400, for comprehensive cover. The Rolls-Royce Enthusiasts Club however runs a special agreed value scheme, without mileage limitation, under which the above cover would only cost £89. The RREC Scheme also includes a very useful salvage clause under which, in the unfortunate event of a total loss, as well as the agreed value, the owner receives the salvage. The difference speaks for itself, as even after paying the RREC membership fee, there is a saving of over £200!

The writer wishes to thank Malcolm Bennett, of the Balmoral Automobile Company Limited for assistance, and Neville Williams and Ray Braithwaite of The Owners Club (UK) Ltd for their help in preparing this article,

Clubs

The Rolls-Royce Enthusiasts Club caters for all Rolls-Royces, including the Silver Shadow, along with Derby and Crewe built Bentleys, including the T Series. As well as the insurance scheme mentioned elsewhere, the club, which has over 5,000 members worldwide, offers social events, local sections, a regular high quality magazine and technical advice and seminars. Further details may be obtained from the Secretary, LT Col. E. Barrass, 6 Montacute Road, Tunbridge Wells, Kent, TN2 5QP.

Bentley T Series owners are also eligible to join the **Bentley Drivers Club**. This is a more sporting-orientated organisation and enquiries should be addressed to Mrs B. Fell, 16 Chearsley Road, Long Crendon, Bucks.

Bargain or bankruptcy? Bargain or bankruptcy? Bargain or bankruptcy? Bargain or bankruptcy? Bargain o

ROLLS-ROYCE SILVER SHADOW

Rolls-Royce Silver Shadow — luxury status symbol or high-speed route to bankruptcy? Prices for early Shadows are now well below the £10,000 mark, but Chris Rees takes a no-holds-barred look at why you *shouldn't* buy one

Shadow of doubt

WHEN ROLLS-ROYCE INtroduced the Silver Shadow in 1965, it cost a whopping £6670 — three times as much as a Jaguar 420G. Today, you can just about find a Shadow for the same money (inflation providing the required distortion). But if you spend £10,000 on a Rolls-Royce, what will you be letting yourself in for? Is the Shadow a dream or a nightmare?

The Silver Shadow marked a definite change of direction in Rolls-Royce design. Not only did the car look different — its straight edges and boxy shape reflecting the trends of the time — but technically it broke entirely new ground. This was the first monocoque model from Crewe, complete with independent suspension and disc brakes all round — borrowing technology from Citroën's hydraulic system — and power assistance for virtually everything.

It sold very well for a Rolls-Royce and spawned many derivatives — including the Corniche and Camargue which used the same floorpan and mechanicals. Older and less well looked-after examples are now available on the second-hand market for under £10,000, a figure which reflects the perceived fortune required to run them.

The Shadow's giant 6.7-litre V8 could hardly be described as economical (between 13 and 16mpg is normal), servicing is not feasible for any but the most determined home mechanic and should anything go wrong and you need parts for the car — well, see our panel for the likely costs. The insurance companies recognise the colossal cost of ownership and rate the Shadow accordingly. All in all, this is not a practical alternative to any other luxury classic car.

On the other hand, the Silver Shadow *is* now undoubtedly a classic and prices have come down so far that many must be tempted to buy one. A good example of a Shadow should, with proper treatment, last almost forever, and prices are bound to escalate given time. ▶▶▶

PHOTOGRAPHY: JAMES MANN

169

This 1979 Shadow II was well gone before restoration began

Sills and rear wheelarches needed replacing, as well as the inner wheelarch filler panels

Offside sill OK, but still checked carefully for rot at front, back and floorpan joint

Front wing lower sections required on both sides — welded, lead loaded then Waxoyled

Lower inner and outer rear wings replaced plus the rear valance

What sort of car are you likely to find for less than £10,000? The car will almost certainly be a Shadow I (ie pre-1977), probably with high mileage and some faults in evidence, particularly in the bodywork and engine departments. At the bottom end of the scale, basket cases can occasionally be found for as little as £5000. But our advice on these cars is to avoid them.

What to look out for

The most important thing to check on a Silver Shadow is its service history — who's looked after it and what's been done to it. Shadows do need regular preventative maintenance such as changing fluids and replacing hoses.

Tell-tale signs of neglect are numerous. The outer bodywork is susceptible to corrosion, especially around the wheelarches. The rubber axle mounts at the rear often go. In the electrics department, the windows and locks can be troublesome and, if a sunroof is fitted, the drip tray will usually be rusted. In the engine, look out for knocking pistons (where the liners have corroded) and oil leaks, which are common and usually occur because of overfilling the sump. High mileage and old age are usually unimportant considerations as long as servicing has been properly, and regularly, completed.

6.7-litre V8 engine striped and rebuilt with new pistons

An example

The car pictured here is a typical restoration project. A 1979 Silver Shadow II, it was brought in to Slough Rolls-Royce specialist, Michael Hibberd, in a rather dilapidated state, unusually so, in fact, for its age. Its bodywork contained holes, the engine was knocking and its interior

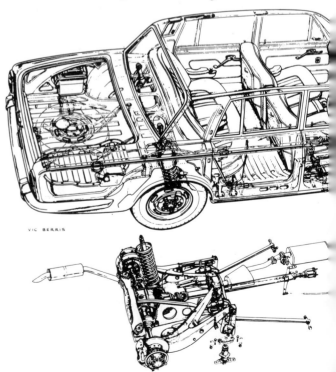

ROLLS-ROYCE SILVER SHADOW

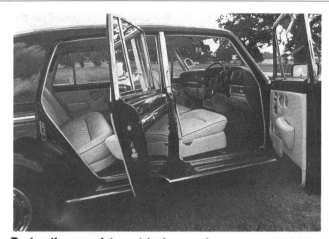

Restoration complete — strip-down and respray cost £8000

PARTS PRICES

Rolls-Royce breakers are few and far between, but parts prices will be less

Prices indicate the approximate cost of parts, in pounds, new from Rolls-Royce and from other sources, where available.

	Rolls-Royce	Other sources
Front wing	1300	650
Rear wing	1300	400
Door (front/rear)	800	400
Bumper (front/rear)	1500	600
Radiator grille	2000	1000
Mascot	300	120
Suspension springs		
front	200	155
rear	300	150
Exhaust	800	500
Engine	11,000	6000
Interior	4000	2000

needed a thorough clean up. In fact, the bodywork was so bad, having been botched up in the past with glassfibre, that an extensive restoration was required.

On the body several panels needed replacing, including front wing lower sections, outer sills, lower rear wings, rear wheelarches, inner lower rear wings and inner wheelarch filler panels; and repairs were carried out on the 'A' post box section and the front and rear valances. Most of those were sourced from remanufacturing specialists.

All the panels were thenfully leaded and Waxoyled.

Because such heavy work had to be done, a complete bare metal strip-down was necessary before a two-pack re-paint was applied.

Inside, two new carpets were installed and all the leatherwork was Connollised (that is, cleaned and redressed). There were no rips in the upholstery, although these can usually be patched up.

The engine — which was knocking — was steam-cleaned and stripped. New pistons and liners were fitted. Having removed the gearbox, its gaskets, seals and filters were all replaced.

On the miscellaneous list were a new set of exhaust clamps, new suspension, shock absorber and engine mounts; lower ball joints, a new alternator belt, scuttle filter and PAS hoses. The dipstick tube needed resealing, the brakes bleeding and the drive shafts greasing.

That may sound like a lot of work — and indeed it was! Michael Hibberd reckoned about the same had been spent on the car as it was worth. The strip down and respray cost £8000 alone! If you're a potential Shadow owner, then that's a sobering idea of what to expect.

Working on a Shadow

The best advice is: don't! The Silver Shadow could in no way be described as a DIY car. The Rolls-Royce Enthusiasts' Club says it is "very unwise" to attempt a restoration yourself. Because the car is technically so complex, the results of even a small mistake could be lethal. Also, the majority of jobs require special tools. For example, the wheel nuts need a 700lb/ft torque wrench.

The owner can undertake routine maintenance, but for effective servicing and tackling larger tasks, it really is best to take it to a specialist.

"Because the Shadow is such a well-built car," explains Michael Hibberd, "it does tend to go on and on. But problems left unattended, or botched, will only reoccur later."

Silver Shadow parts are not cheap. Generally speaking, most parts can be sourced at around 50 per cent of official RR prices and reconditioned items can be had for anything from 20 per cent to 75 per cent off the new RR price. ▶▶▶

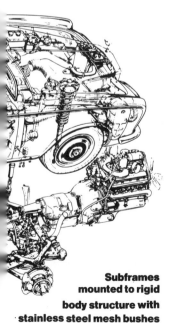

Subframes mounted to rigid body structure with stainless steel mesh bushes

171

ROLLS-ROYCE SILVER SHADOW

PRODUCTION FIGURES

Silver Shadow	16,717
Silver Shadow LWB	2776
James Young 2-dr	35
MPW 2-dr coupé	571
MPW 2-dr convertible	504
Bentley T-series	1712
James Young 2-dr	15
Pininfarina coupé	1
MPW 2-dr coupé & convertible	140
Silver Shadow II	8422
Silver Wraith	2144
Bentley T2	558
Bentley T2 LWB	10
TOTAL	**33,605**

Salvaged parts usually only come from cars written off in accidents, so you seldom find body parts which are usable. Most parts can be overhauled or reconditioned rather than replacing the entire unit, including large items like the engine and braking systems.

Living with a Rolls

You might imagine that life with a Silver Shadow would be as effortless as driving it, but there *are* drawbacks to Rolls-Royce ownership, maybe surprisingly. The incredible cost of just keeping it going is the biggest one, of course, but then you have to have some comeback for owning a car with such sumptous trimmings. It's also a natural target for vandals and 'collectors' of flying lady mascots and hub-caps (cost to replace all four is around £300).

But these are probably easily outweighed by the superb experience of driving it day to day, cossetted in absolute comfort and silence in the most unpleasant of external conditions.

It may seem very tempting to pick up a Silver Shadow for less than £10,000 instead of a MkII Jaguar, or brand new Sierra, but in reality ownership of the flying lady is fraught with actual and potential expense. Moreover, it's not a car on which you can cut corners. Attempting to rebuild anything yourself other than a hinge could lead to disaster.

The Silver Shadow has one of the most complex designs of all time — Citroën included, because Rolls-Royce used Citroën patents in its self-levelling suspension system. This is a car which really does have to be competently serviced by a Rolls-Royce specialist.

Our thanks to Michael Hibberd for his help in compiling of this feature. Specialising in Rolls-Royce, Bentley, Jaguar and other cars, his facility is based at Unit 1d, Middle Green Trading Estate, Middle Green Rd, Langley, Slough, Berks SL3 6DF; tel: 0753 31631.

Driving the Silver Shadow

From its majestic dimensions to the deep lustre of its paintwork, the Silver Shadow is certainly an imposing vehicle. Open the massive doors and you are faced with a sumptuous expanse of leather, deep pile carpets and burr walnut.

The driving position is stately: upright, pampered and spacious. Flicking the key to 'on' produces a hushed whisper somewhere ahead of you and a gentle rush of air from the split-level air conditioning. Grasping the thin steering wheel (strangely reminiscent of an Austin A60's) the slightest pressure brings the servos into action, effortlessly turning the wheels.

Engage 'drive' on the column-mounted gearstick and you're ready to move off. Depressing the accelerator coaxes a sort of muted sound from the engine like the ruffling of skirts. With all the ease of whooshing down a water slide, you are already cruising at 60, and, yes you *can* hear the clock.

As the car negotiates bends and overcomes pot-holes, there is a sensation that you are seaborne, riding the waves of oceans rather than coping with the A308. Servo-assisted braking is superb.

Who would trade this superlative, exquisite machine, with its antique white lettering for each individual electric switch and hand-painted gold coach-lines, for anything less?

SILVER SHADOW HISTORY

1965 — Launch of Silver Shadow and Bentley T-series (only grille was different) originally with 6230cc V8 engine. The first Rolls-Royce with monocoque construction and Citroënesque suspension. All cars fitted with GM400 auto 'box. From 1970, all cars came with the 6750cc V8.
1966 — Coachbuilt two-door James Young saloon introduced. Coupé and convertible, produced by Mulliner Park Ward, became the Corniche in 1971.
1977 — Silver Shadow II and Bentley T2 — facelifted inside and out, with new steering and air conditioning. LWB version became Silver Wraith, with vinyl roof and optional cockpit/passenger division.
1980 — Replaced by Silver Spirit and Bentley Mulsanne.

Pininfarina-styled Bentley coupé (left) never produced. LHD Corniche 'décapotable' (right) built for French market

Nearly 3000 LWB Shadows built (left). Bentley T2 (right) lasted just three years until introduction of Mulsanne